Boundary Layer Flows - Theory, Applications and Numerical Methods

Edited by Vallampati Ramachandra Prasad

Published in London, United Kingdom

IntechOpen

Supporting open minds since 2005

Boundary Layer Flows - Theory, Applications and Numerical Methods
http://dx.doi.org/10.5772/intechopen.78474
Edited by Vallampati Ramachandra Prasad

Contributors
Vu Duc Thai, Bui Van Tung, Volodymyr Voskoboinick, Andriy Voskoboinick, Oleksandr Voskoboinyk, Volodymyr Turick, Vladimir Shalaev, Abdullah Faruque, Ünver Kaynak, Samet Çaka Çakmakçıoğlu, İsmail Hakkı Tuncer, Onur Baş, Asan Omuraliev, Ella Abylaeva, Adrián R. Wittwer, Acir M. Loredo-Souza, Mario E. De Bortoli, Jorge O. Marighetti, Marvin E. Goldstein, Pierre Ricco, Liu Bilong, Zhang Xilong, Kou YiWei, Nabeel S. Dhaidana, Abdalrazzaq K. Abbas

First published in London, United Kingdom, 2020 by IntechOpen
IntechOpen is the global imprint of INTECHOPEN LIMITED, registered in England and Wales, registration number: 11086078, 7th floor, 10 Lower Thames Street, London, EC3R 6AF, United Kingdom
Printed in Croatia

British Library Cataloguing-in-Publication Data
A catalogue record for this book is available from the British Library

Additional hard and PDF copies can be obtained from orders@intechopen.com

Boundary Layer Flows - Theory, Applications and Numerical Methods
Edited by Vallampati Ramachandra Prasad
p. cm.
Print ISBN 978-1-83968-185-1
Online ISBN 978-1-83968-186-8
eBook (PDF) ISBN 978-1-83968-187-5

For EU product safety concerns:
IN TECH d.o.o., Prolaz Marije Krucifikse Kozulić 3, 51000 Rijeka, Croatia, info@intechopen.com or visit our website at intechopen.com.

Meet the editor

Prof. V. Ramachandra Prasad is currently working as a Professor in the Mathematics Department, School Advanced Sciences, Vellore Institute of Technology, Vellore, Tamil Nadu, India. He has 25 years of teaching experience and has authored more than 100 publications in the area of fluid dynamics and heat transfer. Over the past decade he has worked in the areas of multi-physical fluid dynamics and numerical simulation. His work has investigated complex phenomena of interest in mechanical engineering, applied mathematics, chemical engineering, and material processing. He has made fundamental contributions in magnetic fluid dynamics, thermal radiation heat transfer, fluid modeling, micro polar non-Newtonian hydrodynamics, magnetic induction flows, viscoelastic liquid flows, porous media, and boundary flows. He has authored over 100 journal articles in these areas and also numerous conference presentations. He has published six books on radiation heat transfer, thermo-diffusion and diffusion-thermo effects on boundary layer flows, and Viscoelastic flows. More recent work has involved non-Newtonian fluid dynamics, fluid particle suspensions, and lid driven cavities. The numerical methods he has used are the finite difference method, finite element method, network simulation methods, and finite volume codes. I have worked in collaboration with US, Indian, Spanish, British, and Taiwanese researchers. I am presently working as reviewer for over 100 national and international journals. His h-index is 24 and i-10 index is 50.

Contents

Preface

This book covers the boundary layer theory. It presents topics ranging from the fundamentals of fluid mechanics to the mathematical, physical, and experimental advances made in this in recent times. The book includes comprehensive overview of the derivation of n-s equations, flow past simple geometries, laminar flows, laminar turbulent transition, turbulent boundary layers, coupled velocity and thermal boundary layers, three-dimensional boundary layers, natural and mixed convection, flow past bodies (aerodynamics), and numerical methods for laminar and turbulent boundary layers.

Written by experts in the field, this book presents the theory's importance for a wide variety of applications. Applied mathematicians and engineers—especially those working in fluid dynamics—along with engineers of aeronautics, undoubtedly welcome this authoritative, up-to-date book.

Beginning with the basics, the book discusses in detail the techniques and results that reveal the nature of the equations that govern the flow within boundary layers and ultimately describe the laws underlying the motion of fluids with small viscosity. The book is also recommended for undergraduate students.

Vallampati Ramachandra Prasad
Department of Mathematics,
School of Advanced Sciences,
Vellore Institute of Technology,
Vellore, Tamil Nadu, India

Section 1

Three Dimensional Boundary Layers

Chapter 1

3D Boundary Layer Theory

Vladimir Shalaev

Abstract

Some new analytical results in 3D boundary layer theory are reviewed and discussed. It includes the perturbation theory for 3D flows, analyses of 3D boundary layer equation singularities and corresponding real flow structures, investigations of 3D boundary layer distinctive features for hypersonic flows for flat blunted bodies including the heat transfer and the laminar-turbulent transition and influences of these phenomena on flows, and the new approach to the analysis of the symmetric flow instability over thin bodies and studies of the control possibility with the electrical discharge using new model of this phenomenon interaction with the 3D boundary layer. Some new analytical solutions of boundary layer and Navier-Stokes equations are presented. Applications of these results to analyze viscous flow characteristics of real objects such as aircraft wings, fuselages, and other bodies are considered.

Keywords: 3D boundary layer, asymptotic perturbation theory, singularities, flow structures, applications

1. Introduction

Despite the intensive development of computer technologies and numerical methods for the Navier-Stokes and Reynolds equations, problems of the three-dimensional boundary layer are of significant interest in the fluid dynamics. So far these problems have been little studied as a result of objective difficulties related with the large dimensionality and complexity of equations. Therefore, analytic results in this field can play an important role in the depth understanding of fluid dynamics phenomena and their study. In this part, some modern results in the three-dimensional boundary layer theory are discussed.

The small perturbation theory for inviscid flows is well developed and widely applied to estimate aerodynamic characteristics of real flight apparatus. Also it has been attempted to develop such theory for the boundary layer [1]. However, the zero approximation ("flat plate" approximation, zero cross-flow approximation) only given a rational contribution and were used in calculations. Equations for perturbations were complex. They required a numerical solution that was not much simpler than the full equation system. Father investigations of three-dimensional effects in the boundary layer theory became possible only after developments of computers with the enough power, numerical methods, and turbulence models [2].

Another approach was developed on the base of the rational perturbation theory including the first-order approximation [3–10] for some class of flows, such as flows over aircraft wings and fuselages at small angle of attack, which have high importance as for the theory and the practice. In this case, zero-order approximation functions do not depend on the cross coordinate. Equations of the first-order approximation reduce to a two-dimensional system by introducing a new variable. The cross coordinate is included to this system as a parameter. This property of the self-similarity simplifies the solution procedures allowing to apply two-dimensional numerical methods and to reduce computing resources.

The singularity in the solution of 2D steady boundary layer equation is well known as the separation. Singularities arising in solutions of unsteady or 3D laminar boundary layer equations are not related directly with the flow separation and are slightly studied due to difficulties of analytical investigations of complex equations and uncertainty of numerical result treatments. However, this task is of interest for the mathematical physics and for numerical modeling of aerodynamic applications. For the first time, a singularity was found in the solution of 2D unsteady BL equations for the flow around the flat plate impulsively set into motion [12]. The singularity of the similar type was discovered on the side edge of a quarter flat plate in a uniform freestream [13] and at a collision of two jets [14]. In Ref. [15], necessary conditions were formulated for a singularity formation in self-similar solutions of the unsteady model and 3D incompressible laminar boundary layers on a flat surface with pressure gradients. Sufficient conditions and singularity types were not studied, and real flow conditions were not considered. Singularities of numerical solutions (the nonuniqueness or the absence of a solution) were found for the laminar boundary layer in the leeward symmetry plane on a round cone at incidence [16–18]. Similar results were obtained inside the computation region of the 3D turbulent boundary layer on the swept wing [19]. The singular behavior of boundary layer characteristics (the skin friction tends to the infinity in the symmetry plan) was found for the boundary layer on the small span delta wing [8, 10]. The explanation of these phenomena was found on the base of analytical solutions of laminar boundary layer equations on conical surfaces [10, 21–24]. The asymptotic flow structure on the base of Navier-Stocks equations in the singularity vicinity is constructed.

The problem of the flow separation control using plasma actuators on the base of the electrical discharge is assumed as a perspective aerodynamic instrument [26–28]. It is considered as a one method for the control of the separated flow asymmetry near the nose part of aircrafts. The problem was complicated by the absence of an adequate model for the boundary layer-discharge interaction and a criterion for flow asymmetry arising. The use as a criterion numerical results and experimental data is restricted as a result of the high sensitivity of the asymmetry origin to different parameters [29]. Solution of these problems was obtained with the development of new models [30–34].

2. Small perturbation theory for three-dimensional boundary layer

As follows from the cross-flow impulse equation in biorthogonal coordinates [2], the necessary conditions for a small cross velocity ($|w| << 1$) are the relations

$$\frac{1}{\lambda H_2}\frac{\partial p}{\partial z} \sim \cos\theta \sim k_1 \sim \frac{\varepsilon}{\lambda} << 1. \tag{1}$$

The small parameter ε characterizes the gradient of the pressure $p(t,x,z)$ with respect to transverse nondimensional coordinate z; t and x here are dimensionless time and longitudinal coordinate, λ is body span, H_2 is metric coefficient, k_1 is longitudinal coordinate line curvature, and θ is the angle between coordinate lines on the body surface. Using these conditions flow parameters in the 3D boundary layer are presented by asymptotic expansions:

$$v_w = v_{w0}(t,s) + \varepsilon v_{w1}(t,s,z),\ h_w = h_{w0}(t,s) + \varepsilon h_{w1}(t,s,z),\ w = \frac{\varepsilon}{\lambda} w_1(t,s,n,z),$$
$$\mathbf{V} = (u,v,h,\rho,\mu,\kappa) = \mathbf{V}_0(t,s,n) + \varepsilon \mathbf{V}_{10}(t,s,n,z) + \varepsilon_1 \mathbf{V}_{11}(t,s,n,z). \tag{2}$$

Here, $s(t,x,z)$ is a dimensionless length of the coordinate line $z = const$ measured from the critical point $x_c(t,z)$; n is normal coordinate transformed with Dorodnitsyn transformation; $v_w(t,x,z)$ and $h_w(t,x,z)$ are blow (suction) velocity and the surface temperature; u and v, h, ρ, κ and μ are dimensionless longitudinal and normal velocities, enthalpy, density, thermal conductivity, and viscosity. The parameter $\varepsilon_1 < < 1$ is not known a priori, it describes own flow perturbations inside the boundary layer. The following is found from the analysis of equations: for thin wings $\varepsilon_1 = \varepsilon/\lambda^2$, for slightly asymmetric bodies ε characterizes the asymmetry and $\varepsilon_1 = \alpha^*/\lambda$, where λ is relative body thickness [4–10].

To calculate boundary layer characteristics, the equation system for the composite solution incorporated in all terms of asymptotic expansion (2) was derived:

$$\begin{aligned}
&\rho\left[\frac{\partial u}{\partial t} + \left(u - \frac{\beta w}{\lambda H_2}\right)\frac{\partial u}{\partial s} + v\frac{\partial u}{\partial n}\right] + \frac{\partial p}{\partial s} = \frac{\partial}{\partial n}\mu\frac{\partial u}{\partial n},\\
&\rho\left[\frac{\partial h}{\partial t} + \left(u - \frac{\beta w}{\lambda H_2}\right)\frac{\partial h}{\partial s} + v\frac{\partial h}{\partial n}\right] = \frac{\partial}{\partial n}\frac{\kappa}{\mathrm{Pr}}\frac{\partial h}{\partial n}\\
&+(\gamma-1)M^2\left[\frac{\partial p}{\partial t} + \left(u - \frac{\beta w}{\lambda H_2}\right)\frac{\partial p}{\partial s} + \mu\left(\frac{\partial u}{\partial n}\right)^2\right],\\
&\rho\left[\frac{\partial w}{\partial t} + u\frac{\partial w}{\partial s} + v\frac{\partial w}{\partial n} + k_1u^2 - k_2uw\right] = \frac{\partial}{\partial n}\mu\frac{\partial w}{\partial n}\\
&-\frac{1}{\lambda H_2}\left(\frac{\partial p}{\partial z} - \varepsilon\beta\frac{\partial p_1}{\partial s}\right) + \cos\theta\frac{\partial p}{\partial s},\\
&\rho\left[\frac{\partial q}{\partial t} + u\frac{\partial q}{\partial s} + v\frac{\partial q}{\partial n} + \frac{\partial k_1}{\partial z}u^2 - k_2uq\right] = \frac{\partial}{\partial n}\mu\frac{\partial q}{\partial n} -\\
&-\frac{\partial}{\partial z}\left[\frac{1}{\lambda H_2}\left(\frac{\partial p}{\partial z} - \varepsilon\beta\frac{\partial p_1}{\partial s}\right) - \cos\theta\frac{\partial p}{\partial s}\right], q(t,s,n,z) = \frac{\partial w}{\partial z},\\
&\frac{\partial \rho}{\partial t} + \frac{\partial \rho u}{\partial s} + \frac{\partial \rho v}{\partial n} + \frac{1}{\lambda H_2}\left(\rho q - \beta\frac{\partial \rho w}{\partial s}\right) - k_2\rho u = 0,\\
&n = 0: u = w = 0,\ v = v_w(t,x,z),\ h = h_w(t,x,z)\ \left(\frac{\partial h}{\partial n} = 0\right),\\
&n \to \infty:\ u = u_e(t,x,z),\ w = w_e(t,x,z),\ h = h_e(t,x,z).
\end{aligned} \tag{3}$$

Eq. (3) is not true in the vicinity of the wing leading edge, where the pressure perturbation has the singularity. Using the asymptotic theory, singular regions near blunted and sharp leading edges were analyzed. It was found that the boundary layer in these regions is described by equations for the boundary layer on the sweep parabola or wedge. On a body the boundary layer begins in the critical point.

The system (2) was applied to the solution of different problems for wings and bodies [4–10]. To illustrate the developed approach in **Figures 1** and **2**, calculations of displacement thicknesses (**Figure 1**) and skin frictions (**Figure 2**) on the wind tunnel model of the US Air Force fighter TF-8A supercritical wing at Mach numbers M = 0.99 and 0.5 are presented. Solid lines correspond to solutions of Eq. (3) for the wing model ($\mathrm{Re} = 2.246 \cdot 10^6$); dotted lines on **Figure 4** are results for full scale wing ($\mathrm{Re} = 2.58 \cdot 10^7$); symbols present solutions of full 3D boundary layer equations [11].

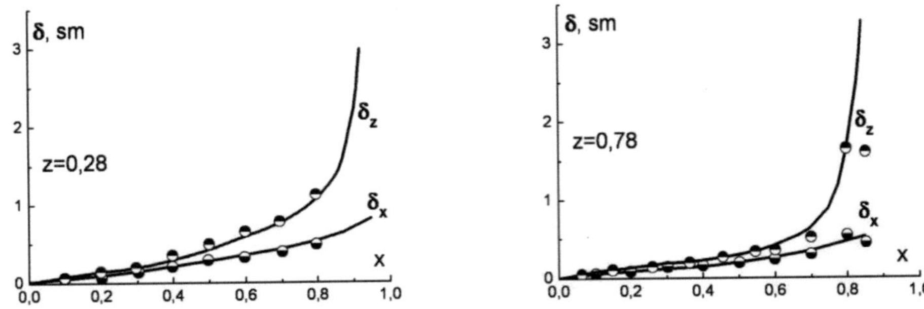

Figure 1.
Displacement thickness distributions on the model of supercritical wing; $M = 0.99$, $Re = 2.246 \cdot 10^6$, *and* $\alpha^* = 3.12^{\circ}$.

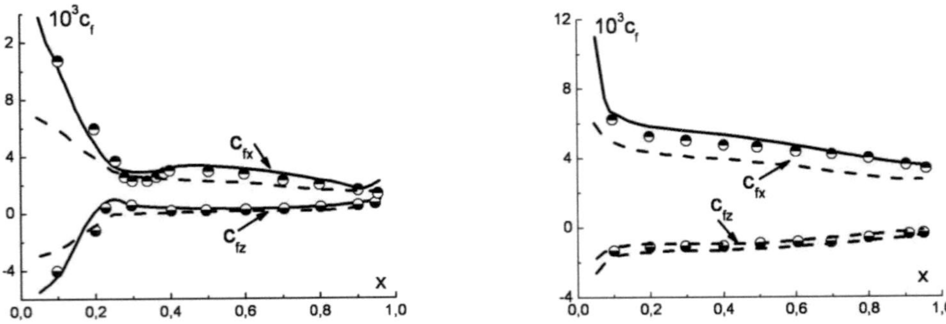

Figure 2.
Skin friction distributions on the model of supercritical wing; $M = 0.5$, $\alpha^* = 12.09^{\circ}$, $Re = 2.246 \cdot 10^6$ *(solid lines), and* $Re = 2.58 \cdot 10^7$ *(dotted lines).*

These figures demonstrate that the asymptotic solution very well reproduce numerical results as for the skin friction and for displacement thicknesses in the large parameter diapason.

3. Singularities in solutions of three-dimensional boundary layer equations

The laminar boundary layer problem on a thin round cone with the half apex angle $\delta_c << 1$ at the angle of attack α^* depends on the parameter $k = 4\alpha^*/(3\delta_c)$ only. Firstly, analytical results about singularities were obtained for outer BL part for a such cone. It is understood from previous works [15–18, 20], the singularity can arise when two subcharacteristic (streamlines) families collided —this is a necessary condition. Such situation arises usually in the leeward symmetry (runoff) plane over a body of revolution at an angle of attack. Unusual properties in numerical solutions of self-similar equations in this plane for a round slender cone in supersonic freestreams were studied in many works due to the practical interest of the heat exchange on flying vehicles head parts [16–18, 20]. In this case, one parameter defines the flow. Two solutions were found in the windward symmetry (attachment) plane and at small angles of attack ($k \leq k_c$) in the leeward symmetry plane. In this plane, no solutions were obtained at moderate angles of attack ($k_c \le k \le 2/3$) and many solutions at larger incidences up to BL separation ($2/3 \le k < 1$). Full BL equation solutions with initial conditions in the windward symmetry plane fixed the violation of symmetry conditions in the runoff plane, a velocity jump through this plane in the angle of attack diapason, when the self-similar solution has been absent [10, 21]. The task for the cone was solved numerically on the base of parabolized Navier-Stokes equations, without the streamwise viscous diffusion [20]. However the problem is retained since the flow structure and reasons of unusual BL properties have not been explained.

Analytical solutions of full equations for the outer BL part on the slender round cone with initial conditions in the windward symmetry plane showed the singularity presence in the leeward symmetry plane of the logarithmic type at $k = 1/3$ and of a power type at $k > 1/3$ [10, 21]. It had been shown numerical solutions provided incorrect results near the singularity due to the accuracy loss. Similar but more complex results were obtained for arbitrary cones; they allow defining the sufficient conditions of the singularity arising [10, 22]. The asymptotic flow structure at large Reynolds number near the singularity on the base of Navier-Stokes equations was constructed, and analytical solutions in different asymptotic regions were obtained, which were matched with BL solutions. The analysis of the viscous-inviscid interaction region, in particular, revealed that the singularity can arise not only in self-similar but in full 3D BL equations [10, 22]. The theory showed that the singularity appearance relates with eigensolutions of the BL equations appearing near the runoff plane; it also explained numerical modeling results on the base of parabolized Navier-Stokes equations.

In the outer BL part, the theory gives the critical angle of attack for the singularity appearance $k_c = 1/3$. However calculations showed that this parameter is a function on numbers of Mach M and Prandtl Pr and the wall temperature h_w, $k_c = k_c(\mathrm{M}_\infty, \mathrm{Pr}, h_w)$ [10]. This indicates that a singularity can arise in the near-wall region. The series decomposition of the near-wall solution in the runoff plane showed the presence of a parameter α, the linear combination of skin friction components, and the sign change of which leads to the change of the physical flow

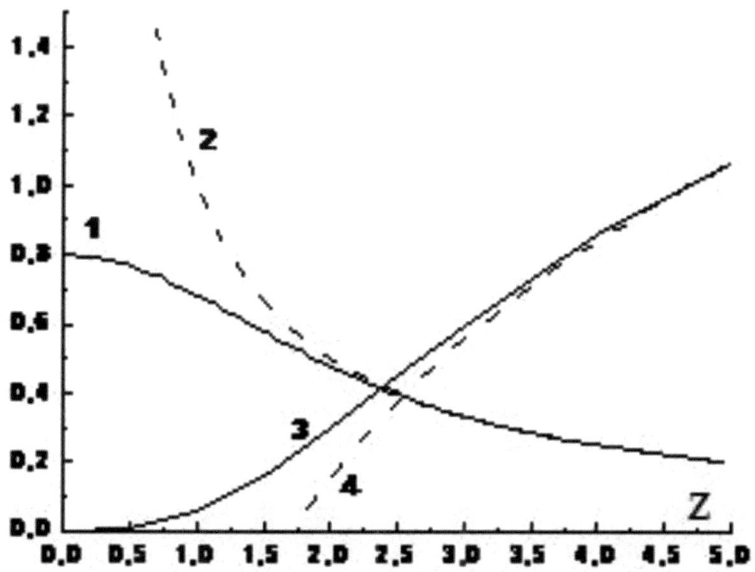

Figure 3.
Solutions of boundary layer equations (dotted lines) and parabolized Navier-Stokes equations (solid lines).

topology near this plane [24]. The analysis of BL equations in the near-wall region showed that $\alpha = 0$ corresponds to the critical value k_c, and it was confirmed by all published numerical calculations [16–18, 20]. In the runoff plane, the new power type singularity in solutions of full BL equations revealed that it is related with the eigensolutions appearing near this plane. Calculation results for BL on delta wing confirm the singularity presence.

3.1 Self-similar boundary layer on a cone

The 3D laminar boundary layer on a conical surface in the orthogonal coordinate system $xy\varphi$ (**Figure 3**) is described by following self-similar equations and boundary conditions [10, 22]:

$$
\begin{gathered}
u_{yy} = Awu_\varphi + vu_y + A_1 w(u - w), \\
w_{yy} = Aww_\varphi + vw_y + w\left(\frac{2}{3}u + Kw\right) - h\left(\frac{2}{3} + K\right), \\
h_{yy} = Awh_\varphi + vh_y - M_e\left(u_y^2 + \frac{3}{2}A_1 w_y^2\right), \ \rho h = 1, \\
y = \varepsilon\sqrt{\frac{3\rho_e u_e}{2x\mu_e}}\int_0^{y^*}\rho d\frac{y^*}{l}, \ \mathrm{Re} = \varepsilon^{-2} = \frac{\rho_\infty u_\infty l}{\mu_\infty}, \\
f_y = u, \ g_y = w, \ v = -f - \left[K - \frac{1}{2}A(\ln(\rho_e\mu_e/u_e))_\varphi\right]g - Ag_\varphi, \\
y = 0: \ u = v = w = 0, \ h = h_w \ (h_y = 0); \ y = \infty: u = w = h = 1.
\end{gathered}
\tag{4}
$$

Equation coefficients are defined by expressions

$$
\begin{gathered}
M_e(\varphi) = (\gamma - 1)M_\infty^2\frac{u_e^2}{h_e}, \ K(\varphi) = \frac{2w_{e\varphi}}{3Ru_e}, \\
A(\varphi) = \frac{2w_e}{3Ru_e}, \ A_1(\varphi) = \frac{2}{3}\left(\frac{w_e}{u_e}\right)^2
\end{gathered}
\tag{5}
$$

In these equations, to reduce formulas, $\mathrm{Pr} = 1$ and the linear dependence of the viscosity on the temperature ($\rho\mu = 1$) are assumed. Indexes y and φ denote derivatives with respect to the corresponding variables; x is the distance from the body nose along the generator referenced to the body length l; y is the Dorodnitsyn variable; y^* is normal to the body surface; φ is the transversal coordinate, and it can be the polar angle for a round cone; $f(y,\varphi)$ and $g(y,\varphi)$ are longitudinal and transverse stream functions; $v(y,\varphi)$ is transformed normal velocity; and $R(\varphi)$ is the metric coefficient. The density ρ, the enthalpy h, the viscosity μ, the longitudinal u, and transversal w velocities are referenced to the values at the outer boundary indexed by e, which are normalized to their freestream values indexed by ∞; they are functions of φ only. The transversal velocity on the outer boundary layer edge $w_e = 0$ in the initial value plane (the attachment plane) $\varphi = 0$, in which $K(0) > 0$, and in the runoff plane $\varphi = \varphi_1$, in which $K(\varphi_1) = -k < 0$, and two boundary layer parts that came from different sides of the attachment plane collided. For the round cone, $\varphi_1 = \pi$.

Eq. (4) is simplified for slender bodies since in this case, $u_e = \rho_e = \mu_e = 1$, $A_1 << 1$. Neglecting proportional to A_1 terms in (4), we obtain the Crocco integral for the enthalpy and momentum equations in the form

$$
\begin{aligned}
&h = h_w + h_r u - \frac{1}{2} M_e u^2,\ h_r = 1 - h_w + \frac{1}{2} M_e, \\
&M_e = (\gamma - 1) M^2,\ v = -\left(f + Kg + Ag_\varphi\right), \\
&u_{yy} = Awu_\varphi + vu_y, \\
&w_{yy} = Aww_\varphi + vw_y + w\left(\frac{2}{3}u + Kw\right) - h\left(\frac{2}{3} + K\right).
\end{aligned} \tag{6}
$$

For the slender round cone with the apex half angle $\delta_c << 1$ at the angle of attack α^*, simple expressions for outer functions are

$$
w_e = 2\alpha^* \sin\varphi,\ K(\varphi) = k\cos\varphi,\ A(\varphi) = k\sin\varphi,\ k = \frac{4\alpha^*}{3\delta_c} \tag{7}
$$

3.2 Singularities in the outer boundary layer region

In the outer boundary layer region, $y \gg 1$, (y is the Dorodnitsyn variable normal to the wall), flow functions are represented as [17, 36]

$$
\begin{aligned}
&u = 1 + U(\eta, \varphi),\ w = 1 + W(\eta, \varphi),\ \eta = (y - \delta)/\sqrt{a(\varphi)}, \\
&h = 1 + H = 1 - \left(\frac{1}{2} M_e + h_w - 1\right) U - \frac{1}{2} M_e U^2
\end{aligned} \tag{8}
$$

Here $\delta(\varphi)$ is the displacement thickness defined by the equation of F. Moore [6], the function $a(\varphi)$ is found from the local self-similarity condition, and $U << 1$ and $W << 1$ are velocity perturbations with respect to boundary conditions, which in the first-order approximation satisfy to equations [10, 21]

$$
U_{\eta\eta} + \eta U_\eta - aAU_\varphi = 0,\ W_{\eta\eta} + \eta W_\eta - \frac{2}{3} a\left[\frac{3}{2} AW_\varphi + (1 + 3K)W\right] = \frac{2}{3} ap(\varphi)U. \tag{9}
$$

These equations have solutions:

$$
\begin{aligned}
&U(\eta, \varphi) = C_1 \operatorname{erfc}(\eta/\sqrt{2}) \\
&W(\eta, \varphi) = -b(\varphi)U,\ W_1(\eta, \varphi) = -b(\varphi)U + B_1(k)V(\eta, \varphi)
\end{aligned} \tag{10}
$$

Constants C_1 and B_1 are calculated from matching condition with a numerical solution inside the boundary layer. These solutions satisfy to initial conditions in the attachment plane and must tend to zero at $\eta \to \infty$. The function $V(\eta, \varphi)$ is the solution of the homogeneous equation for the cross-velocity perturbation, when the right-hand side equals to zero; it is expressed by Veber-Hermite functions [21]. The coefficient $B_1 \sim 1/K(0)$ has the singularity at $K(0) \to 0$. For the round cone this limit corresponds to zero angle of attack; in this case, the analytical expression for $W_1(\eta, \varphi)$ shows the presence of the power type singularity in the leeward plane $\varphi = \varphi_1$ [10, 21]. The first solution $W(\eta, \varphi)$ is regular in this limit, and its behavior is defined by functions $a(\varphi)$ and $b(\varphi)$, which satisfy to equations [10, 21, 22]

$$
\begin{aligned}
&w_e b_\varphi + 2(1 + M) w_{e\varphi} b = 2pMw_{e\varphi},\ p(\varphi) = 1 + \left(1 + \frac{3}{2}K\right)\left(\frac{1}{2}M_e + h_w - 1\right), \\
&w_e a_\varphi + 2(N + 1) w_{e\varphi} a = 2Nw_{e\varphi},\ N(\varphi) = 3M(\varphi) = K^{-1}.
\end{aligned} \tag{11}
$$

Solutions of these equations with initial conditions in the attachment plane are represented in integral forms in the general case and have analytical expressions for the round cone [10, 21]. Their properties near the leeward plane, at $\zeta = \varphi_1 - \varphi << 1$, are represented by expressions

$$w_e = \frac{3}{2}kR\zeta,\ k = -K(\varphi_1),\ R = R(\varphi_1),\ p_1 = p(\varphi_1),\ n = 3m = -K^{-1}(\varphi_1),$$

$$m \neq 1 : b = \frac{mp_1}{m-1} - b_m\zeta^{2(m-1)},\ m = 1 : b = -2p_1 \ln\zeta + b_1, \tag{12}$$

$$n \neq 1 :\ a = \frac{n}{n-1} + a_n\zeta^{2(n-1)},\ n = 1 :\ a = -2\ln\zeta + a_1$$

Here a_n and b_m are known coefficients [10, 21]. These formulas are true for non-slender bodies also [10, 22, 23].

These results show the presence in the outer BL part of two singularity types in the leeward plane related with properties of functions $a(\varphi)$ and $b(\varphi)$. For $k<1$ the function $U(\eta,\zeta)$ exists at $\zeta = 0$ but reaches this limit irregularly; its behavior is studied analytically in details for the slender round cone [10, 21]. For $k \geq 1$ the function $U(\eta,\zeta)$ is singular at $\zeta \to 0$ since $a(\zeta) \to \infty$ and the BL thickness tend to infinity as $\sqrt{a(\zeta)}$: the logarithmic singularity type takes place at $k = 1$, and it is of the power type at $k > 1$. At $k \geq 1$ the flow separation is observed in experimental and numerical studies; this phenomenon changes not only the outer part but also the inner boundary layer structure. It should be noted that such behavior of velocity viscous perturbations near the BL outer part at the separation development is a new property in the comparison with the 2D flows.

The function $W(\eta,\zeta)$ has irregular but finite limit in the leeward plane for $\zeta \to 0$ at $k < 1/3$. This limit is singular at $k \geq 1/3$: the singularity has the logarithmic or power type, if $k = 1/3$ or $k > 1/3$. At $1/3 \leq k < 1$ the singularity is related with the behavior of cross-flow velocity only. This singularity leads to the longitudinal vortex component strengthening in the outer part of the viscous region. The singularity takes place, if the pressure gradient is negative ($k \leq 2/3$) or positive ($k > 2/3$). It is formed by BL proper solutions, which have homogeneous conditions on both boundaries and arise near the runoff plane. The critical value $k_c = 1/3$ for the outer BL part is undependable on the wall temperature and Mach and Prandtl numbers; however the considered singularities define the real flow structure near the leeward plane at $k \geq 1/3$ [17, 36, 37].

3.3 Asymptotic flow structure near the singularity

Due to the irregularity of solutions already at $k \geq 1/6$ ($m \leq 2$), the vortex boundary region near the runoff plane is formed with transverse dimension $\zeta \sim \varepsilon^{\frac{1}{2-m}}$; at $m \sim 1$ this value is of the order of the BL thickness ε. In this region, the transverse diffusion is the effect of the first order, and to describe it we introduce the following variables:

$$\begin{gathered}\varepsilon_1 = \left[\tfrac{3}{2}\mathrm{Re}\rho_e(\varphi_1)u_e(\varphi_1)/\mu_e(\varphi_1)\right]^{-\frac{1}{2}}. \\ z = \sqrt{kx}R\zeta/\varepsilon_1,\ u = u(y,z),\ h = h(y,z),\ w = w(y,z)\end{gathered} \tag{13}$$

Using these variables from Navier-Stokes equations at $\zeta \sim \varepsilon_1 << 1$ for this region, we derive self-similar equations, which in its outer part, at $y \gg 1$, reduce to the form

$$U_{yy} + kU_{zz} + (1-k)yU_y + kzU_z = 0,$$
$$W_{yy} + kW_{zz} + (1-k)yW_y + \left(\frac{2}{z} + kz\right)W_z + 2k(m-1)W + \frac{2}{3}p_1U = 0 \tag{14}$$

For $k < 1$ these equations have the solution corresponding to the regular at $k \to 0$ solution of BL equations:

$$U(y,z) = C_1\, erfc\left(y\sqrt{(1-k)/2}\right) erf\left(z/\sqrt{2}\right),\; W = -B(z)C_1 \mathrm{erfc}\left(y\sqrt{(1-k)/2}\right),$$
$$B_{zz} + \left(\frac{2}{z} + z\right)B_z - 2(m-1)B = -2mp_1F(z),\; F(z) = \mathrm{erf}\left(z/\sqrt{2}\right) \tag{15}$$

The function $B(z)$ is expressed by Kummer's function $\Phi(a, b, x)$ [10, 22, 23]:

$$B = mp_1B_0(z) + B_m\Phi\left(1-m, \frac{3}{2}, -\frac{1}{2}z^2\right),\; B_m = b_m\left(R\sqrt{kx}/\varepsilon_1\right)^{2(1-m)}. \tag{16}$$

$B_0(z)$ is a particular solution of the inhomogeneous equation; the coefficient B_m is determined from matching condition.

In **Figure 4**, comparisons of solutions of BL (dotted lines) and Navier-Stokes (solid lines) equations for $m = 1/2$ (curves 1 and 2) and $m = 1$ (curves 3 and 4) are presented. It is seen that regular solutions of Navier-Stokes equations are converged quickly to singular solutions of BL equations.

Another effect generated by the singularity at $k \geq 1/3$ due to the BL growth at $\zeta \to 0$ is the viscous-inviscid interaction. This effect is important in the region, where the inviscid and induced cross velocities have same orders; this condition defines the transverse dimension of the region $\Delta\varphi$ and the velocity scale:

$$\Delta\varphi \sim \sqrt{\varepsilon}x^{-\frac{1}{4}},\quad w_e \sim kRu_e\sqrt{\varepsilon}x^{-\frac{1}{4}}. \tag{17}$$

In this region, the flow has the two-layer structure. Assuming the potential flow in the outer inviscid region, the solution here is presented by the improper integral

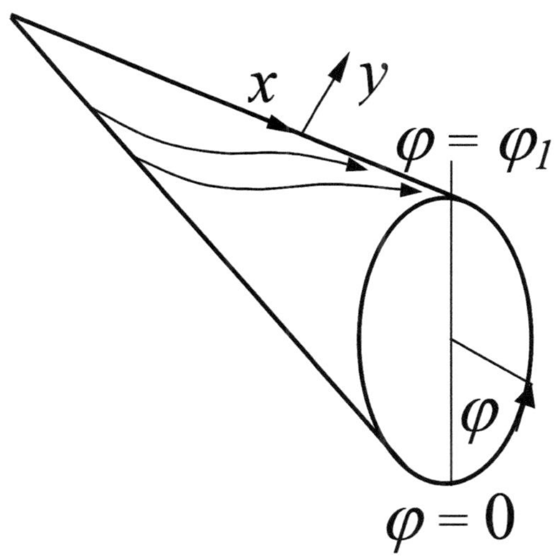

Figure 4.
The general flow scheme and the coordinate system for a cone.

from the displacement thickness $\delta(x,s)$. In the boundary layer, the flow is described by full 3D equations:

$$s = \frac{R\zeta}{\sqrt{\varepsilon}},\ w_e = \frac{3}{2}u_e\sqrt{\varepsilon}W_e(x,s)\ W_e(x,s) = -ks[1+r],\ r = \frac{4m}{\pi}\frac{\partial}{\partial x}\int_0^\infty \frac{\delta(x,t)dt}{s^2-t^2}$$

$$v = f + Kg + Ag_s + \frac{2}{3}xf_x,\ h = h_w + h_r u - \frac{1}{2}M_e(\varphi_1)u^2 \tag{18}$$

$$u_{yy} = W_e w u_s + v u_y + \frac{2}{3}x u u_x$$

$$w_{yy} = W_e w w_s + v w_y + w\left(\frac{2}{3}u + W_{es}w\right) - h\left(\frac{2}{3} + W_{es}\right) + \frac{2}{3}x u w_x$$

For these equations boundary conditions have the form (1). A solution of these equations will be matched with the boundary layer solution at $s \to \infty$. Initial conditions are needed at some streamwise location $x = x_0$, which can be obtained from a solution of Navier-Stokes equations near the body nose; this feature does the problem more complicated. Obtained equations allow a self-similar solution for hypersonic flows at some additional assumptions.

The solution in the outer boundary layer part, at $y \gg 1$, is described by formulas

$$\begin{aligned} &t = y/\sqrt{d(x,s)},\ u = 1 + U(x,t,s),\ w = 1 - c(x,s)U \\ &U = C_1\,erfc(t/\sqrt{2}),\ p_0 = \frac{3}{2}\left(\frac{1}{2}M_0 + h_w - 1\right) \\ &(1+r)sd_s - 2mxd_x - 2(n-1-r_s)d = -2n \\ &(1+r)sc_s - 2mxc_x - 2(m-1-r_s)c = -2m(p_1 - qp_0) \end{aligned} \tag{19}$$

Along characteristics $\xi(x,s) = const$, which are streamlines of the inviscid flow, the equations for functions $d = d(\xi,s)$ and $c = c(\xi,s)$ are integrated. At $s \to 0$ these functions are represented in the form

$$\begin{aligned} &c = Cs^L + \frac{m(p_1 + p_0 r_s)}{m-1-r},\ L(\xi,s) = \frac{m-1-r_s}{1+r};\ d = Ds^I + \frac{n}{n-1-r_s}, \\ &I(\xi,s) = \frac{n-1-r_s}{1+r},\ C = b_m\varepsilon^{m-1},\ D = a_n\varepsilon^{n-1}. \end{aligned} \tag{20}$$

Coefficients C and D are obtained by matching $d(\xi,s)$ and $c(\xi,s)$ at $s \to \infty$ in relation to $a(\zeta)$ and $b(\zeta)$ at $\zeta \to 0$ [17, 36]. The logarithmic singularity appears in these functions at $I = 0$ or $L = 0$. At $L(\xi,0) < 0$ or $I(\xi,0) < 0$, the singularity is of the power type.

Following from presented results, in contrast with the 2D separation, the viscous-inviscid interaction does not eliminate the singularity in 3D boundary layer; this effect moves only the critical value of k_c.

3.4 Singularities in the boundary layer near-wall region

The singularity in the outer BL part gives the critical value $k_c = 1/3$, although calculations show $k_c = k_c(\mathrm{M}_\infty, \mathrm{Pr}, h_w)$. This indicates on the possibility of singularity arising in the near-wall region. To study this possibility, at the first, we study the solution behavior of Eq. (6) at $y < < 1$ in the runoff plane $\varphi = \varphi_1$ where the solution is presented in the form

$$\tau_0 = \frac{du_0(0)}{dy}, \ \theta_0 = \frac{dw_0(0)}{dy},$$
$$u_0 = \tau_0 y + U_0(y), \ w_0 = \theta_0 y + W_0(y), \ v_0 = -\alpha y^2 - F_0 + kG_0, \ \alpha = \frac{1}{2}(\tau_0 - k\theta_0) \tag{21}$$

Second terms of these decompositions can be presented by series

$$U_0 = F_{0y} = \sum_{i=0} \frac{\alpha_i y^{i+4}}{(i+4)!}, \ F_0 = \sum_{i=0} \frac{\alpha_i y^{i+5}}{(i+5)!},$$
$$W_0 = G_{0y} = \sum_{i=0} \frac{\beta_i y^{i+2}}{(i+2)!}, \ G_0 = \sum_{i=0} \frac{\beta_i y^{i+3}}{(i+3)!}. \tag{22}$$

First three coefficients of these series are defined by relations

$$\alpha_0 = -2\tau_0\alpha, \ \alpha_1 = k\tau_0\beta_0, \ \alpha_2 = k\tau_0\beta_1$$
$$\beta_0 = -ph_w, \ \beta_1 = -p\tau_0 h_r, \ \beta_2 = \frac{1}{3}(\tau_0 - 3k\theta_0)\theta_0 + 2pM_e\tau_0^2 \tag{23}$$

Using these decompositions we can study qualitatively a dependence of the flow structure near the runoff plane from parameters by analyzing the subcharacteristic behavior. The transformed normal to the body surface v and transverse w velocities at $\zeta < < 1$ and $y < < 1$ in the first-order approximation are represented in the form

$$v = v_0 = -\left(\alpha y^2 - \frac{1}{6}k\beta_0 y^3\right) = -\frac{1}{6}k\beta_0 y^2(y + y_c)$$
$$y_c = -\frac{6\alpha}{k\beta_0} = \frac{6\alpha}{kph_w}, \ w = -w_0 = -k\theta_0\zeta y \tag{24}$$

In the plane $\zeta = 0$, the cross-flow velocity $w = 0$ due to symmetry conditions. Here two critical points, in which $v = 0$, can be. The first point locates on the cone surface $y = 0$, and the second one $y = -y_c$ appears in the physical space at $\alpha < 0$, if $p > 0$ ($k < 2/3$), that corresponds to small angles of attack for the round cone and at $\alpha > 0$, if $p < 0$. Commonly, the critical value of the cross-flow velocity gradient $k_c \le 1/3$ corresponds to the negative cross-flow pressure gradient $p > 0$, the transverse skin friction in this region $\theta_0 > 0$.

Using these expressions, the equation for the subcharacteristics is obtained in the form

$$\frac{y_c dy}{y(y + y_c)} = \beta\frac{d\zeta}{\zeta}, \ \beta = \frac{\alpha}{k\theta_0}; \ \alpha \ne 0 : y = \frac{y_c y_0 s^\beta}{y_c + y_0(1 - s^\beta)}, \ s = \left|\frac{\zeta}{\zeta_0}\right|,$$
$$\alpha = 0 : y = \frac{y_0}{1 - y_0 d\ln s}, \ d = \frac{ph_w}{6\theta_0} \tag{25}$$

Here y_0 and z_0 define the initial point in the cross-plane.

The subcharacteristic behavior is shown in **Figure 5a** and **b** for $p > 0$. At $\alpha > 0$ velocities $v < 0$ and $w < 0$; the only critical point node is in the coordinate origin, and subcharacteristics go to it from the region $\zeta \ne 0$ (**Figure 5a**). At $\alpha = 0$ $y_c = 0$ and the point $\zeta = y = 0$ is double critical point of the type saddle node: the saddle is

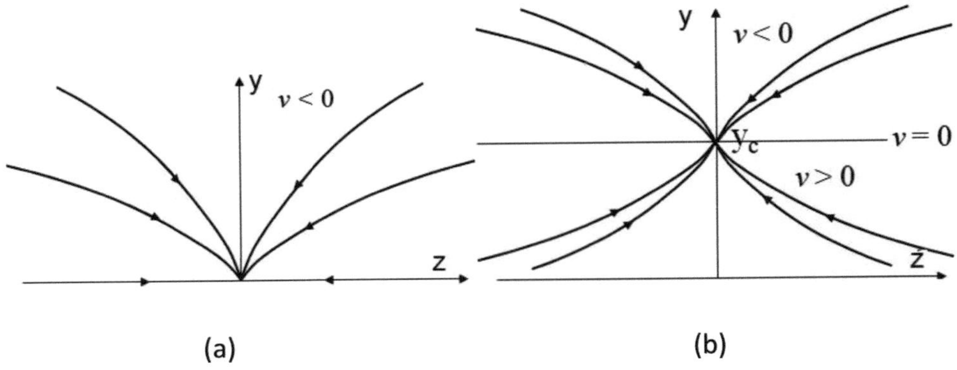

Figure 5.
Subcharacteristics in the cross-plane at $\alpha \geq 0$ (a) and $\alpha < 0$ (b); $p > 0$.

in the lower half-plane, i.e., out of the physical space. The node is in the upper half-plane, and the subcharacteristic pattern retains the same as at $\alpha > 0$. At $\alpha < 0$ the node drifts in the point $\zeta = 0, y = y_c > 0$, and the coordinate origin becomes by the saddle point (**Figure 5b**). In this case, at $y > y_c$ the normal velocity $v < 0$ and at $0 < y < y_c$ $v > 0$; $v = 0$ on the line $y = y_c$.

This analysis shows that at the parameter α sign change, the physical flow structure varies qualitatively, and the value $\alpha = 0$ is a criterion of the new flow property appearance. It should be noted that in solutions of Navier-Stokes equations for similar problems near the coordinate origin $z = y = 0$ in the leeward symmetry plane, the streamwise-oriented vortex arises, and the flow is not described by the BL theory since the viscous diffusion inside the vortex is distributed along the radius from its axis, but not along the normal to the body surface. On the base of this qualitative analysis, it is supposed that the critical value $k_c(h_w, \mathrm{M})$ is defined by the relation

$$2\alpha(k_c) = \tau_0(k_c) - k_c\theta_0(k_c) = 0 \tag{26}$$

To support this hypothesis, equations for functions $U_0(y)$ and $W_0(y)$ are analyzed by substituting near-wall decompositions to Eq. (6). Considering functions $U_0(y)$ and $W_0(y)$ as perturbations, we can linearize resulting equations and obtain in the first-order approximation:

$$\begin{aligned}
&U_{0yy} + \alpha y^2 U_{0y} + \tau_0(F_0 - kG_0) = -\alpha\tau_0 y^2,\\
&W_{0yy} + \alpha y^2 W_{0y} - \frac{2}{3}(\tau_0 - 3\theta_0)yW_0 + \theta_0(F_0 - kG_0) =\\
&\beta_0 + \beta_1 y + \frac{1}{2}\beta_2 y^2 + \left[\frac{2}{3}\theta_0 y - p(h_r - 2M_e\tau_0 y)\right]U_0
\end{aligned} \tag{27}$$

At $y \to 0$ $U_0(y)$ and $W_0(y)$ are expressed by above series, and in order to match them with the solution of full Eq. (6) in the main BL part, it is required that these functions will grow at $y \to \infty$ not faster than a power function. To study their solution behavior at $y \to \infty$ and $\alpha \neq 0$, we introduce the new variable:

$$\xi = -\alpha y^3/3, \; y = -(3\xi/\alpha)^{\frac{1}{3}}. \tag{28}$$

At the limit $\xi \to \infty$, previous equations are reduced in the first-order approximation to the form

$$\xi\frac{\partial^2 U_0}{\partial\xi^2}+\left(\frac{2}{3}-\xi\right)\frac{\partial U_0}{\partial\xi}=-\frac{\tau_0}{3}\left(\frac{3\xi}{\alpha}\right)^{\frac{1}{3}},\ c=2\frac{\tau_0-3k\theta_0}{9\alpha}$$
$$\xi\frac{\partial^2 W_0}{\partial\xi^2}+\left(\frac{2}{3}-\xi\right)\frac{\partial W_0}{\partial\xi}+cW_0=-\frac{\beta_1}{3\alpha}+\frac{\beta_2}{6\alpha}\left(\frac{3\xi}{\alpha}\right)^{\frac{1}{3}}-\frac{2}{9\alpha}(\theta_0+3M_e\tau_0)U_0 \tag{29}$$

Solutions of these equations can be represented as

$$U_0=A_{00}\int_0^y e^{-\frac{1}{3}\alpha s^3}ds+\tau_0\left(\frac{3\xi}{\alpha}\right)^{\frac{1}{3}},$$
$$W_0=B_{00}\Phi\left(-c,\frac{2}{3},\xi\right)+B_{01}\xi^{1/3}\Phi\left(\frac{1}{3}-c,\frac{4}{3},\xi\right)-\frac{3\beta_1}{2(\tau_0-3k\theta_0)}$$
$$-\frac{3\beta_2}{\tau_0-9k\theta_0}\left(\frac{3\xi}{\alpha}\right)^{\frac{1}{3}}-\frac{\theta_0+3M_e\tau_0}{\tau_0-3k\theta_0}U_{00} \tag{30}$$

First terms of these expressions are solutions of homogeneous equations, with zero right-hand sides; A_{00}, B_{00}, and B_{01} are constants; $\Phi(a,b,x)$ is Kummer's degenerate hypergeometric function, which has asymptotes at $\xi\to\infty$:

$$\alpha>0,\xi<0:\ \ \Phi\sim(-\xi)^c;\ \alpha<0,\xi>0:\ \ \Phi\sim e^{\xi}\xi^{c-2/3} \tag{31}$$

Solutions grow exponentially at $\alpha<0$ and $p>0$; they cannot be matched with the solution in the main BL part. Therefore, at these conditions a solution of BL equations cannot exist. This conclusion and also the criterion (26) for the boundary of the existing leeward symmetry plane solution are confirmed by numerical calculations for the slender round cone at an angle of attack [25–32, 37], a part of which is presented in **Figure 6**. In this figure, symbols correspond to calculations of limit

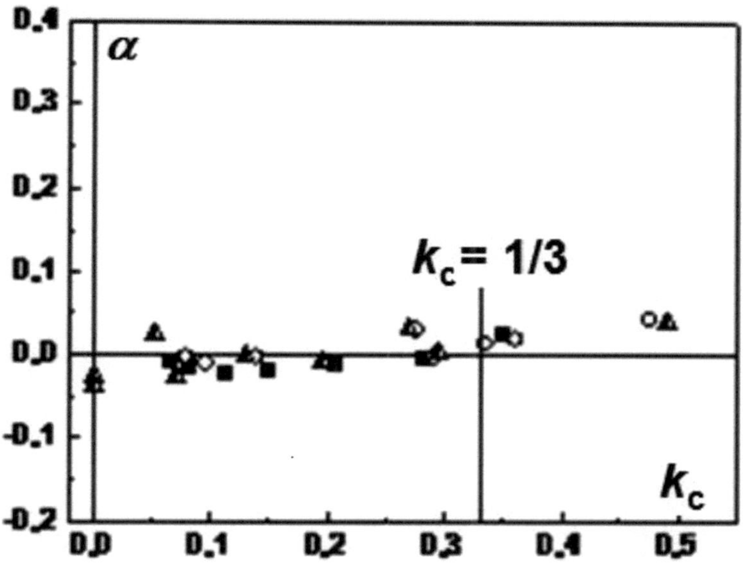

Figure 6.
The boundary of the solution existing in the leeward symmetry plane of the slender round cone at the angle of attack and Pr = 1 *in the dependence of the critical value k_c: ▲, [28]; ■, [29]; and ○, [37].*

values $\alpha(k_c)$ for the solution existing at different boundary conditions in the diapason of Mach numbers from 2 to ∞ at the Prandtl number 1 for different surface temperatures. At $k < 1/3$ data are grouped near the value $\alpha = 0$ in accordance with the criterion (26). The data scatter is, apparently, due to the decrease of the calculation accuracy at the approach to the critical value k_c and also with errors of data copying from papers. At $k > 1/3$, all calculations are finished with $\alpha > 0$, since the solution existing in this region is determined by singularities in the outer BL part, but not in the near-wall region.

Then we consider the solution behavior of full BL equations in the near-wall region beside the runoff plane at $\zeta < < 1$. 3D BL equations have the parabolic type, and their solution before the runoff plane knows nothing about the solution in this plane; however, in order for the first solution to move smoothly into the last one at $\alpha > 0$, the first will be locally self-similar. Due to this condition, the streamwise $\tau(\zeta)$ and cross-flow $\theta(\zeta)$ friction stresses and the self-similar variable η at $\zeta < < 1$ will be defined by expressions

$$\tau(\zeta) = \frac{\tau_0}{a(\zeta)},\ \theta(z) = \frac{\theta_0}{a(\zeta)},\ \eta = \frac{y}{a(\zeta)} \tag{32}$$

The function $a(z)$ at $\alpha \geq 0$ will satisfy to the condition $a(0) = 1$. In this case, flow functions in the boundary layer near the wall can be represented in the form

$$\begin{aligned}
&f(\eta,\zeta) = a(\zeta)\left[\tau_0\frac{\eta^2}{2} + F(\eta,\zeta)\right],\ u(\eta,\zeta) = f_\eta = \tau_0\eta + U(\eta,\zeta)\\
&g(\eta,\zeta) = a(\zeta)\left[\theta_0\frac{\eta^2}{2} + G(\eta,\zeta)\right],\ w(\eta,\zeta) = g_\eta = \theta_0\eta + W(\eta,\zeta)\\
&v = a\left[\left(\alpha - \frac{1}{2}\theta_0 k\zeta\frac{a_\zeta}{a}\right)\eta^2 + F - kG\left(1 + k\zeta\frac{a_\zeta}{a}\right) - k\zeta G_\zeta - k\zeta\eta_\zeta W\right]
\end{aligned} \tag{33}$$

Substituting these expressions to Eq. (6) and linearizing the result with respect to disturbances, we obtain the first-order approximation for the flow in the near-wall region beside the runoff plane:

$$\begin{aligned}
&U_{\eta\eta} + \alpha\eta^2 U_\eta + a^2\left\{k\theta_0\zeta\eta U_\zeta + \tau_0\left[F - kG\left(1 + \frac{a_\zeta}{a}\right) - k\zeta G_\zeta\right]\right\} = -\alpha\tau_0\eta^2\\
&W_{\eta\eta} + \alpha\eta^2 W_\eta + a^2\left\{k\theta_0\zeta\eta W_\zeta + \theta_0\left[F - kG\left(1 + \frac{\zeta a_\zeta}{a}\right) - k\zeta G_\zeta\right] - 3\alpha c\eta W\right\} =\\
&-\alpha\theta_0\eta^2 + a^2\left\{\beta_0 + \beta_1\eta + \frac{1}{2}\beta_3\eta^2 + \left[\frac{2}{3}(\theta_0 + 3pM_e\tau_0)\eta - ph_r\right]U\right\}
\end{aligned} \tag{34}$$

Here $\beta_3 = \frac{2}{3}\tau_0\theta_0 - k\theta_0^2 + pM_e\tau_0^2$. Due to local self-similarity at $\alpha \geq 0$, we define the function $a(\zeta)$ as

$$\alpha a^2 - \frac{1}{2}k\theta_0\zeta a a_\zeta = \alpha,\ a^2 = 1 + C\zeta^q,\ q = \frac{4\alpha}{k\theta_0} \tag{35}$$

The constant C is found from a comparison with numerical calculations. It follows from this relation at $\alpha \geq 0$ and $q < 2$ the solution of Eq. (6) in the near-wall region at $\zeta < < 1$ can find in the form of the series:

$$F(\eta,\zeta) = F_0(\eta) + \zeta^q F_q(\eta) + \ldots,\ U(\eta,\zeta) = U_0(\eta) + \zeta^q U_q(\eta) + \ldots,$$
$$G(\eta,\zeta) = G_0(\eta) + \zeta^q G_q(\eta) + \ldots,\ W(\eta,\zeta) = W_0(\eta) + \zeta^q W_q(\eta) + \ldots, \quad (36)$$

The first term of this expansion is the solution for the runoff plane but depends on the self-similar variable. Second terms define the proper solution of BL Eq. (6) at $\zeta << 1$, which at $\eta \to \infty$, has the form [37]

$$\xi = -\frac{\alpha\eta^3}{3},\ U_q(\xi) = A_{q0}\Phi\left(\frac{4}{3},\frac{2}{3},\xi\right) + A_{q1}\xi^{\frac{1}{3}}\Phi\left(\frac{5}{3},\frac{4}{3},\xi\right)$$
$$W_q(\xi) = B_{q0}\Phi\left(\frac{4}{3}-c,\frac{2}{3},\xi\right) + B_{q1}\xi^{1/3}\Phi\left(\frac{5}{3}-c,\frac{4}{3},\xi\right) + \frac{9\beta_1}{2\tau_0} + \frac{3\beta_2}{\frac{11}{3}\tau_0 - k\theta_0}\left(\frac{3\xi}{\alpha}\right)^{\frac{1}{3}}$$
$$+\frac{\theta_0 + 3M_e\tau_0}{2\tau_0}U_0 - \frac{\theta_0 + 3M_e\tau_0}{\tau_0 - 3k\theta_0}U_q. \quad (37)$$

Here A_{q0}, A_{q1}, B_{q0}, and B_{q1} are constants. These relations show that the proper solution in near-wall BL region near the runoff plane is nonzero. It is irregular at $\alpha \geq 0$ and it is singular at $\alpha < 0$. The logarithmic singularity is not in this case, and the solution of BL equations exists at the critical value k_c in contrast to the outer region.

In the work of [15], at the analysis of perturbations in the boundary layer related with the angle of attack, it was found that they lead to infinite disturbances in the symmetry plane, although equations have no visible singularities contained. In this case, the first-order approximation is described by the Blasius solution for the delta flat plate. In **Figure 7**, dimensionless longitudinal and transverse skin friction distributions $f_1''(z)$ and $g_1''(z)$, induced by the second order BL approximation (**Figure 7a**) and the angle of attack (**Figure 7b**) are presented in dependence on transverse coordinate $z = 1 - Z/X$, where X and Z are Cartesian streamwise and transverse coordinates. By approaching the symmetry plane ($z = 1$), skin friction perturbations infinitely grow. Detailed investigation of equations for these functions showed that in these cases singularities take place as in the near-wall and outer BL parts. In the outer part, the singularity corresponds to values of the parameter

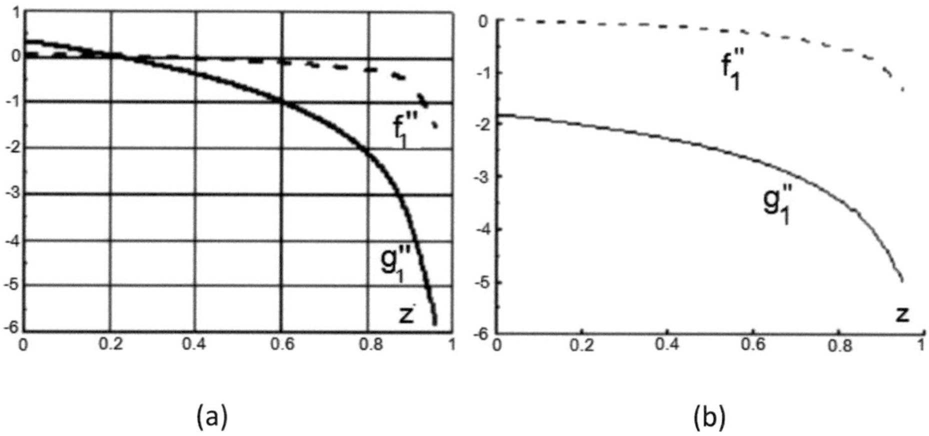

Figure 7.
Skin friction distributions on the small aspect ratio delta wing at M = 2 related with (a) second BL approximation and (b) angle of attack.

m = 3/4 and 7/8 in relation to cases a and b, respectively. The longitudinal velocity perturbation singularity is related only with the near-wall singularity.

Near-wall singularities generate the flow structure including three asymptotic sublayers describing the viscous-inviscid interaction similar as near the 2D separation point. However, the viscous-inviscid interaction is not enough to remove the singularity of the obtained type. Near the wall sublayer close to the symmetry plane the fourth region is formed, in which the flow is described by the parabolized Navier-Stocks equations similar to the above case of the outer singularity.

4. Studies of the symmetric flow instability over thin bodies and the control possibility on the base of the interaction model of 3D boundary layer with the electrical discharge

The electric discharge is considered as one of effective methods for control of the flow asymmetry over bodies [23–27]. However, to select optimal control parameters, it needs to have a reasonable criterion for the asymmetry origin and a possibility for fast estimation of the control effect. For the second problem, the model of the boundary layer and discharge interaction is proposed. The scheme of this model is shown in **Figure 8** [28–31, 37].

It is assumed the plasma discharge effect can be modeled by the heat source in the boundary layer. The effect of gas ionization is neglected since the ionization coefficient is of the order of 10^{-5}. This source in the energy equation is presented by formulas:

$$Q = \frac{Q^* xl}{h_\infty u_\infty} = Q_0 y^2 \exp\left[-\frac{(y - y_c(\varphi))^2}{\sigma}\right], \quad y_c = 2y_0\sqrt{|(\varphi - \varphi_1)(\varphi_2 - \varphi)|} \quad (38)$$

Here Q^* is a dimensional source intensity; Q_0 is a maximum of dimensionless heat-release intensity; σ characterizes the discharge width; $y_c(\varphi)$ is a centerline of the discharge that is approximated by the parabola; y_0 is a maximum distance from the discharge centerline to the wall; and the angles φ_1 and φ_2 determine the electrode locations.

Calculations of the turbulent boundary layer characteristics were conducted using the method [10] for a slender cone of half-apex angle $\delta_c = 5°$ at the angle of attack $\alpha = \alpha^*/\delta_c = 3.15$. Other parameters are: $l = 1$ m, $T_\infty = 288$ K, $u_\infty = 10$ m/s, $\sigma = 1$, and $y_0 = 1$; the center between electrodes is located at $\phi_0 = 0.5(\phi_1 + \phi_2) =$ 1.714 rad (98.25°), $\phi_1 = \phi_0 - 3\Delta\phi$, and $\phi_2 = \phi_0 + 3\Delta\phi$, where $\Delta\phi = 0.0314159$ is the integration step of the finite-difference approximation.

In **Figure 9**, the dimensionless enthalpy (**Figure 9a**) and circumferential velocity (**Figure 9b**) profiles across the boundary layer are shown as functions of η for $Q_0 = 200$ and for different polar angles φ. These profiles are similar to the source

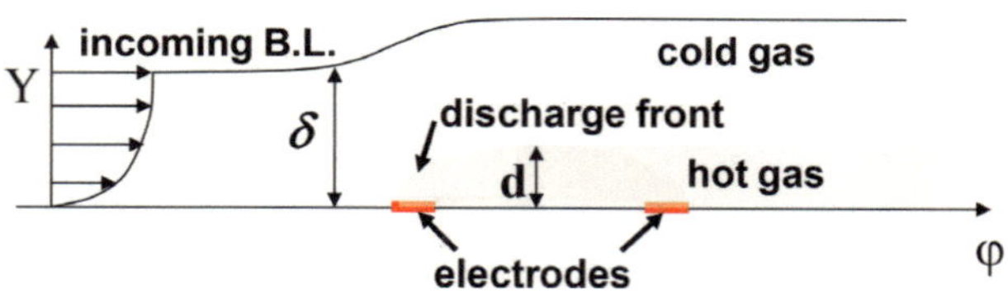

Figure 8.
A scheme of discharge interaction with the boundary layer.

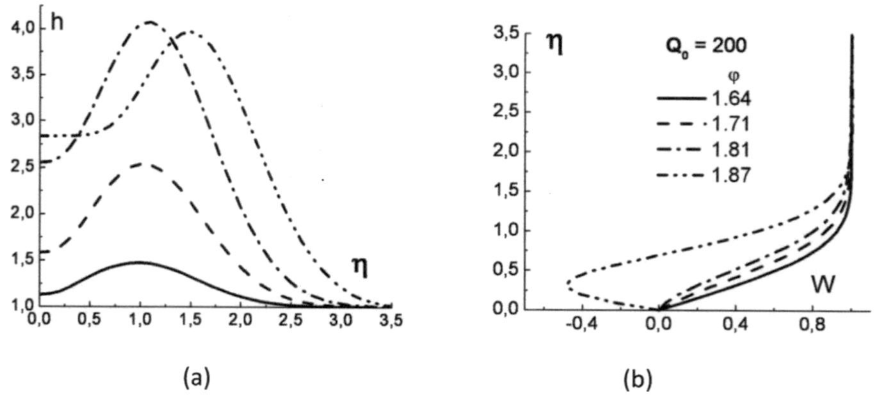

Figure 9.
Profiles of temperature (a) u and circumferential velocity (b) across the boundary layer near the heat-release region.

heat intensity distribution across the boundary layer. The temperature reaches the maximum value near the rear electrode, $\varphi = \varphi_2 = 1.809$. Behind the heat source region, the temperature maximum decreases and moves toward the upper boundary-layer edge due to the heat diffusion. The station $\varphi = 1.87$ is located just after the separation point

Figure 10a demonstrates the plasma discharge effect on the separation point. As the heat source intensity increases from 0 to 400, the separation angle, φ_s, decreases from 133° to about 105°. It is seen that the plasma heating is more effective in the range $Q_0 < 100$, where the slope $d\varphi_s/dQ_0$ is relatively large.

Figure 10b illustrates feasibility of the vortex structure control using a local boundary-layer heating on the base of the developed criterion of symmetric flow stability (solid line). Due to the heat release, the flow configuration changes from the initial asymmetric state ($\varphi_s \approx 133°$, symbol 1) to the symmetric state with $\theta_s \approx 120°$ (symbol 2). This requires a nondimensional heat source intensity $Q_0 \approx 30$ that corresponds to the total power which is approximately equal to 480 W. This example indicates that the method is feasible for practical applications of the global flow structure control.

The method of the global flow stability was developed [27–31] using the asymptotic approach for the flow over slender cones, the separated inviscid flow model [34] and the stability theory of autonomous dynamical systems [35]. Comparison of the calculated criteria for different elliptic slender cones with experimental data for laminar and turbulent boundary layers sowed its efficiency.

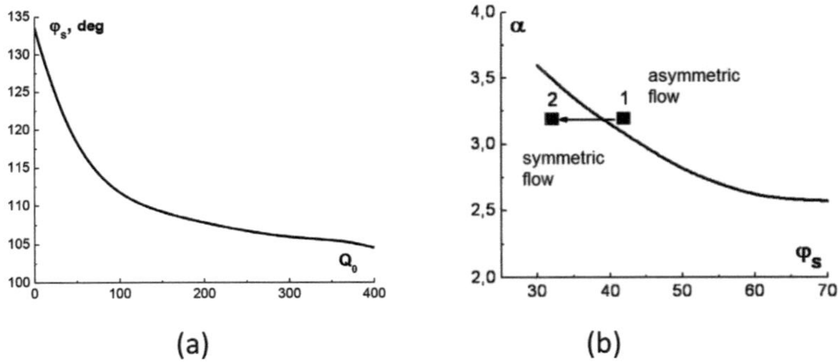

Figure 10.
Discharge effect on the separation angle (a) and flow state (b).

5. Investigations of abnormal features of the heat transfer and the laminar-turbulent transition for hypersonic flows around flat delta wing with blunted leading edges

Although found in the experimental zones of abnormal high heat fluxes on the windward flat surface of the half cone with blunted nose and delta wings with blunted leading edges, the phenomenon of the early laminar-turbulent transition [38–46] cannot be explained in frameworks of the boundary layer theory and on the base of solutions of parabolized Navier-Stocks equations. Only detailed flow simulations using full Navier-Stocks equations allowed to find reasons of such anomalies [46–48].

Figure 8 shows the comparison of calculated (the upper part) and experimental (the lower part) heat flux distributions on the delta wing with the leading edge sweep angle $\chi = 75°$, the bluntness radius of cylindrical edges and the spherical nose R = 8 at the angles of attack $\alpha = 0°$, M = 6, unit Reynolds numbers $Re_1 = 1.1556 \times 10^6\ m^{-1}$ [47, 48]. Similar patterns were obtained in numerical simulations for different Reynolds numbers and Mach numbers up to 10.5 [46]. At moderate Mach numbers, a flow on such simple surface outside the nose and leading edge regions is described very well by the flat plate approximation and has no anomalies.

At hypersonic speeds, high heat flux regions, which is present in **Figure 11**, are observed in the middle wing span and near the symmetry plane. It is seen that the experimental middle high heat flux streak is finished by the turbulent wedge. Calculations were conducted only for the laminar flow.

To understand the reason for the heat flux anomaly, the cross-flow pattern helps (**Figure 9**). Three longitudinal vortexes are in this flow. The largest vortex is in the inviscid region above shock (the dark layer) and boundary (the light layer) layers. Vortex near the symmetry plane and in the middle of the span occupies both layers. Its mutual location depends on the blunt radius, Mach, and Reynolds numbers [43, 46]. For the considered case, the middle vortex is above the high heat flux region that is shown below the cross-flow pattern (**Figure 12**).

The analysis shows that high heat flux streaks are formed by the convective transfer of heat gas from the shock layer to the wing surface by the gas rotation inside the vortex. In the considered case, the middle vortex is formed before the symmetry plane vortex near the nose in the narrowing flow region between the head shock and the leading edge due to the cross-flow acceleration near the leading edge and the induced pressure gradient related with the domed flow structure near the symmetry plane.

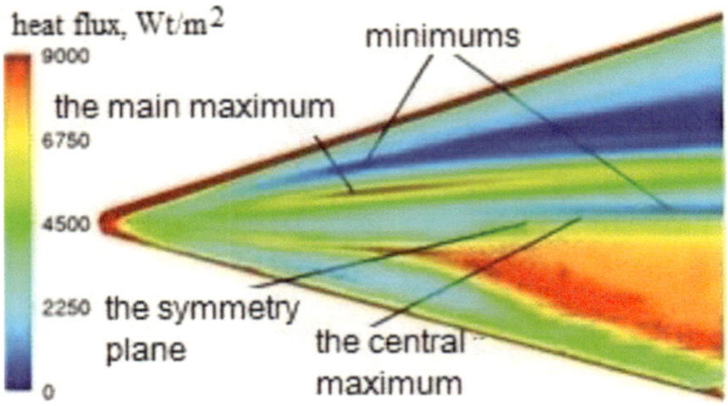

Figure 11.
Comparison of numerical (the upper part) and experimental (the lower part) specific heat flux distribution on the wing surface.

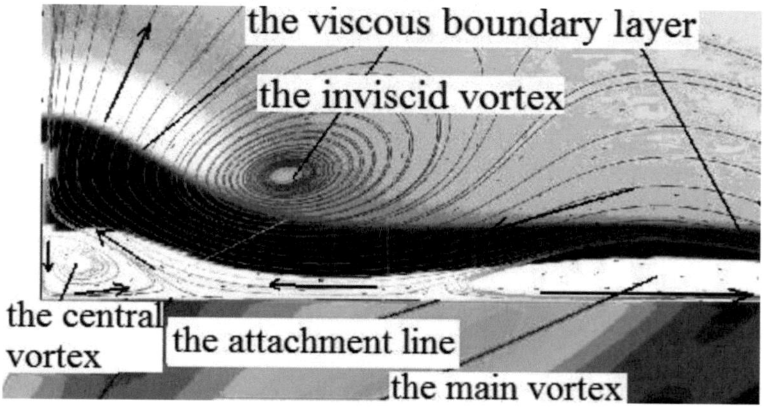

Figure 12.
The cross-flow structure above the wing in the section X = 0.1 m.

In considered conditions, the middle vortex also is the reason for the laminar-turbulent transition. Formed along the vortex center, streamwise velocity profiles have inflection points that lead to the Rayleigh instability development. Transverse velocity profiles along this line have the S-shaped form that leads to the cross-flow instability. Both these processes result to the more early transition than Tollmien-Schlichting wave evolution.

6. Conclusions

In this work, the short review of researches on the study of BL equation singularities, which are formed when two streamline families are collided, is presented. This phenomenon can arise only in unsteady and 3D problems and has no analogue in 2D flows. A typical example of such problem is the flow around a slender cone in the vicinity of the runoff plane. In this case, solutions are found in the analytical form that allows to analyze explicitly the singularity character.

The analysis of solutions for the outer flow part revealed two singularity types. One type is in streamwise and cross-velocity viscous perturbations; it arises at values of relative cross pressure gradient $k \geq 1$ and leads to the exponential disturbance growth as the runoff plane is approached. At $k = 1$ the singularity is logarithmic and at $k > 1$ it is power; its appearance is correlated with the BL separation appearance. Another singularity type at smaller values of $k \geq 1/3$ in the first-order approximation leads to the infinite growth of transverse velocity perturbations only and is not related directly with the flow separation; at $k = 1/3$ the singularity is logarithmic, and at $k > 1/3$ it is power. These BL singularities correspond to some asymptotic flow structure at $\mathrm{Re} \gg 1$. This structure includes the boundary region with the dimension of the order of the BL thickness, in which the viscous transverse diffusion effect smoothes the singularity. The comparison of obtained parabolized Navier-Stokes equation solutions describing the flow in the boundary region with BL equations solutions confirms this conclusion. Second region induced by the viscous-inviscid interaction effect has the transverse dimension of the order of square root from the BL thickness and the two-layer structure. For the potential flow in the outer inviscid subregion, the integral solution representation is found on the base of the slender wing theory. The inner subregion is described by full 3D BL equations, the solution of which is obtained for the outer viscous subregion part. It was shown that the viscous-inviscid interaction does not eliminate the singularity but drifts it in the parametric space. To eliminate the irregularity, the boundary region is needed.

To find the dependence of the critical parameter of the singularity appearance k_c on Mach and Prandtl numbers and the wall temperature BL equations, solutions are studied in the near-wall region beside the runoff plane. Equation subcharacteristic (streamlines) analysis showed the presence of one parameter α, the sing of which defines the qualitative change of the streamline topology and, consequently, the physical flow structure. It is shown and confirmed by comparison with all available calculations that the boundary of the solution which exists in the runoff plane corresponds to the criterion $\alpha(k_c) = 0$. The analysis of BL equation solutions near the runoff plane revealed the presence at $\alpha \geq 0$ of irregular and at $\alpha < 0$ singular proper solutions. This is confirmed by numerical calculations of the flow around slender delta wing with the small aspect ratio. Singularities in the near-wall region generate the some flow structure in its vicinity, the study of which is out of this paper framework. Presented results do not depend on outer boundary conditions and are true for the full freestream velocity diapason including hypersonic flows.

Presented research allows concluding that the flow in symmetry planes, for example, on wings, has the complex structure, which is needed to take into account the numerical modeling in order to eliminate the accuracy loss. Regular flow function decompositions commonly used at solutions of BL equations are not applied near this plane, and it cannot be considered as a boundary condition plane due to a possible solution disappearance.

Author details

Vladimir Shalaev
Moscow Institute of Physics and Technology, Zhukovsky, Russia

*Address all correspondence to: vi.shalaev@yandex.ru

References

[1] Mager A. Three-dimensional boundary layer with small cross-flow. Journal of the Aeronautical Sciences. 1954;**21**:835-845

[2] Cebeci T, Cousteix J. Modeling and Computations of Boundary Layer Flows. Long Beach; Berlin, Heidelberg, N. Y: Horizons Publishing Inc.; Springer; 2005, 502p

[3] Lunev VV, Senkevich EA. Method of meridional sections in problems of a three-dimensional boundary layer. Fluid Dynamics. 1986;**21**:394-400

[4] Shalaev VI. Boundary layer on slender wings in a supersonic gas flow. In: Numerical Methods of Continuum Mechanics. Vol. 17. Novosibirsk: Siberian Branch of the Academy of Sciences of the USSR; ITPM; 1986. pp. 127-134, N 5: [in Russian]

[5] Shalaev VI. Three-dimensional boundary layer on slender wings and bodies at small angles of attack. In: Modeling in Continuum Mechanics. Novosibirsk: Siberian Branch of the Academy of Sciences of the USSR; ITPM; 1988. pp. 148-153. [in Russian]

[6] Khonkin AD, Shalaev VI. Three-dimensional boundary layer on slender wings of finite span. Reports of USSR Academy of Sciences. 1989;**307**(2): 312-315. [in Russian]

[7] Khonkin AD, Shalaev VI. Three-dimensional boundary layer on bodies with slight cross-section asymmetry at small angles of attack. Reports of USSR Academy of Sciences. 1990:**313**(5): 1067-1071 . [in Russian]

[8] Shalaev VI. Boundary layer on slender small span wing. Journal of Applied Mechanics and Technical Physics. 1992;**1**:71-78

[9] Shalaev VI. Unsteady boundary layers with small cross flows. Fluid Dynamics. 2007;**42**(3):398-409

[10] Shalaev VI. Application of Analytical Methods in Modern Aeromechanics. Part. 1. Boundary Layer Theory. Moscow: MIPT; 2010. 300p. (In Russian)

[11] Cebeci T, Kaups X, Ramsey JA. A General Method for Calculating Three-Dimensional Compressible Laminar and Turbulent Boundary Layers on Arbitrary Wings. NASA CP; 1977. N 2777

[12] Brown SN, Stewartson K. On the propagation of disturbances in a laminar boundary layer I. Mathematical Proceedings of the Cambridge Philosophical Society. 1973;**73**:493-503

[13] Stewartson K. The Theory of Laminar Boundary Layers in Compressible Fluids. Oxford: Clarendon Press; 1964

[14] Stewartson K, Simpson CJ. On a singularity initiating a boundary layer collision. Quarterly Journal of Mechanics and Applied Mathematics. 1982;**35**:1-16

[15] Williams JC. Singularities in solution of three-dimensional boundary layer equations. Journal of Fluid Mechanics. 1985;**160**:257-279

[16] Bashkin VA. On uniqueness of self-similar solutions of three-dimensional laminar boundary layer equations. Izvestia AS USSR. MZhG. 1968;**5**:35-41

[17] Wu P, Libby PA. Laminar boundary layer on a cone near a plane of symmetry. AIAA Journal. 1973;**11**(3): 326-333

[18] Murdock JW. The solution of sharp cone boundary layer equations in the plane of symmetry. Journal of Fluid Mechanics. 1972;**54**:665-678. Pt 4

[19] Cousteix J. Houdeville R. Singularities in three-dimensional

turbulent boundary-layer calculations and separation phenomena. AIAA Paper. 1981;1201

[20] Rubin SG, Lin TC, Tarulli F. Symmetry plane viscous layer on a sharp cone. AIAA Journal. 1977;15(2): 204-211

[21] Shalaev VI. Singularities in the boundary layer on a cone at incidence. Fluid Dynamics. 1993;**6**:25-33

[22] Shalaev VI. Singularities of 3D laminar boundary layer equations and flow structure near a sink plane on conical bodies. Fluid Dynamics. 2007; **42**(4):560-570

[23] Shalaev VI. Singularities in laminar boundary layer and flow structure near sink plane on conical bodies. In: 29th ICAS Congress, St. Petersburg, September 7—12, 2014. Available from: http://www.icas.org/ICAS_ARCHIVE/ICAS2014/data/papers/2014_0675_paper.pdf

[24] Shalaev VI. Singularities of 3D laminar boundary layer equations and flow structure in their vicinity on conical bodies. In: International Conference on the Methods of Aerophysical Research. AIP Conference Proceedings. Vol. 1770. 2016. pp. 030055-1-030055-14

[25] Stewartson K. Is the singularity at separation removable? Journal of Fluid Mechanics. 1970;**44**:347-364, part 2

[26] Lowson MV, Ponton AJC. Symmetric breaking in vortex flows on conical bodies. AIAA Journal. 1992; **30**(6):1576-1583

[27] Kumar A, Hefner JN. Future challenges and opportunities in aerodynamics. In: 22th ICAS Congress, Harrogate, 28 August–1 September, 2000. pp. 0.2.1-0.2.14

[28] Meng X, Wang J, Cai J, Liu F, Luo S. Effect of plasma actuation on asymmetric vortex flow over a slender conical forebody. AIAA Paper. 2012;**0287**

[29] Séraulic A, Pailhas G. Implementation of DBD plasma actuators to control boundary layers in subsonic flows. International Journal of Aerodynamics. 2013;**3**(1):3-25

[30] Shalaev V, Fedorov A, Malmuth N, Zharov V, Shalaev I. Plasma control of forebody nose symmetry breaking. AIAA Paper. 2003;**0034**

[31] Shalaev V, Fedorov A, Malmuth N, Shalaev I. Plasma control of forebody nose symmetry breaking. AIAA Paper. 2004;**0842**

[32] Shalaev VI, Shalaev IV. A stability criterion of symmetric separated flow over slender bodies and flow control by volumetric and surface gas heating. In: XV International Conference on the Methods of Aerophysical Research (ICMAR). Abstracts. Vol. 1. Novosibirsk: Parallel; 2010. pp. 223-224

[33] Shalaev VI, Shalaev IV. On stability of symmetric vortex structures at flow over bodies. Bulletin of Nizhny Novgorod University. 2011;**4**:1265-1266, part 3

[34] Reijasse P, Knight D, Ivanov M, Lipatov I, editors. EUCASS book series. Progress in flight physics. Chapter 4. Vortex, wakes and base flows. Shalaev VI, Shalaev IV. Stability of Symmetric Vortex Flow over Slender Bodies and Possibility of Control by Local Gas Heating. Paris: EDP Sciences. 2013;**5**: 155-168. Doi:10.1051/eucass/201305155

[35] Dyer DE, Fiddes SP, Smith JHB. Asymmetric vortex formation from cone at incidence—A simple inviscid model. Aeronautical Quarterly. 1982; **3**(6):293-312

[36] Gilmore R. Catastrophe Theory for Scientists and Engineers. New York: Wiley; 1981: 350p

[37] Maslov A, Zanin B, Sidorenko A, Malmuth N, et al. Plasma control of separated flow asymmetry on a cone at high angle of attack. AIAA Paper. 2004; **0843**

[38] Borovoi VY, Davlet-Kildeev RZ, Ryzhkova MV. On heat exchange peculiarities on the surface of some lifting bodies at large supersonic speeds. Fluid Dynamics. 1968;**1**

[39] Kondratiev IA, Yushin AY. On local heat flux increasing on the windward surface of a delta wing with blunted leading edges. In: Aerothermodynamics of Aerospace Systems; Report Collection of TsAGI School-Seminar "Mechanics of Liquid and Gas". Vol. 1. Zhukovsky: TsAGI; 1990. pp. 167-175

[40] Gubanova OI, Zemlyansky BA, Lesin AB, Lunev VV, Nikulin AN, Syusin AV. Abnormal heat exchange on the winward side of the delta wing with the blunted nose at hypersonic speeds. In: Aerothermodynamics of Aerospace Systems; Report Collection of TsAGI School-Seminar "Mechanics of Liquid and Gas". Vol. 1. Zhukovsky: TsAGI; 1990. pp. 188-196

[41] Kovaleva NA, Kolina NP, Yushin AY. Experimental investigation of heat flux and laminar-turbulent transition on half delta wing models with blunted leading edge in supersonic flow. Uchenye Zapiski Tsentralnogo Aero-Gidrodinamicheskogo Instituta (TsAGI). 1993;**24**(3):46-52

[42] Lesin AB, Lunev VV. On peak heat fluxes on delta flat plate with blunted nose in hyersonic flow. Fluid Dynamics. 1994;**2**:131-137

[43] Bragko VN, Vaganov AV, Dudin GN, Kovaleva NA, Lipatov II, Skuratov AS. Experimental investigation of delta wing aerodynamic heating pecularities at high Mach numbers. Proceedings of MIPT. 2009;**1**(3):57-66

[44] Vaganov AV, Yermolaev YG, Kosinov AD, Semenov NV, Shalaev VI. Experimental investigation of flow structure and transition in boundary layer on delta wing with blunted leading edges at Mach numbers 2, 2, 5 and 4. Proceedings of MIPT. 2013;**3**(19):164-173

[45] Vlasov VI, Gorshkov AB, Kovalev RV, Lunev VV. Thin delta flat plat with blunted nose in viscous hypersonic flow. Fluid Dynamics. 2009;**4**:134-145

[46] Bragko VN, Vaganov AV, Neyland VY, Starodubtsev MA, Shalaev VI. Simulation of flow peculiarities on delta wing windward side with blunted leading edges on the base of Navier-Stocks equation numerical solution. Proceedings of MIPT. 2013;**5**(2):13-22

[47] Aleksandrov SV, Vaganov AV, Shalaev VI. Physical mechanisms of longitudinal vortexes formation, appearance of zones with high heat fluxes and early transition in hypersonic flow over delta wing with blunted leading edges. In: International Conference on the Methods of Aerophysical Research, Permision from AIP Conference Proceedings. Vol. 1770. 2016. pp. 020011-1-020011-10. DOI: 10.1063/1.4963934

[48] Aleksandrov SV, Vaganov AV, Shalaev VI. Physical mechanisms of inc reased heat flux zone appearance and la minar-turbulent transition in the boundary layer on blunted delta wings at high freestream velocities. In: Proceedings of 7th European Conference for Aeronautics and Aerospace Sciences. Milan; 2017. p. 082. DOI: 10.13009/EUCASS2017-81

Chapter 2

TBL-Induced Structural Vibration and Noise

Zhang Xilong, Kou YiWei and Liu Bilong

Abstract

One of most import noise sources in a jet powered aircraft is turbulent boundary layer (TBL) induced structural vibration. In this chapter, the general model for the prediction of TBL-induced plate vibration and noise is described in detail. Then numerical examples for a typical plate are illustrated. Comparisons of plate vibration and radiated noise between numerical results and wind tunnel test are presented. The effects of structural parameters on modal-averaged radiation efficiency and therefore the radiated noise are discussed. The result indicates that an increment of flow velocity will increase the acoustic radiation efficiency below the hydrodynamic coincidence frequency range. The main reason for this phenomenon is that a higher convection velocity will coincide with lower order modes which have higher radiation efficiencies.

Keywords: turbulent boundary layer, plate vibration, radiated noise, modal radiation efficiency

1. Introduction

The interior noise level in a jet aircraft is mainly depend on noise which generated by turbulent boundary layers (TBL), if the rest of noise sources such as ventilation systems, fans, hydraulic systems, etc. have been appropriately acoustically treated. When the aircraft passes through the atmosphere, the turbulent boundary layer creates pressure fluctuations on the fuselage. These pressure fluctuations cause the aircraft fuselage to vibrate. The noise generated by the vibration is then transmitted to the cabin.

The noise emitted by the aircraft fuselage depends on the speed of the vibrating plate, which in turn depends on the speed of the aircraft, the geometry and size of the plates, and the loss or damping of the plates. It is obvious that the acoustic performances of the internal system, trim panels etc., will also affect the noise inside the aircraft. Graham [1] came up with a model in aircraft plates to predict TBL induced noise, in which the modal excitation terms were calculated by an analytical expression. In Graham's another research [2], the advantages of various models describing the cross power spectral density induced by a flow or TBL across a structure was discussed. Han et al. [3] tried to use energy flow analysis to predict the noise induced by TBL. The method can better predict the response caused by the TBL excitation. However, the noise radiation caused by the flat panel cannot be predicted well. To avoid this deficiency, Liu et al. [4–6] described a model to predict TBL induced noise for aircraft plates. In their work, the modal excitation terms and acoustical radiation efficiency can be predicted properly and the predicted results are also compared with that of the wind tunnel and in-flight test. Rocha and Palumbo [7] further investigated the sensitivity of sound power radiated by aircraft panels to TBL parameters, and discussed the findings by Liu [4] that ring stiffeners may increase TBL induced noise radiation significantly.

The radiation efficiency of a plate plays an important role in vibro-acoustic problems. In recent related research, the sound medium around the fuselage of the aircraft is often considered to be stationary. Under this assumption, Cremer and Heckl [8] used a more concise formula to predict the acoustic radiation efficiency of an infinite plate. Wallace [9] derived an integral formula based on far-field acoustic radiation power to calculate the modal acoustic radiation efficiency of a finite plate. Kou et al. [10] proposed modifications to the classical formulas given by Cremer and Leppington, regarding the influence of structural damping on the radiation efficiency.

A comparison of the acoustic radiation of the plate with stationary fluid and convective fluid-loaded can be found in [11–13]. Graham [11] and Frampton [12] studied the influence of the mean flow on the modal radiation efficiency of a rectangular plate. They found that at high speeds, as the modal wave moves upstream, the increase of flow velocity would reduce the modal critical frequency. As a consequence, the acoustics radiation efficiency under the critical frequency of the plate would be higher. Kou et al. [13] also conducted a research for the effect of convection velocity in the TBL on the radiation efficiency. Kou et al. found that the modal averaged radiation efficiency will increase with the increase of the convection velocity below the hydrodynamic coincidence frequency. The study also showed that the increase of the structural loss factor could increase the modal average radiation efficiency at the subcritical frequencies, and the damping effect increases with the increase of the flow velocity.

For a plate subjected to a TBL fluctuation, although a large amount of research work used experimental and computational methods for the vibro-acoustical properties of plates, it is worth a chapter to introduce the prediction model and summarize recent findings for TBL induced plate vibrations and noise radiations. The following sections begins with a description of models for the wavenumber-frequency spectrum of TBL, and then a specific presentation of the calculation of vibro-acoustic responses of the wall plate excited by TBL is followed. In the end, the effect of flow velocity (M_c) and structural damping on the modal averaged radiation efficiency is discussed.

2. Models for the wavenumber-frequency spectrum of turbulent boundary layer fluctuating pressure

As for the research about wavenumber-frequency spectrum of turbulent boundary layer, Corcos [14], Efimtsov [15], Smolyakov-Tkachenko [16], Williams [17], Chase [18, 19] and other researchers put up with a series of widely used of wavenumber-frequency spectrum model. The models are established according to a large number of experimental data and statistical theory of turbulence. The following parts introduce some typical wavenumber-frequency spectrum models.

2.1 The Corcos model

The model proposed by Corcos during the last few decades has been widely used for many different types of problems. The model is applicable in the immediate neighborhood of the so-called convective ridge [20], as long as $\omega\delta/U_\infty > 1$. In this expression δ is the thickness of the boundary layer and U_∞ the velocity of the flow well away from the structure. The flat-plate boundary layer is taken to lie in the x-y plane of a Cartesian coordinate system, with mean flow in the direction of the x-axis. Corcos assumes that the cross power spectral density, between the pressures at two different positions separated by the vector n can be expressed as

$$S_{pp}\left(\xi_x, \xi_y, \omega\right) = \Phi_{pp}(\omega) \exp\left(-\gamma_1 k_c |\xi_x|\right) \exp\left(-\gamma_3 k_c \left|\xi_y\right|\right) \exp\left(-jk_c \xi_x\right) \tag{1}$$

where $\Phi_{pp}(\omega)$ is the auto-power spectral density of turbulent boundary layer fluctuating pressure, $k_c = \omega/U_c$ is the convection wave number. γ_1 and γ_3 can be obtained by fitting experimental data, γ_1 and γ_3 are 0.11–0.12 and 0.7–0.12 respectively for smooth rigid siding.

The Fourier Transform of ξ_x and ξ_y can obtain wavenumber-frequency spectrum

$$\begin{aligned} S_{pp}\left(k_x, k_y, \omega\right) &= \iint S_{pp}\left(\xi_x, \xi_y, \omega\right) \exp\left[j\left(k_x \xi_x + k_y \xi_y\right)\right] d\xi_x d\xi_y \\ &= \Phi_{pp}(\omega) \frac{2\gamma_1 k_c}{\left(k_x - k_c\right)^2 + \left(\gamma_1 k_c\right)^2} \cdot \frac{2\gamma_3 k_c}{k_y^2 + \left(\gamma_3 k_c\right)^2} \end{aligned} \tag{2}$$

So, the normalized wavenumber-frequency spectrum in wavenumber domain is

$$\begin{aligned} \hat{S}_{pp}\left(k_x, k_y, \omega\right) &= \frac{k_c^2}{\Phi_{pp}(\omega)} S_{pp}\left(k_x, k_y, \omega\right) \\ &= \frac{2\gamma_1}{\left(k_x/k_c - 1\right)^2 + \gamma_1^2} \cdot \frac{2\gamma_3}{\left(k_y/k_c\right)^2 + \gamma_3^2} \end{aligned} \tag{3}$$

2.2 The generalized Corcos model

Caiazzo and Desmet [21] proposed a generalized model which based on the Corcos model. The model uses butterworth filter to replace exponential decay of x and y direction in the Corcos model. It can make the wavenumber-frequency spectrum attenuation rapidly near the convection wave number by adjusting the parameters. Expression of this model is as follows

$$\begin{aligned} S_{pp}\left(\xi_x, \xi_y, \omega\right) &= -\Phi_{pp}(\omega) \sin(\pi/2P) \sin(\pi/2Q) \exp\left(-jk_c \xi_x\right) \\ &\times \sum_{p=0}^{P-1} \exp\left[j\left(\theta_p + \gamma_1 k_c |\xi_x|\right)\right] \times \sum_{q=0}^{Q-1} \exp\left[j\left(\theta_q + \gamma_1 k_c |\xi_x|\right)\right] \end{aligned} \tag{4}$$

where $\theta_p = (\pi/2P)\cdot(1 + 2p)$, $\theta_q = (\pi/2Q)\cdot(1 + 2q)$. When $P = Q = 1$, Eq. (4) is equal to the Corcos model.

Analogously, the normalized wavenumber-frequency spectrum in wavenumber domain is

$$\begin{aligned} \hat{S}_{pp}\left(k_x, k_y, \omega\right) &= -\frac{k_c^2}{\pi^2} \frac{PQ\left(\gamma_1 k_c\right)^{2P-1}}{\left[\left(k_x - k_c\right)^{2P} + \left(\gamma_1 k_c\right)^{2P}\right] \sum_{p=0}^{P-1} e^{j\theta_p}} \\ &\times \frac{Q\left(\gamma_3 k_c\right)^{2Q-1}}{\left[\left(k_y\right)^{2Q} + \left(\gamma_3 k_c\right)^{2Q}\right] \sum_{q=0}^{Q-1} e^{j\theta_q}} \end{aligned} \tag{5}$$

2.3 The Efimtsov model

The Efimtsov model assumes, as in the Corcos model, that the lateral and the longitudinal effects of the TBL can be separated. However, in the Efimtsov model the dependence of spatial correlation on boundary layer thickness, δ, as well as spatial separation is taken into account. Correlation length $1/\gamma_1 k_c$ and $1/\gamma_3 k_c$ in Corcos model are replaced with Λ_x and Λ_y. The Efimtsov model gives the cross power spectral density of the pressure at two different positions separated by the vector ξ as

$$S_{pp}\left(\xi_x, \xi_y, \omega\right) = \Phi_{pp}(\omega) \exp\left(-|\xi_x|/\Lambda_x\right) \exp\left(-\left|\xi_y\right|/\Lambda_y\right) \exp\left(-jk_c\xi_x\right) \tag{6}$$

where

$$\Lambda_x = \delta\left[\left(\frac{a_1 Sh}{U_c/U_\tau}\right)^2 + \frac{a_2^2}{Sh^2 + (a_2/a_3)^2}\right]^{-1/2} \tag{7}$$

$$\Lambda_y = \begin{cases} \delta\left[\left(\frac{a_4 Sh}{U_c/U_\tau}\right)^2 + \frac{a_5^2}{Sh^2+(a_5/a_6)^2}\right]^{-1/2}, & M_\infty < 0.75 \\ \delta\left[\left(\frac{a_4 Sh}{U_c/U_\tau}\right)^2 + a_7^2\right]^{-1/2}, & M_\infty > 0.9 \end{cases} \tag{8}$$

where Sh is the Strouhal number and equal to $Sh = \omega\delta/U_\tau$ and U_τ the friction velocity which varies with the Reynolds number but is typically of the order $0.03\ U_\infty$–$0.04\ U_\infty$. At high frequencies these expressions correspond to a Corcos model with $\gamma_1 = 0.1$ and $\gamma_3 = 0.77$. Coefficient a_1_a_7 are 0.1, 72.8, 1.54, 0.77, 548, 13.5 and 5.66 respectively. When $0.75 < M_\infty < 0.9$, the Λ_y can be determined by numerical interpolation. At high frequency, the Efimtsov model and the Corcos model are equal while $\gamma_1 = 0.10$ and $\gamma_3 = 0.77$.

The normalized wavenumber-frequency spectrum is

$$\hat{S}_{pp}\left(k_x, k_y, \omega\right) = \frac{2\Lambda_x^{-1}}{(k_x/k_c - 1)^2 + (\Lambda_x k_c)^{-2}} \cdot \frac{2\Lambda_y^{-1}}{k_y^2 + \Lambda_y^{-2}} \tag{9}$$

2.4 The Smolyakov-Tkachenko model

Like Efimtsov model, Smolyakov-Tkachenko model also takes the boundary layer thickness and scale space separation of boundary layer effect of fluctuating pressure into account. Based on the experimental results, the difference is that the Smolyakov-Tkachenko model amend the space scale function index $\exp\left[-\left(|\xi_x|/\Lambda_x + \left|\xi_y\right|/\Lambda_y\right)\right]$ to $\exp\left[-\sqrt{\left(\xi_x^2/\Lambda_x^2 + \xi_y^2/\Lambda_y^2\right)}\right]$, in order to make the computing result is consistent with the experimental results.

The normalized wavenumber-frequency spectrum is

$$\hat{S}_{pp}\left(k_x, k_y, \omega\right) = 0.974 A(\omega) h(\omega)\left[F\left(k_x, k_y, \omega\right) - \Delta F\left(k_x, k_y, \omega\right)\right] \tag{10}$$

where

$$A(\omega) = 0.124\left[1 - \frac{1}{4k_c\delta^*} + \left(\frac{1}{4k_c\delta^*}\right)^2\right]^{1/2} \tag{11}$$

$$h(\omega) = \left[1 - \frac{m_1 A}{6.515\sqrt{G}}\right]^{-1} \tag{12}$$

$$m_1 = \frac{1 + A^2}{1.025 + A^2} \tag{13}$$

$$G = 1 + A^2 - 1.005m_1 \tag{14}$$

$$F(k_x, k_y, \omega) = \left[A^2 + (1 - k_x/k_c)^2 + \left(\frac{k_y/k_c}{6.45}\right)\right]^{-3/2} \tag{15}$$

$$\Delta F(k_x, k_y, \omega) = 0.995\left[1 + A^2 + \frac{1.005}{m_1}\left\{(m_1 - k_x/k_c)^2 + (k_y/k_c)^2 - m_1^2\right\}\right]^{-3/2} \tag{16}$$

where δ^* is the thickness of boundary layer, which is also set as $\delta^* = \delta/8$.

2.5 The Ffowcs-Williams model

Ffowcs-Williams using Lighthill acoustic analogy theory to deduce a frequency-wave spectrum model, in which the speed of the pneumatic equation is set as the source term by Corcos form. A number of parameters in the model and function need further experiments to determine, which is not widely used at present. Hwang and Geib [22] ignore compression factor of the influence of this model to put forward a simplified model. The normalized wavenumber-frequency spectrum is

$$\hat{S}_{pp}(k_x, k_y, \omega) = \left(\frac{|\boldsymbol{k}|}{k_c}\right)^2 \frac{2\gamma_1}{(k_x/k_c - 1)^2 + \gamma_1^2} \cdot \frac{2\gamma_3}{(k_y/k_c)^2 + \gamma_3^2} \tag{17}$$

2.6 The Chase model

Chase's model is another model commonly used and believed to describe the low-wavenumber domain better than Corcos's model, which has the same starting point with the Ffowcs-Williams model. The normalized wavenumber-frequency spectrum can be described as

$$\hat{S}_{pp}(k_x, k_y, \omega) = \frac{(2\pi)^3 \rho k_c^2 U_\tau^3}{\Phi(\omega)}\left(C_M k_x^2 K_M^{-5} + C_T |\boldsymbol{k}|^2 K_T^{-5}\right) \tag{18}$$

where

$$K_M^2 = \frac{(\omega - U_c k_x)^2}{h^2 U_\tau^2} + |\boldsymbol{k}|^2 + (b_M \delta)^{-2} \tag{19}$$

$$K_T^2 = \frac{(\omega - U_c k_x)^2}{h^2 U_\tau^2} + |\boldsymbol{k}|^2 + (b_T \delta)^{-2} \tag{20}$$

$$\Phi(\omega) = \frac{(2\pi)^2 \rho^2 h U_\tau^4}{3\omega(1 + \mu^2)}(C_M F_M + C_T F_T) \tag{21}$$

$$F_M = \left[1 + \mu^2 \alpha_M^2 + \mu^4(\alpha_M^2 - 1)\right] / \left[\alpha_M^2 + \mu^2(\alpha_M^2 - 1)\right]^{3/2} \tag{22}$$

$$F_T = \left[1 + \alpha_T^2 + \mu^2(3\alpha_T^2 - 1) + 2\mu^4(\alpha_T^2 - 1)\right] / \left[\alpha_T^2 + \mu^2(\alpha_T^2 - 1)\right]^{3/2} \tag{23}$$

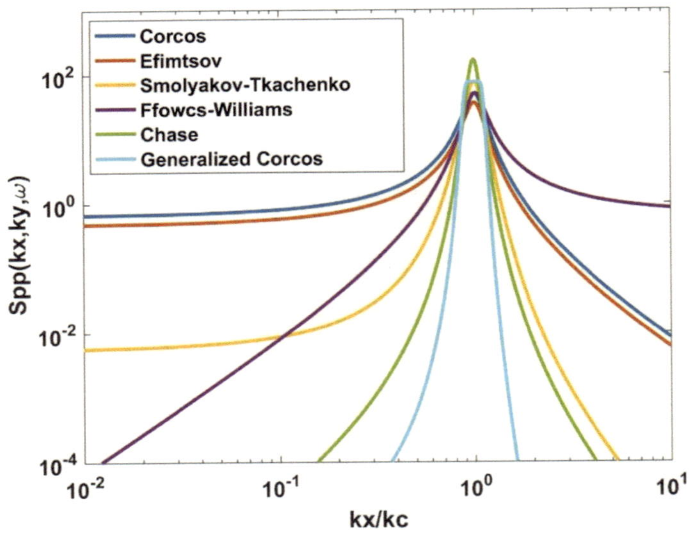

Figure 1.
A comparison of models for different wavenumber-frequency spectrum of turbulent boundary layer fluctuating pressure, reproduced from Ref. [23].

$$\alpha_M^2 = 1 + (b_M k_c \delta)^{-2}, \quad \alpha_T^2 = 1 + (b_T k_c \delta)^{-2} \tag{24}$$

$$\mu = hU_\tau / U_c \tag{25}$$

$$C_M = 0.0745, C_T = 0.0475, b_M = 0.756, \quad b_T = 0.378, \quad h = 3.0 \tag{26}$$

2.7 Comparison of models

Figure 1 shows the comparison of the above models. In the figure, the parameters used by the Corcos model are $\gamma_1 = 0.116$, $\gamma_3 = 0.77$, the order of Generalized Corcos model is ($P = 1$, $Q = 4$). From the comparison among those models, it can be seen that the Generalized Corcos model attenuates quickly in the vicinity of the convective wave number, and its order is adjustable, which can effectively control the computational accuracy. The model can obtain more accurate prediction results by adjusting parameters. In addition, the Chase model is considered to be able to better describe the pressure characteristics of TBL pulsation at low wave number segment, while other models have some defects at low wave number segment. However, Corcos model is the most commonly used in practical application. Because the model is simple in form and has clear physical significance, a simple calculation formula can usually be obtained when solving the structural vibration and sound response induced by turbulent boundary layer. It should be noted that the structure radiated sound predicted by Corcos model tends to be larger at low wave number.

3. Calculation of vibro-acoustic responses of the wall plate excited by TBL

Consider a simply supported thin rectangular plate excited by TBL, as shown in **Figure 2**. In the figure, U_c is turbulent flow velocity, and the direction of the incoming flow is parallel to the X-axis. In this chapter, vibro-acoustic responses are solved by modal superposition method [23].

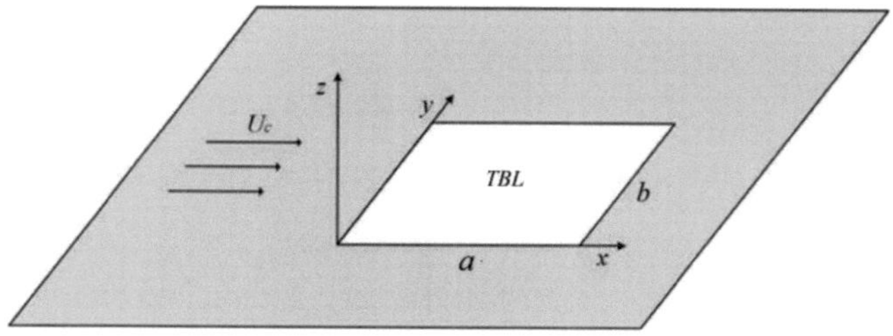

Figure 2.
Schematic diagram of simply supported thin rectangular plate excited by TBL.

Assume that point s on the plate is excited by a normal force F *at points*, and the vibration displacement response at point $\boldsymbol{r}$can be calculated by

$$W(\boldsymbol{r}, \omega) = H(\boldsymbol{r}, \boldsymbol{s}, \omega) \cdot F(\boldsymbol{s}, \omega) \tag{27}$$

where $\boldsymbol{s} = (x_o, y_0)$, $\boldsymbol{r} = (x, y)$.

The impulse response H satisfies the following governing equation

$$\left[D(1 + j\eta)\nabla^4 - m_s\omega^2\right]H(\boldsymbol{r}, \boldsymbol{s}, \omega) = \delta(\boldsymbol{r} - \boldsymbol{s}) \tag{28}$$

The impulse response can be expanded as

$$H(\boldsymbol{r}, \boldsymbol{s}, \omega) = \sum_{m=1}^{M} \sum_{n=1}^{N} H_{mn}(\omega)\Psi_{mn}(\boldsymbol{r})\Psi_{mn}(\boldsymbol{s}) \tag{29}$$

The modal amplitude of impulse response by using the Galerkin method can be described as

$$H_{mn}(\omega) = \frac{1}{DK_{mn}(1 + j\eta) - m_s\omega^2} \tag{30}$$

3.1 Vibro-acoustic responses of plate solved by spatial domain integration

Cross spectral density of displacement response for any two points on the plate can be defined as

$$\begin{aligned} S_{WW}(\boldsymbol{r}_1, \boldsymbol{r}_2, \omega) &= \textstyle\int_S\int_S S_{pp}(\boldsymbol{s}_1 - \boldsymbol{s}_2, \omega)H^*(\boldsymbol{r}_1, \boldsymbol{s}_1, \omega)H(\boldsymbol{r}_2, \boldsymbol{s}_2, \omega)d\boldsymbol{s}_1 d\boldsymbol{s}_2 \\ &= \Phi_{pp}(\omega)\sum_{m=1}^{M}\sum_{n=1}^{N}|H_{mn}(\omega)|^2\Psi_{mn}(\boldsymbol{r}_1)\Psi_{mn}(\boldsymbol{r}_2)J_{\mathrm{mn}}(\omega) \end{aligned} \tag{31}$$

where

$$J_{\mathrm{mn}}(\omega) = \int_S\int_S S_{pp}(\boldsymbol{s}_1 - \boldsymbol{s}_2)\Psi_{mn}(\boldsymbol{s}_1)\Psi_{mn}(\boldsymbol{s}_2)\, d\boldsymbol{s}_1 d\boldsymbol{s}_2 \tag{32}$$

In the above equation, $J_{mn}(\omega)$ is called modal excitation term.

When using the Corcos model, the coordinate transformation of the quadruple integral in the modal excitation term can be obtained

$$J_{\mathrm{mn}}(\omega) = \frac{4}{S}\left(\frac{1}{k_m k_n}J^1_{mn} + J^2_{mn} + \frac{1}{k_m}J^3_{mn} + \frac{1}{k_n}J^4_{mn}\right) \tag{33}$$

Where

$$\begin{Bmatrix} J_{mn}^1 \\ J_{mn}^2 \\ J_{mn}^3 \\ J_{mn}^4 \end{Bmatrix} = \int_0^b \int_0^a \begin{Bmatrix} 1 \\ (a-x)(b-y) \\ (b-y) \\ (a-x) \end{Bmatrix} \times \begin{Bmatrix} \sin k_m x \cdot \sin k_n y \\ \cos k_m x \cdot \cos k_n y \\ \sin k_m x \cdot \cos k_n y \\ \cos k_m x \cdot \sin k_n y \end{Bmatrix} \tilde{S}_{pp}(x, y, \omega) dxdy \tag{34}$$

$$\tilde{S}_{pp}(x, y, \omega) = \exp(-\gamma_1 k_c x) \exp(-\gamma_3 k_c y) \cos(k_c x) \tag{35}$$

When $\boldsymbol{r}_1 = \boldsymbol{r}_2$, the auto-spectral density of displacement response can be obtained as

$$S_{WW}(\boldsymbol{r}, \omega) = \Phi_{pp}(\omega) \sum_{m=1}^{M} \sum_{n=1}^{N} |H_{mn}(\omega)|^2 \Psi_{mn}^2(\boldsymbol{r}) J_{\mathrm{mn}}(\omega) \tag{36}$$

As for vibration ($V = j\omega W$) the auto-spectral density is

$$\begin{aligned} S_{VV}(\boldsymbol{r}, \omega) &= \omega^2 S_{WW}(\boldsymbol{r}, \omega) \\ &= \omega^2 \Phi_{pp}(\omega) \sum_{m=1}^{M} \sum_{n=1}^{N} |H_{mn}(\omega)|^2 \Psi_{mn}^2(\boldsymbol{r}) J_{\mathrm{mn}}(\omega) \end{aligned} \tag{37}$$

So, vibration energy and acoustic radiation energy can be expressed as

$$\begin{aligned} \langle V^2 \rangle &= \frac{1}{S} \iint S_{VV}(x, y, \omega) dS \\ &= \frac{1}{S} \omega^2 \Phi_{pp}(\omega) \sum_{m=1}^{M} \sum_{n=1}^{N} J_{mn}(\omega) |H_{mn}(\omega)|^2 \end{aligned} \tag{38}$$

$$\Pi^r = \rho_0 c_0 \omega^2 \Phi_{pp}(\omega) \sum_{m=1}^{M} \sum_{n=1}^{N} \sigma_{mn} J_{mn}(\omega) |H_{mn}(\omega)|^2 \tag{39}$$

According to the definition, the modal average acoustic radiation efficiency excited by TBL of the thin plate is

$$\sigma = \frac{\sum_{m=1}^{M} \sum_{n=1}^{N} \sigma_{mn} J_{mn}(\omega) |H_{mn}(\omega)|^2}{\sum_{m=1}^{M} \sum_{n=1}^{N} J_{mn}(\omega) |H_{mn}(\omega)|^2} \tag{40}$$

3.2 Vibro-acoustic responses of plate solved by wavenumber domain integration

Another approach to obtain the cross spectral density of vibration response is to solve it directly by using the separable integral property of some turbulent boundary layer pulsating pressure models in the wavenumber domain [24].

The wavenumber-frequency spectrum of TBL satisfies the following relationship

$$\begin{aligned} S_{pp}(\boldsymbol{s_1} - \boldsymbol{s_2}, \omega) &= \frac{1}{(2\pi)^2} \int S_{pp}(\boldsymbol{k}, \omega) \exp[-j\boldsymbol{k}(\boldsymbol{s_1} - \boldsymbol{s_2})] d\boldsymbol{k} \\ &= \frac{1}{(2\pi)^2} \iint S_{pp}(k_x, k_y, \omega) \exp\left[-j\left(k_x \xi_x + k_y \xi_y\right)\right] dk_x dk_y \end{aligned} \tag{41}$$

where $s_1 - s_2 = \left(\xi_x, \xi_y\right)$, $k = \left(k_x, k_y\right)$.

The formula can be obtained by substituting the cross spectral density of the vibration response

$$\begin{aligned}S_{WW}(r_1, r_2, \omega) &= \iint S_{pp}(s_1, s_2, \omega) H^*(r_1, s_1, \omega) H(r_2, s_2, \omega) ds_1 ds_2 \\ &= \frac{1}{(2\pi)^2} \int S_{pp}(k, \omega) \exp\left[-jk(s_1 - s_2)\right] dk \iint H^*(r_1, s_1, \omega) H(r_2, s_2, \omega) ds_1 ds_2 \\ &= \frac{1}{(2\pi)^2} \int S_{pp}(k, \omega) dk \int H^*(r_1, s_1, \omega) \exp(-jks_1) ds_1 \int H(r_2, s_2, \omega) \exp(jks_2) ds_2 \\ &= \frac{1}{(2\pi)^2} \int S_{pp}(k, \omega) G^*(r_1, k, \omega) G(r_2, k, \omega) dk\end{aligned} \tag{42}$$

where

$$\begin{aligned}G(r, k, \omega) &= \int H(r, s, \omega) \exp(jks) ds \\ &= \sum_{m=1}^{M} \sum_{n=1}^{N} H_{mn}(\omega) \Psi_{mn}(r) \int \Psi_{mn}(s) \exp(jks) ds \\ &= \sum_{m=1}^{M} \sum_{n=1}^{N} H_{mn}(\omega) \Psi_{mn}(r) I_{mn}(k)\end{aligned} \tag{43}$$

$$\begin{aligned}I_{mn}(k) &= \int \Psi_{mn}(s) \exp(jks) ds \\ &= \frac{2}{\sqrt{ab}} \int_0^b \int_0^a \sin(k_m x) \sin(k_n y) \exp\left[j\left(k_x x + k_y y\right)\right] dx dy \\ &= \frac{2}{\sqrt{ab}} \cdot \frac{k_m\left[1 - \cos(m\pi) \exp\left(jk_x a\right)\right]}{k_x^2 - k_m^2} \cdot \frac{k_n\left[1 - \cos(n\pi) \exp\left(jk_y b\right)\right]}{k_y^2 - k_n^2}\end{aligned} \tag{44}$$

Similarly, the spectral density of the vibration velocity can be obtained as

$$\begin{aligned}S_{VV}(r, \omega) &= \frac{\omega^2}{(2\pi)^2} \int S_{pp}(k, \omega) |G(r, k, \omega)|^2 dk \\ &= \frac{\omega^2}{(2\pi)^2} \sum_{m=1}^{M} \sum_{n=1}^{N} \Psi_{mn}^2(r) |H_{mn}(\omega)|^2 \int S_{pp}(k, \omega) |I_{mn}(k)|^2 dk\end{aligned} \tag{45}$$

As for the Corcos model, we can obtain that

$$\int S_{pp}(k, \omega) |I_{mn}(k)|^2 dk = \frac{4}{S} \Phi_{pp}(\omega) \left[2\gamma_1 k_c \Lambda_m(\omega)\right] \left[2\gamma_3 k_c \Gamma_n(\omega)\right] \tag{46}$$

where

$$\Lambda_m(\omega) = 2k_m^2 \int_{-\infty}^{\infty} \frac{1 - \cos(m\pi) \cos(k_x a)}{\left(k_x^2 - k_m^2\right)^2 \left[(k_x - k_c)^2 + (\gamma_1 k_c)^2\right]} dk_x \tag{47}$$

$$\Gamma_n(\omega) = 2k_n^2 \int_{-\infty}^{\infty} \frac{1 - \cos(n\pi)\cos(k_y b)}{\left(k_y^2 - k_n^2\right)^2 \left[k_y^2 + (\gamma_3 k_c)^2\right]} dk_y \tag{48}$$

According to the residue theorem, $\Lambda_m(\omega)$ and $\Gamma_n(\omega)$ can be further simplified as

$$\begin{aligned}\Lambda_m(\omega) &= 2k_m^2 \int_{-\infty}^{\infty} \frac{1 - \cos(m\pi)\cos(k_x a)}{(k_x^2 - k_m^2)^2 \left[(k_x - k_c)^2 + (\gamma_1 k_c)^2\right]} dk_x \\ &= 2\pi k_m^2 \left\{ \frac{a}{4k_m^2\left[(k_m + k_c)^2 + (\gamma_1 k_c)^2\right]} + \frac{a}{4k_m^2\left[(k_m - k_c)^2 + (\gamma_1 k_c)^2\right]} \right. \\ &\left. + \frac{1 - \cos(m\pi)\exp\left[-(j+\gamma_1)k_c a\right]}{(2\gamma_1 k_c)\left[k_c^2(1 - j\gamma_1)^2 - k_m^2\right]^2} + \frac{1 - \cos(m\pi)\exp\left[(j-\gamma_1)k_c a\right]}{(2\gamma_1 k_c)\left[k_c^2(1 + j\gamma_1)^2 - k_m^2\right]^2} \right\}\end{aligned} \tag{49}$$

$$\begin{aligned}\Gamma_n(\omega) &= 2k_n^2 \int_{-\infty}^{\infty} \frac{1 - \cos(n\pi)\cos(k_y b)}{\left(k_y^2 - k_n^2\right)^2 \left[k_y^2 + (\gamma_3 k_c)^2\right]} dk_y \\ &= 2\pi k_n^2 \left\{ \frac{b}{2k_n^2\left[k_n^2 + (\gamma_3 k_c)^2\right]} + \frac{1 - \cos(n\pi)\exp(-\gamma_3 k_c b)}{(\gamma_3 k_c)\left[k_n^2 + (\gamma_3 k_c)^2\right]^2} \right\}\end{aligned} \tag{50}$$

Vibration energy and sound radiation energy are

$$\begin{aligned}\langle V^2 \rangle &= \frac{\omega^2}{(2\pi)^2 S} \int\int S_{pp}(\boldsymbol{k}, \omega) |G(\boldsymbol{r}, \boldsymbol{k}, \omega)|^2 d\boldsymbol{k} d\boldsymbol{r} \\ &= \frac{\omega^2}{(2\pi)^2 S} \sum_{m=1}^{M} \sum_{n=1}^{N} |H_{mn}(\omega)|^2 \int S_{pp}(\boldsymbol{k}, \omega) |I_{mn}(\boldsymbol{k})|^2 d\boldsymbol{k} \\ &= \frac{1}{S} \omega^2 \Phi_{pp}(\omega) \left(\frac{4}{S} \frac{\gamma_1 k_c}{\pi} \frac{\gamma_3 k_c}{\pi} \right) \sum_{m=1}^{M} \sum_{n=1}^{N} \Lambda_m(\omega) \Gamma_n(\omega) |H_{mn}(\omega)|^2\end{aligned} \tag{51}$$

$$\begin{aligned}\Pi^r &= \frac{1}{(2\pi)^2} \rho_0 c_0 \omega^2 \sum_{m=1}^{M} \sum_{n=1}^{N} \sigma_{mn} |H_{mn}(\omega)|^2 \int S_{pp}(\boldsymbol{k}, \omega) |I_{mn}(\boldsymbol{k})|^2 d\boldsymbol{k} \\ &= \rho_0 c_0 \omega^2 \Phi_{pp}(\omega) \left(\frac{4}{S} \frac{\gamma_1 k_c}{\pi} \frac{\gamma_3 k_c}{\pi} \right) \sum_{m=1}^{M} \sum_{n=1}^{N} \sigma_{mn} \Lambda_m(\omega) \Gamma_n(\omega) |H_{mn}(\omega)|^2\end{aligned} \tag{52}$$

Compare the above two equations, it can be seen that

$$J_{mn}(\omega) = \frac{4}{S} \left[\frac{\gamma_1 k_c}{\pi} \Lambda_m(\omega) \right] \times \left[\frac{\gamma_3 k_c}{\pi} \Gamma_n(\omega) \right] \tag{53}$$

Finally, the modal average acoustic radiation efficiency can be obtained as

$$\sigma = \frac{\sum_{m=1}^{M} \sum_{n=1}^{N} \sigma_{mn} \Lambda_m(\omega) \Gamma_n(\omega) |H_{mn}(\omega)|^2}{\sum_{m=1}^{M} \sum_{n=1}^{N} \Lambda_m(\omega) \Gamma_n(\omega) |H_{mn}(\omega)|^2} \tag{54}$$

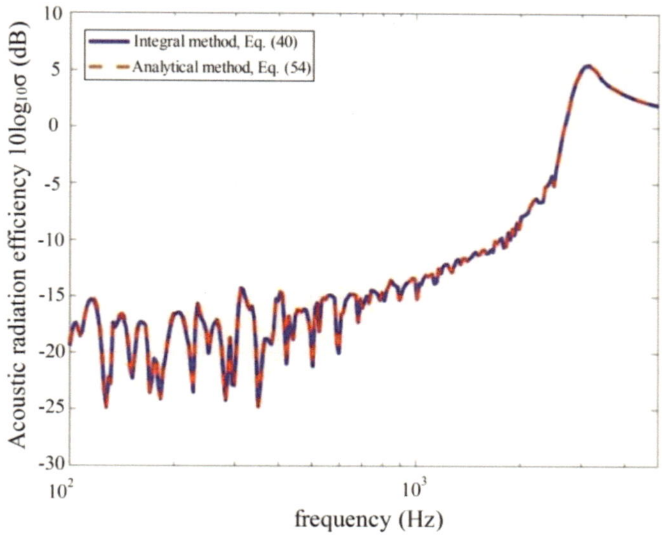

Figure 3.
Comparison of calculation methods of the modal averaged radiation efficiency excited by TBL. Reproduced from Ref. [23].

By observing the above equation, it can be found that only the modal excitation term in the modal averaged radiation efficiency is related to turbulence.

Figure 3 shows the comparison of two methods for calculating the modal averaged radiation efficiency excited by TBL. The size of the plate is 1.25 × 1.1 m, and the thickness is 4 mm, structural loss factor of aluminum plate is 1%, mach number is 0.5. Obviously, the accuracy of the two methods is equal. Computation speed of analytical method is much faster than integral method, but its range of application has limitations. Only the Corcos model and Efimtsov model can be used to separate integrals in the wave number domain.

The comparison of measured and predicted velocity spectral density and the radiated sound intensity of a plate ($a \times b = 0.62 \times 0.3$ m, and the thickness is 1.1 mm) is shown in **Figure 4**, which is only compared in narrow band. In this study, the loss factor of the plate assumes as 1.5%. The measured and predicted results for radiated sound intensity and auto spectrum of velocity have a good agreement with the frequency ranges from 100 to 3500 Hz. The agreement of the two type curves provides solid verification to test measured and predicted results.

3.3 Characteristic frequency in hydrodynamic coincidence

When the velocity of bending wave in the wall plate is close to the sound velocity in the air, the sound radiation efficiency reaches the maximum value. The corresponding frequency is the so-called critical frequency, and its expression is

$$f_c = \frac{c_0^2}{2\pi}\sqrt{\frac{m_s}{D}} \tag{55}$$

In the case of flow, when the velocity of flexural wave propagation in the wall plate is close to the turbulent convection velocity, the wall plate is most excited by the fluctuating pressure of TBL. The corresponding frequency is defined as the hydrodynamic coincidence frequency

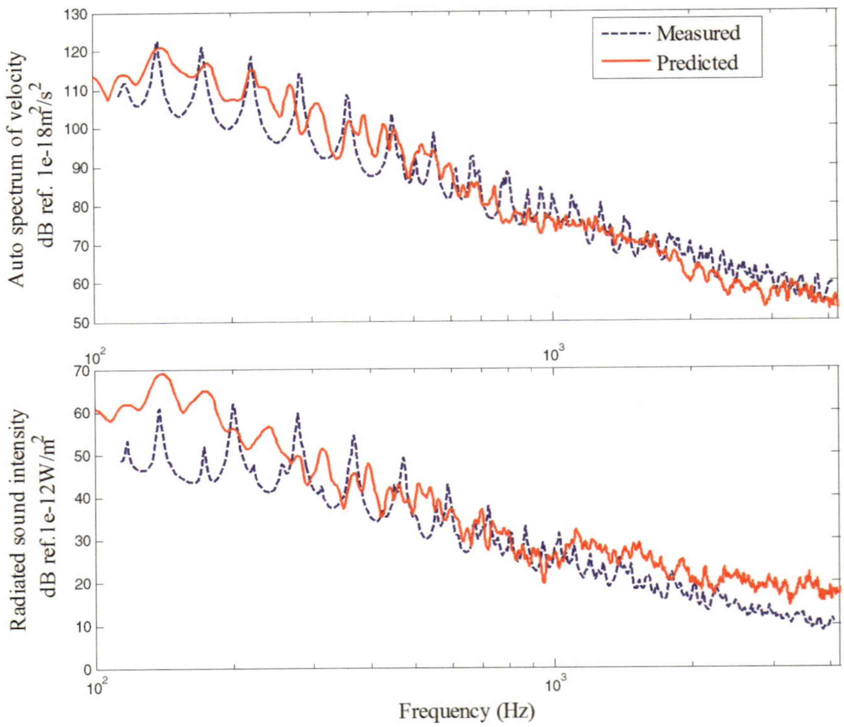

Figure 4.
Measured and predicted velocity auto spectrum and the radiated sound intensity of the plate with the size of a × b = 0.62 × 0.3 *m. Narrow band analysis in per Hz. Flow speed 86 m/s.*

$$f_h = \frac{U_c^2}{2\pi}\sqrt{\frac{m_s}{D}} \tag{56}$$

Similarly, for order (m, n) mode, its critical frequency and hydrodynamic coincidence frequency are

$$f_{c,mn} = \frac{c_0}{2\pi}k_{mn} \tag{57}$$

$$f_{h,mn} = \frac{U_c}{2\pi}k_{mn} \tag{58}$$

In conclusion, the relationship between critical frequency and hydrodynamic coincidence frequency can be summarized as follows

$$f_h = M_c^2 \cdot f_c \tag{59}$$

$$f_{h,mn} = M_c \cdot f_{c,mn} \tag{60}$$

In the above two equations, $M_c = U_c/c_0$ is mach number. Subsonic turbulence is generally considered, so the hydrodynamic coincidence frequency is always less than the critical frequency of the plate. It is important to note that the characteristics of frequency is a reference value which is based on the infinite plate hypothesis. Actually, the characteristics frequency of the limited plate slightly higher than a reference value. In addition, for the transverse flow problem, modal power line frequency can be thought of only related to the transverse mode. That is to say, $f_{h,mn} \approx U_c k_m/2\pi$, where $k_m = m\pi/a$ is lateral modal wave number.

4. Effect of flow velocity and structural damping on the acoustic radiation efficiency

4.1 Effect of convection velocity on the modal averaged radiation efficiency

The specific parameters and dimensions used in the calculation are listed in **Table 1**.

The increment of vibration power and acoustic radiation energy are different with the increase of the velocity, which indicates that the changing of velocity can affect the modal averaged radiation efficiency. The modal averaged radiation efficiency of the aluminum plate at three flow velocities (M_c = 0.5; 0.7; 0.9) is shown in **Figure 5**. It can be seen that when the M_c increases from 0.5 to 0.9, the modal averaged radiation efficiency will increase by 3–7 dB below the hydrodynamic coincidence frequency. And the corresponding hydrodynamic coincidence frequencies (f_h) are 1482, 2905, and 4802 Hz, respectively. The results show that the modal averaged radiation efficiency increases in the frequency range below the hydrodynamic coincidence frequency. The increase of the modal averaged radiation efficiency indicates that with the increase of flow velocity, the increment of the radiated sound power is larger than that of the mean square velocity.

The phenomenon that the modal averaged radiation efficiency increases with the flow velocity can be explained by the hydrodynamic coincidence effect. For the lateral incoming flow problem, the hydrodynamic coincidence is mainly

Plate length	a	1.25 m
Plate width	b	1.1 m
Plate thickness	h	0.002 m
Plate surface density	m_s	5.4 kg/m^2
Plate bending stiffness	D	52 Nm
Air density	ρ_0	1.21 kg/m^3
Sound speed	c_0	340 m/s

Table 1.
Parameters used in calculation.

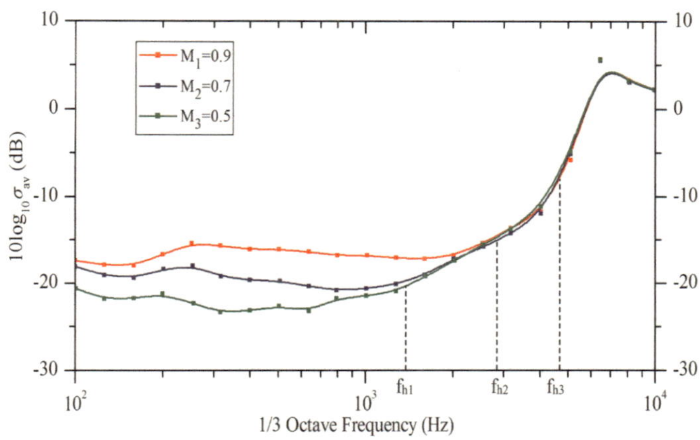

Figure 5.
Effect of the convective Mach number on the modal averaged radiation efficiency of the finite aluminum plate. Reproduced from Ref. [23].

determined by the lateral modal trace speed and the convection velocity. When the bending wave velocity of the lateral mode is the same as the turbulent flow velocity ($U_c = 2\pi f/k_m$), the corresponding hydrodynamic coincidence frequency is $f = mU_c/2a$. Thus a higher convection velocity at the same frequency will lead the TBL excitation to coincide with a lower order lateral mode.

The reason for above phenomenon may be further explored through the modal excitation terms. As illustrated in **Figure 6**, the lateral modal excitation term ($10\log_{10}\Lambda_m(\omega)$) is plotted with the lateral mode number (m) and frequency for different flow velocity (M_c). In the figure, the peak of the lateral mode excitation term corresponds to the maximum excitation and its position depends on the hydrodynamic coincidence frequency. The black bold lines in the two sub graphs are the positions where the hydrodynamic coincidence occurs. It can be seen that the slope of the hydrodynamic coincidence line is inversely proportional to the flow velocity, and the higher the velocity is, the lower the order of a certain frequency is. In addition, the lateral modes near the hydrodynamic coincidence line are all strongly excited. As the frequency increases, the number of these modes increases, but the amplitude of their corresponding mode excitation term decreases. Below the

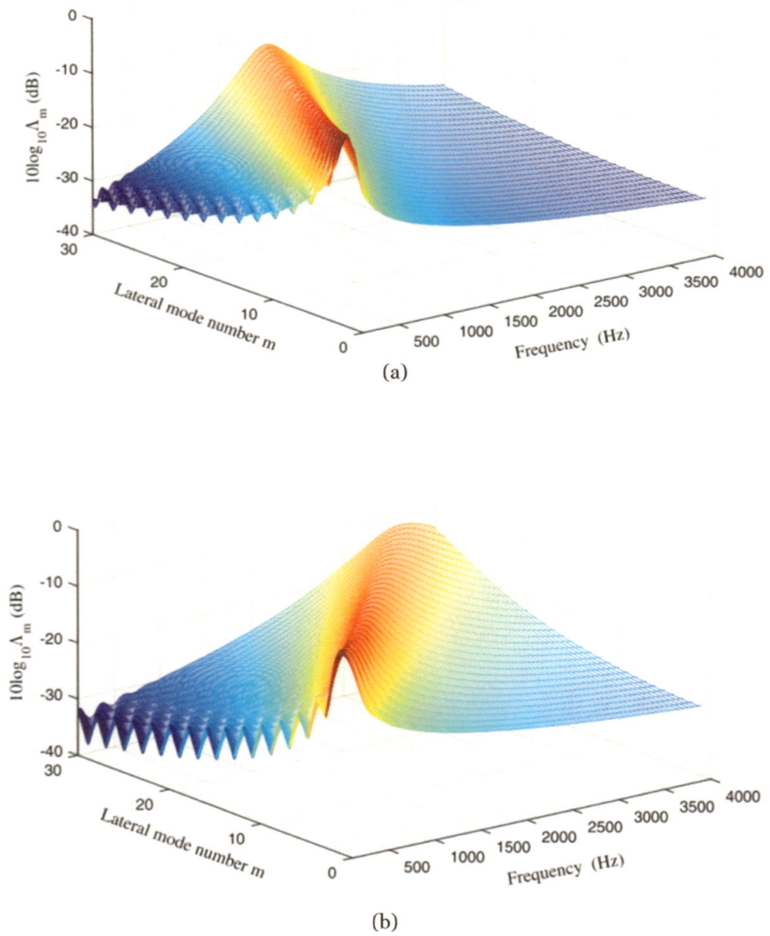

Figure 6.
Variation of the lateral modal excitation term with the lateral mode number and the frequency of a finite aluminum plate. (a) Convective $M_c = 0.5$ *and (b) convective* $M_c = 0.9$. *Reproduced from Ref. [13].*

critical frequency, a lower order lateral mode always has higher modal averaged radiation efficiency than that of a higher order lateral mode with the same n, since the modal critical frequency moves to lower frequency. So plate with higher flow velocity is supposed to have higher modal averaged radiation efficiency.

As an example, the hydrodynamic coincidence lines for different flow velocity (M_c) and the modal radiation efficiencies of mode $(m, 1)$ are illustrated in **Figure 7**. The black solid lines in the figure are the hydrodynamic coincidence line corresponding to the mode order and frequency. It can be seen that at a certain frequency, the modal averaged radiation efficiency of the hydrodynamic coincidence modes at higher velocity is always greater than that of the low velocity. In a word, an increase of the flow velocity will increase the modal radiation efficiency of the coincided mode, and then results in the increase of the modal averaged radiation efficiency. Besides, owing to the low pass property of the modal excitation term, the increase of the modal radiation efficiency is restrained above the hydrodynamic coincidence frequency. As a consequence, the modal averaged radiation efficiency is great affected by the flow velocity which only occurs below the hydrodynamic coincidence frequency.

4.2 Effect of structural damping on modal averaged radiation efficiency

The modal averaged radiation efficiency changes with structural loss factors for different flow velocity (M_c), as shown in **Figure 8**. The reference value is calculated according to Leppington's formula [25]. Though Leppington's formula is widely used in statistical energy analysis, it does not take the flow and structural damping into account. **Figure 8** indicates that an increase of the structural loss factor will increase the modal averaged radiation efficiency under the critical frequency, but the increments are different for different flow velocity. It is found that the modal averaged radiation efficiency is not sensitive to the change of structure loss factor at low Mach number. For example, for a typical high-speed train (M_c = 0.25), the increased modal averaged radiation efficiency is less than 2 dB in the frequency band below the critical frequency when the structural loss factor increases from 1 to 4%. In the case of high flow velocity, the effect of structure loss factor on the modal averaged radiation efficiency is much obvious. When M_c = 0.7, the modal averaged radiation efficiency will increase by about 5 dB if the structural loss factor has the same increment. The results show that the influence of structural damping on the

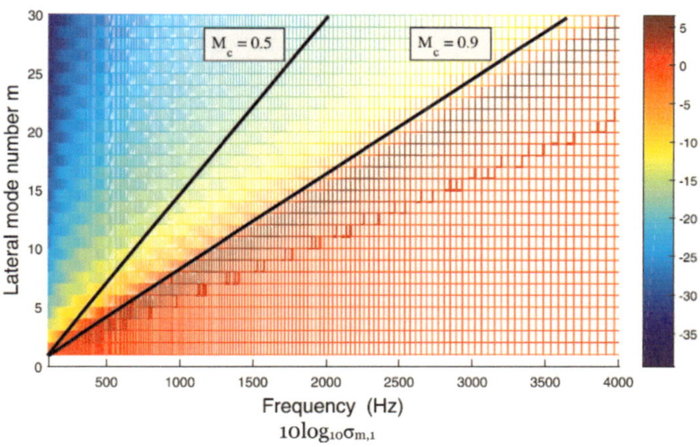

Figure 7.
Hydrodynamic coincidence lines and variation of the modal radiation efficiency with the lateral mode number and the frequency of a finite aluminum plate. m *varies,* n = 1. *Reproduced from Ref. [13].*

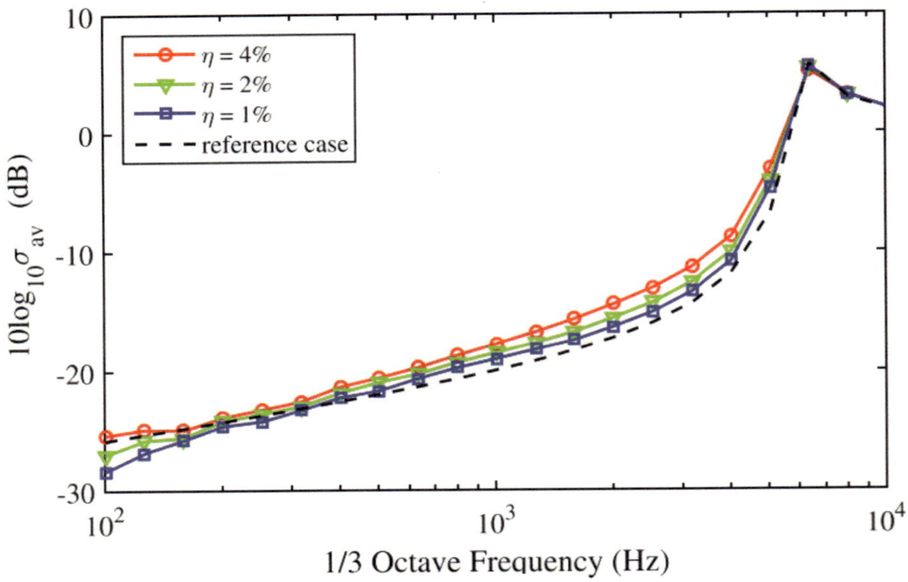

(a)

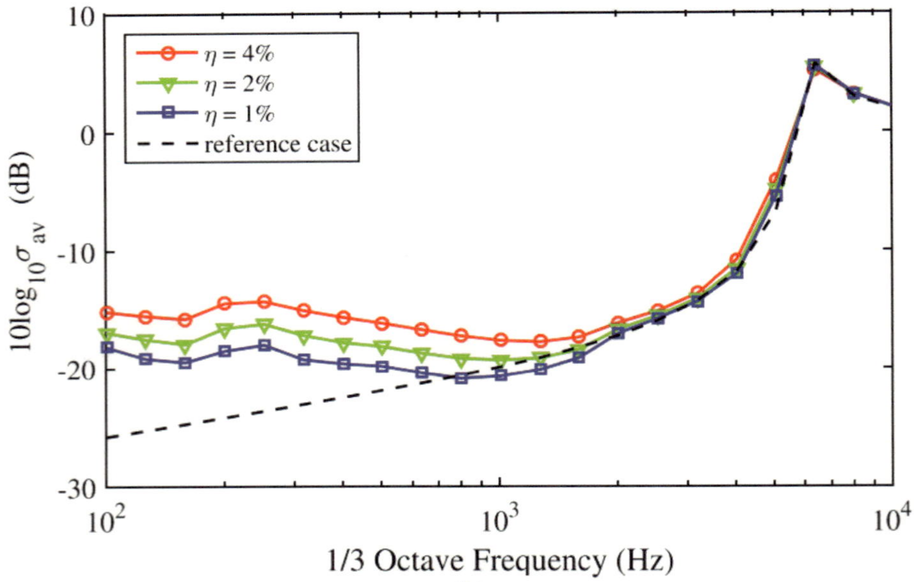

(b)

Figure 8.
Effect of the structural loss factor on the modal averaged radiation efficiency of a finite aluminum plate. (a) Convective $M_c = 0.25$ *and (b) convective* $M_c = 0.7$*. Reproduced from Ref. [13].*

modal averaged radiation efficiency is related to the flow velocity, and the influence of structural damping can be enhanced by increasing the flow velocity.

The effect of structural damping on the modal averaged radiation efficiency can be qualitatively explained by Eq. (61)

$$\sigma_{av} = \frac{\prod^t}{\rho_0 c_0 S\langle V^2\rangle} = \frac{\sum_{m=1}^{\infty}\sum_{n=1}^{\infty}\sigma_{mn}(\omega)J_{mn}(\omega)|V_{mn}(\omega)|^2}{\sum_{m=1}^{\infty}\sum_{n=1}^{\infty}J_{mn}(\omega)|V_{mn}(\omega)|^2} \tag{61}$$

Eq. (61) shows that the modal averaged radiation efficiency is equivalent to the weighted average function of the modal velocity response, and the weighted coefficient is the modal averaged radiation efficiency. In the frequency band below the critical frequency, the radiation efficiency of each mode varies in the range from 0 to 1. Due to this weighted effect of Eq. (61), the vibration energy (denominator in the equation) decreases more effectively than the acoustic radiation power (molecule in the equation). Thus the radiation efficiency increases in the frequency band below the critical frequency. However, the phenomenon that the radiation efficiency of a damped plate is enlarged with increment of flow velocity has not yet been clearly interpreted.

Moreover, it is observed that the effect of structural damping on modal averaged radiation efficiency has a good agreement with the research of Kou [23] at low flow velocity. In their work, it is shown that the modal averaged radiation efficiency of heavily damped structures is sensitive to the change of structural loss factor without turbulent flow. It also implies that Leppington's equation is not applicable to the prediction of modal averaged radiation efficiency of damped structures at high flow velocity.

5. Conclusion

This chapter studies the vibro-acoustic characteristics of the wall plate structure excited by turbulent boundary layer (TBL). Based on the modal expansion and Corcos model, the formulas for calculating the modal averaged radiation efficiency are derived. The results indicate that an increment of flow rate will increase the vibration energy and the radiated sound energy of the structure. However, the amplitude of two cases varies with the velocity are not the same, and when the velocity increases, the acoustic radiation efficiency will increase below the hydrodynamic coincidence frequency range. The main reason for this phenomenon is that a higher convection velocity will coincide with lower order modes which have higher radiation efficiencies.

The modal averaged radiation efficiency increases with the increase of structural damping below the critical frequency band. The larger the flow rate, the more significant the effect of structural damping on acoustic radiation efficiency. In the case of low flow velocity, the modal averaged radiation efficiency is not sensitive to the change of structural damping. The structural damping increases from 1 to 4%, and the increase of modal averaged radiation efficiency less than 2 dB. In the case of high flow rate, the modal averaged radiation efficiency will increase by 5 dB when the increment of the structural damping is from 1 to 4%.

Acknowledgements

Thanks to the financial support by the Taishan Scholar Program of Shandong (no. ts201712054).

Conflict of interest

Figures 6–8 in this chapter are reproduced from an AIP Publishing journal paper written by the second and third authors, and all the figures are cited in this text.

According to AIP webpage for Copyright and Permission to Reuse AIP materials, AIP Publishing allows authors to retain their copyrights (https://publishing.aip.org/

authors/copyright-reuse). Each Copyright Owner retains the nonexclusive rights to "Reprint portions of the Work (excerpts, figures, tables) in future works created by the Author, in keeping with professional publication ethics", without obtaining permission from AIP Publishing, and provided clear credit is given to its first publication in an AIP Publishing journal. Any reuse must include a full credit line acknowledging AIP Publishing's publication and a link to the VOR on AIP Publishing's site.

Author details

Zhang Xilong[1], Kou YiWei[2] and Liu Bilong[1*]

1 School of Mechanical and Automotive Engineering, Qingdao University of Technology, Qingdao, China

2 Key Laboratory of Noise and Vibration Research, Institute of Acoustics, Chinese Academy of Sciences, Beijing, China

*Address all correspondence to: liubilong@qut.edu.cn

References

[1] Graham WR. Boundary layer induced noise in aircraft, part I: The flat plate model. Journal of Sound and Vibration. 1996;**192**(1):101-120. DOI: 10.1006/jsvi.1996.0178

[2] Graham WR. A comparison of models for the wavenumber-frequency spectrum of turbulent boundary layer pressures. Journal of Sound and Vibration. 1997;**206**:541-565. DOI: 10.1006/jsvi.1997.1114

[3] Han F, Bernhard RJ, Mongeau LG. A model for the vibro-acoustic response of plates excited by complex flows. Journal of Sound and Vibration. 2001; **246**(5):901-926. DOI: 10.1006/jsvi.2001.3699

[4] Liu BL. Noise radiation of aircraft panels subjected to boundary layer pressure fluctuations. Journal of Sound and Vibration. 2008;**314**:693-711. DOI: 10.1016/j.jsv.2008.01.045

[5] Liu BL, Feng LP, Nilsson A, Aversano M. Predicted and measured plate velocities induced by turbulent boundary layers. Journal of Sound and Vibration. 2012;**331**(24):5309-5325. DOI: 10.1016/j.jsv.2012.07.012

[6] Liu BL, Zhang H, Qian ZC, Chang DQ, Yan Q, Huang WC. Influence of stiffeners on plate vibration and radiated noise excited by turbulent boundary layers. Applied Acoustics. 2014;**80**:28-35. DOI: 10.1016/j.apacoust.2014.01.007

[7] Rocha J, Palumbo D. On the sensitivity of sound power radiated by aircraft panels to turbulent boundary layer parameters. Journal of Sound and Vibration. 2012;**331**:4785-4806. DOI: 10.1016/j.jsv.2012.05.030

[8] Cremer L, Heckl M. Structure-Borne Sound: Structural Vibrations and Sound Radiation at Audio Frequencies. 3nd ed. Berlin: Springer-Verlag; 1988. pp. 502-505. DOI: 10.1121/1.2060712

[9] Wallace CE. Radiation-resistance of a rectangular panel. The Journal of the Acoustical Society of America. 1972;**51**: 946-952. DOI: 10.1121/1.1912943

[10] Kou YW, Liu BL, Tian J. Radiation efficiency of damped plates. The Journal of the Acoustical Society of America. 2015;**137**:1032-1035. DOI: 10.1121/1.4906186

[11] Graham WR. The effect of mean flow on the radiation efficiency of rectangular plates. Proceedings of the Royal Society of London. Series A. 1998;**454**:111-137. DOI: 10.1098/rspa.1998.0149

[12] Frampton KD. Radiation efficiency of convected fluid-loaded plates. The Journal of the Acoustical Society of America. 2003;**113**:2663-2673. DOI: 10.1121/1.1559173

[13] Kou YW, Liu BL, Chang D. Radiation efficiency of plates subjected to turbulent boundary layer fluctuations. The Journal of the Acoustical Society of America. 2016;**139**: 2766-2771. DOI: 10.1121/1.4949021

[14] Corcos GM. Resolution of pressure in turbulence. The Journal of the Acoustical Society of America. 1963; **35**(2):192-199

[15] Efimtsov BM. Characteristics of the field of turbulent wall pressure-fluctuations at large Reynolds-numbers. Soviet Physics Acoustics. 1982;**28**(4): 289-292

[16] Smolyakov AV, Tkachenko VM. Model of a field of pseudosonic turbulent wall pressures and experimental-data. Soviet Physics Acoustics. 1991;**37**(6):627-631. DOI: 10.1002/chin.200334260

[17] Williams JF. Boundary-layer pressures and the Corcos model: A development to incorporate low-wavenumber constraints. Journal of Fluid Mechanics. 1982;**125**:9-25. DOI: 10.1017/S0022112082003218

[18] Chase DM. Modeling the wavevector-frequency spectrum of turbulent boundary-layer wall pressure. Journal of Sound and Vibration. 1980; **70**(1):29-67. DOI: 10.1121/1.2017510.

[19] Chase DM. The character of the turbulent wall pressure spectrum at subconvective wave-numbers and a suggested comprehensive model. Journal of Sound and Vibration. 1987; **112**(1):125-147. DOI: 10.1016/S0022-460X(87)80098-6

[20] Howe MS, Feit D. Acoustics of fluid–structure interactions. Physics Today. 1999;**52**(12):64-64. DOI: 10.1063/1.882913

[21] Caiazzo A, Desmet W. A generalized Corcos model for modelling turbulent boundary layer wall pressure fluctuations. Journal of Sound and Vibration. 2016;**372**(23):192-210. DOI: 10.1016/j.jsv.2016.02.036

[22] Hwang Y, Geib F. Estimation of the wavevector-frequency spectrum of turbulent boundary layer wall pressure by multiple linear regression. Journal of Vibration, Acoustics, Stress, and Reliability in Design. 1984;**106**(3): 334-342. DOI: 10.1115/1.3269199

[23] Kou YW. Investigation on sound and vibration characteristics of aircraft plates subjected to aerodynamic loads [thesis]. Beijing: University of Chinese Academy of Sciences; 2016 (in Chinese)

[24] Nilsson AC, Liu BL. Vibro-Acoustics, Vol. 2. Berlin Heidelberg: Springer-Verlag; 2016

[25] Leppington FG, Broadbent EG, Heron KH. The acoustic radiation efficiency of rectangular panels. Proceedings of the Royal Society of London. Series A. 1982;**382**:245-271. DOI: 10.1098/rspa.1982.0100

Section 2

Turbulent Boundary Layers

Chapter 3

Roughness Effects on Turbulence Characteristics in an Open Channel Flow

Abdullah Faruque

Abstract

A comprehensive study was carried out to understand the effects of roughness on the turbulence characteristics of flow in an open channel and would be presented in this chapter. Tests were conducted with four different types of bed surface conditions (an impermeable smooth bed, impermeable rough bed, permeable sand bed and an impermeable bed with distributed roughness) and at two different Reynolds number (R_e = 47,500 and 31,000). The variables of interest include the mean velocity, turbulence intensity, Reynolds shear stress, shear stress correlation and higher-order moments. Quadrant decomposition was also used to extract the magnitude of the Reynolds shear stress from the turbulent bursting events. The effect of bed roughness on the turbulence characteristics can be seen throughout the depth of flow and thus dispute the 'wall similarity hypothesis'. In comparison to other roughness, distributed roughness shows the greatest effect on both streamwise and vertical turbulence intensities. Velocity triple products that reflects the transportation of turbulent kinetic energy is also seen to be affected by roughness of the channel bed with a variation of 200–300% compared to the flow over smooth bed. To analyze the turbulent bursting events, quadrant decomposition tools were used and found that the roughness affected heavily in the production of extreme turbulent events. The increases of the intensity and frequency of this turbulent burst causes the increase of instantaneous Reynolds shear stress. Transport of the sediment, pollutant suspension from the channel bed, changing the composition of the nutrient in the flow, sustainability of the benthic organisms, entrainment and exchange of energy and momentum are all influenced by this change of Reynolds shear stress. The sand used to form the various bed roughness conditions is same but found that the effect on different turbulence characteristics are different for different roughness. This is a strong indication that the geometric formation of the roughness is the cause of the differences in turbulence characteristics for different roughness formed by the same sand grain.

Keywords: turbulence, open channel flow, roughness, Reynolds shear stress, quadrant analysis, higher-order moment

1. Introduction

1.1 Open channel flow: general

Open channel flow comprises a sheared boundary layer like flow [1]. It is in utmost interest for the engineers and researchers to understand the structure and dynamics of the open channel flow. Numerical modeling and laboratory experiments are two tools used by the researchers to explain the sediment transport, resuspension, formation of channel bed, entrainment in the flow and the exchange of energy and momentum in an open channel flow. Turbulence affects the horizontal and vertical transfer of energy and momentum and causes disruption to nutrition/oxygen utilization rates of some benthic organisms. Turbulent mixing increases with the increment of current speed and enhances the transport of phytoplankton. There were lot of studies with the explanation of mechanism of the above-mentioned phenomena but there are still a lot of unanswered questions and dispute. As indicated by [2] that a significant modulation of turbulence can be the result of average bed particle volume fractions as low as 10^{-4}. The other contribution factors to the modulation of turbulence are the shape, size and arrangement of bed particles. The research in open channel turbulent flow is much less comparing the vast amount of research done on turbulent boundary layer and pipe flow. Although there are significance in engineering application for the flow over rough surfaces but research study on turbulent flow over smooth surfaces [3–9] in both form of experimental and numerical since 1970 superseded the research on flow over rough surfaces. As research grows on the flows over rough surfaces but remains to be the Achilles heel of turbulent research [10]. There are basic differences between the flow in open channel and boundary layer due to the presence of the free surface and channel aspect ratio in an open channel flow and always debatable among researchers to use turbulent boundary layer data for modeling open channel flow [11]. Formation and enhancement of secondary currents occur due to the presence of the free surface and the side walls of the open channel. Free surface also dampens the vertical velocity fluctuations.

1.2 Open channel flow: effect of roughness

The flow progression from a developing state to a fully developed condition was studied by [12]. They have observed that for the case of a section with fully developed flow and the aspect ratio $b/d \geq 3$, the boundary layer extends to the surface of the water. At the channel centerline and near free surface, the velocity profile does not dip even for channel aspect ratio as low as $b/d = 3$. As discussed earlier about the differences between the flow in open channel and turbulent boundary due to the existence of free surface, [13] observed similarity on the velocity field due to the effect of roughness in a zero-pressure gradient turbulent boundary layer. The formation of secondary currents in an open channel flow is related to the aspect ratio (width/depth ratio of flow, b/d) and [7] noted the velocity-dip phenomenon for $b/d < 5$ where the measurement of maximum velocity on the centerline of a flume are seen to be below the free surface. In Ref. [14] indicated that the streamwise mean velocity profiles follow the well-known logarithmic law for the smooth surface, and with an appropriate shift, for the rough surface. In Ref. [15] observed that wall roughness led to higher turbulence levels in the outer region of the boundary layer. In Ref. [13] noted that roughness enhances the levels of the turbulence intensities over most of the flow.

Particle motion near a solid boundary causing sediment deposition and entrainment is influenced by the coherent structures near the wall as noted by [16] in their study of the particle behavior in the turbulent boundary layer. The generation of high-speed regions by vortices in the viscous layer sweeping along the wall causes particles pushing out of the way [16]. In Ref. [17] reported that for locations above the roughness sublayer, the distributions of the second-order turbulent stresses are similar to the smooth-wall distributions. In Ref. [13] noted that roughness enhances the levels of the Reynolds shear stress over most of the flow. The specific geometry of the roughness elements causes significant enhancement to the levels of the Reynolds stresses as stated by [18]. The enhancement to the levels of the Reynolds stresses does not contain near the bed only but progresses over most of the flow creating a stronger interaction between the regions of flow (inner and outer) than would be implied by the wall similarity hypothesis.

In case of the three-dimensional flow (when b/d ≤ 5) [7] predicted a reversal of the sign of the Reynolds shear stress ($-\overline{uv}$) from positive to negative at the location closer to the free surface. Correlation coefficient of the Reynolds shear stress is defined as $\left(R = \frac{-\overline{uv}}{u \times v}\right)$and is an indicator of the degree of similarity of turbulence-uv. The turbulence intensity in streamwise direction (u) and normal to the bed (v) is used to non-dimensionalize the Reynolds shear stress ($-\overline{uv}$). The correlation coefficient of the Reynolds stress only required the turbulence in streamwise and normal to the bed direction and [7] emphasized that the correlation coefficient of the Reynolds stress is very important because the estimation of the friction velocity is not required. The variation of R as stated by [7] is that after monotonous increment with respect to y/d in the region closer to the bed (y/d < 0.1) decreases as one moves away from near bed to the free-surface region. R attains a near constant value in the range of 0.4–0.5 for the middle portion of the flow depth (0.1 ≤ y/d ≤ 0.6). Properties of the mean flow in an open channel and the bed roughness have no effect on the value of R as noted by [7] and called the distribution of R universal. For an open channel flow [19] noted that the value of Reynolds shear stress increases to a maximum at the location closer to the bed and decreases after that. The researchers [19] explained that for the flow over smooth wall, the above mentioned variation of Reynolds shear stress is the effect of viscosity, whereas for the flow over rough bed, the emerging mechanisms for momentum extraction in the existing roughness sublayer is responsible. They blamed the lower value of aspect ratio that created secondary currents for the contradiction in the characteristics of the Reynolds stress with respect to the Reynolds number variation.

In Ref. [17] reported that the relative contributions of sweep and ejection events within the sublayer showed that sweep events provide the dominant contribution to the Reynolds shear stress within this region. In Ref. [13] noted that triple correlations and turbulence diffusion were strongly modified by the surface roughness. In Ref. [18] noted that surface roughness significantly enhances the levels of the turbulence kinetic energy, and turbulence diffusion in a way that depends on the specific geometry of the roughness elements. In Ref. [8] showed that the wall condition affects the variation of the triple products and the effects are not restrained to the near wall but extended to the full depth of flow. Ejection events shown clear dominance over other events for the full depth of the flow as noted by [8] and they also noted significant variation of ejection events with respect to bed roughness. To compare the effect of rough wall with smooth wall on the magnitude of the extreme events, they did the quadrant decomposition of the instantaneous velocity and found much higher magnitude for the flow over rough bed compared to the smooth wall flow. This is an indication of the effect of roughness propagating into the full depth of flow and not constraint to the region closer to the bed. Quadrant analysis is also done by [11] to compare the turbulent structures of open channel flow with the same in boundary layer flow. They found that the turbulent structures are very similar if all turbulent events are included in the analysis but found very significant difference if only the extreme events are used in analysis.

2. Experimental setup

2.1 Open channel flume

A 9-m long open channel flume at the University of Windsor with a rectangular cross-section dimension of 1100 mm × 920 mm is used to perform the experiment. **Figure 1** shows the schematic of the experimental setup with open channel flume. A squire cross-section dimension of 1.2 m and depth of 3 m header tank is placed at the beginning of the flume. The depth of flow for this series of experiments are kept to 100 mm, eventually achieving the aspect ratio (width/depth ratio of flow, b/d) of 11. Choice of this aspect ratio is based on the expectation that the generation of the secondary current will be minimum and the flow can be a representation of two-dimensional flow [7]. Two centrifugal pumps of 15 horsepower capacity each are used to recirculate the water. Tempered transparent glass materials are used to build the sidewalls and bottom of the flume and will enable the LDA (laser Doppler anemometer) to measure the instantaneous velocity. There were many previous studies [20–21] confirmed the quality of the flow of this permanent facility. The flume has an adjustable slope mechanism at the bottom but was kept horizontal for this series of test. 720 and 450 GPM are the two constant flow rate used for the tests.

2.2 Test conditions to study the effect of roughness

One hydraulically smooth and three characteristically different rough surfaces are used in this study to capture and understand the open channel flow characteristics. **Figure 2a** shows the hydraulically smooth bed condition made up by a polished aluminum plate spanning full width of the flume. Sand composed of uniform particles with gradation characteristics as shown in **Table 1** is used to create the three different rough surfaces. Four different types of bed surface conditions were used in this study. **Figure 2b** shows the 'distributed roughness' rough surface, **Figure 2c** shows the 'continuous roughness' rough surface and **Figure 3** shows the 'natural sand bed' rough surface. A 18 mm wide sand strip is glued on top of the polished aluminum plate spanning full width of the flume alternate by a 19 mm wide smooth strip to generate the distributed roughness. The same sand grain is glued on top of the entire polished aluminum plate spanning full width of the flume to generate the continuous roughness. Natural sand bed condition is consist of 3.7 m long 200 mm thick uniform sand of the same characteristics

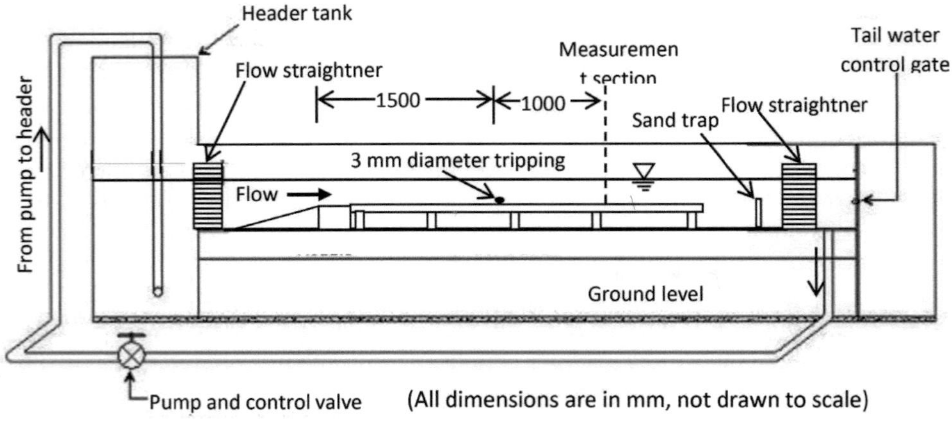

Figure 1.
Schematic of the open channel flume and experimental setup.

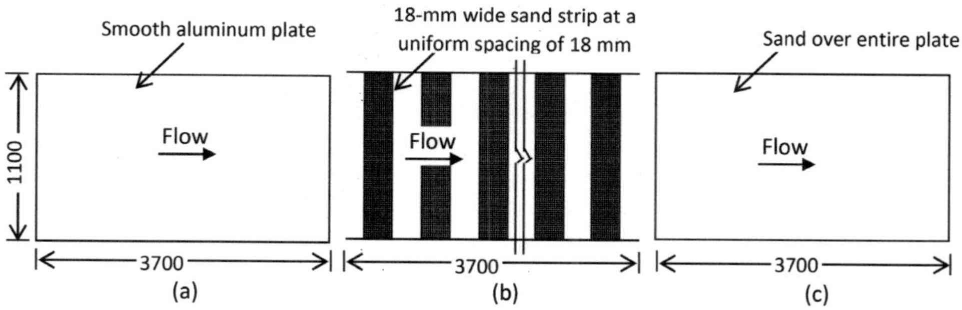

Figure 2.
Plan view of different fixed bed condition. (a) Hydraulically smooth surface, (b) Distributed roughness surface, (c) Continuous roughness surface.

d_{50} (mm)	2.46
d_{95}/d_5	1.91
d_{95}/d_{50}	1.34
d_{84}/d_{50}	1.26
$\sigma_g = \sqrt{d_{84}/d_{16}}$	1.24
$C_z = d_{30}^2/(d_{60}d_{10})$	1.00

Table 1.
Gradation measurements of the sand.

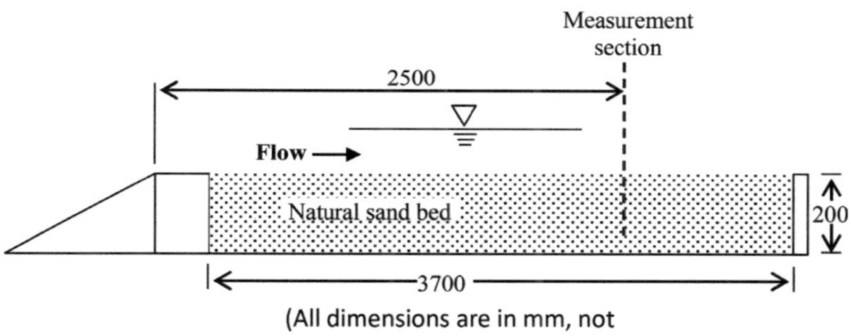

Figure 3.
Section of natural sand bed.

spanning full width of the flume. Special care had been taken in maintaining the flow condition in such a way that there were no sand movement in any period of time of running the test. As a precautionary measure of accidental sand movement and sand entering into the pipe/pump system causing damage to the pump, a sand trap is constructed at the end of the flume.

Two different flow Reynolds numbers ($R_e = U_{avg}d/\nu \approx 47{,}500$ and 31,000) correspondence to two different Froude numbers ($F_r = U_{avg}/(gd)^{0.5} \approx 0.40$ and 0.24) respectively are used for each four bed surface conditions. The parameters used for Reynolds and Froude number calculations are the average streamwise velocity (U_{avg}), nominal depth of flow (d), kinematic viscosity of the fluid (ν) and gravitational acceleration g. The flow conditions are maintained to be subcritical (i.e., Froude numbers less than unity) and choose the flow Reynolds numbers accordingly. The variation of water surface elevation were measured for the test section and there are less than 1 mm variation of surface water for a streamwise distance of 600 mm proves that pressure gradient is negligible. In order of conditioning the flow, two sets of flow straighteners are placed at the beginning and end of the flume. A turbulent boundary layer presence is ensured by tripping the flow using a 3 mm diameter rod at the upstream of the measurement section as shown in **Figure 1**. Shape factor of the boundary layer (the ratio of displacement to momentum thickness) for the flow over smooth bed is found for this case as 1.3 and flow can be considered as fully developed turbulent flow [22]. The instantaneous velocity measurement is carried out on top of the 60th sand strip for the flow over distributed roughness bed. To minimize the effect of secondary current, measurements for all flow test conditions are carried out along the flume centerline. Preliminary tests for all bed conditions are carried out to confirm that the flow condition is fully developed. **Table 2** presents the summary of various test conditions.

2.3 The laser Doppler anemometry

Velocity measurements were done using A commercial two-component fiber-optic LDA system (Dantec Inc.) which is powered by a 300-mW Argon-Ion laser. Details of this is avoided for brevity because using the same system in several previous studies [20, 23, 9]. A Bragg cell and a focusing lens of 500 mm with beam spacing of 38 mm are the optical elements of the LDA system. A large amount of data collected (10,000 validated samples at each and every measurement location) to minimize the uncertainty of the data collection. The data rate varied widely based on the location of the measurement and ranges from 4 Hz to 65 Hz. The water used in the test is seeded with hollow spheres with density of 1.13 g/cc with mean particle size of 12 microns after filtering the water for many days and it is done prior to the start of the measurement. The seeded particles can stuck on the flume side wall and can cause extraneous scattered light distributed throughout the illuminating beams. The glass side wall around the measurement region were cleaned before each set of measurement to avoid the erroneous data collection due to the scattered light. Due to the measurement location at the flume centerline, two scattered beams of the present two-component LDA system measuring the vertical component of the velocity cannot reach at very close to the bed or very close to the free surface but measurement of streamwise one-component velocity were carried out for full depth of flow. Following the footsteps of other researchers [16, 6] who have successfully tilted the probe by 3° and 2°, respectively, in their pursuit to collect two dimensional velocity data closer to the wall, the LDA probe for the present tests was tilted 2° towards the bottom wall to capture data for two-component velocity measurements at near proximity of the wall.

Test	Bed condition	d (mm)	R_e	F_r
1	Smooth bed	~100	~47,500	~0.40
2		~100	~31,000	~0.24
3	Distributed roughness	~100	~47,500	~0.40
4		~100	~31,000	~0.24
5	Continuous roughness	~100	~47,500	~0.40
6		~100	~31,000	~0.24
7	Natural sand bed	~100	~47,500	~0.40
8		~100	~31,000	~0.24

Table 2.
Summary of test conditions to study the effect of roughness.

3. Results and discussions

The purpose of the present study is to explain how the roughness and Reynolds number affect flow characteristics in an open channel flow (OCF). Tests were conducted with four different types of bed surface conditions (an impermeable smooth bed, impermeable rough bed, permeable sand bed and an impermeable bed with distributed roughness) and at two different Reynolds number (R_e = 47,500 and 31,000) for each and every bed surface. Instantaneous velocity components are used to analyze the streamwise mean velocity, turbulence intensity in both streamwise and vertical direction, Reynolds shear stress including shear stress correlation and higher-order moments including vertical flux of the turbulent kinetic energy. Quadrant decomposition was also used to extract the magnitude of the Reynolds shear stress from the turbulent bursting events.

3.1 Mean velocity profiles

3.1.1 Outer coordinates

Figure 4 shows the variation of streamwise component of the velocity with respect to the depth of flow in outer coordinates. The mean velocity (U) is non-dimensionalize by the maximum velocity (U_e) and the wall normal distance (y) is non-dimensionalize by the maximum flow depth (d). As one can see in the inset in **Figure 4** that the velocity profiles of every flow conditions show a slight dip in the outer region where the location of maximum velocity happened to be occurred below the free surface with $dU/d\partial y$ is negative in the location close to the free surface. Velocity dip is different with different rough bed conditions with flow over natural sand bed showing the biggest dip followed by distributed roughness and continuous roughness bed. However, the flow over smooth surface shows the dip higher than the flow over distributed roughness and continuous roughness bed. Effect of bed roughness is very evident at the location close to the bed with velocity profile for the flow over smooth wall is fuller compared the flow over different rough beds. The same phenomenon was also observed by [15] and blamed it to the increment of surface drug due to the effect bed roughness. Comparing the effect of various type of bed roughness on the streamwise velocity component as one can see from **Figure 4a** that distributed roughness profile has the biggest deviation from smooth bed profile with continuous roughness and natural sand bed shows identical deviation. The variation of streamwise component of the velocity with respect to the depth of flow in outer coordinates with respect to the lower Reynolds number is shown in **Figure 4b**. The velocity profile characteristics are very similar for the lower Reynolds number flow compared to the flow for higher Reynolds number with the exception of flow over natural sand bed, which shows much higher deviation than flow over the bed of continuous roughness. One can correlate this with the interchange of fluid and momentum across the boundary, which is permeable like the flow over the bed of natural sand. The subsequent momentum/energy loss due to the effect of infiltration and corresponding differences on mean velocity reduces with the increment of Reynolds stress.

3.1.2 Inner coordinates

Figure 5 shows the variation of streamwise component of the velocity with respect to the depth of flow in inner coordinates. The Clauser method was used to calculate the friction velocity for flow over smooth and rough bed conditions by

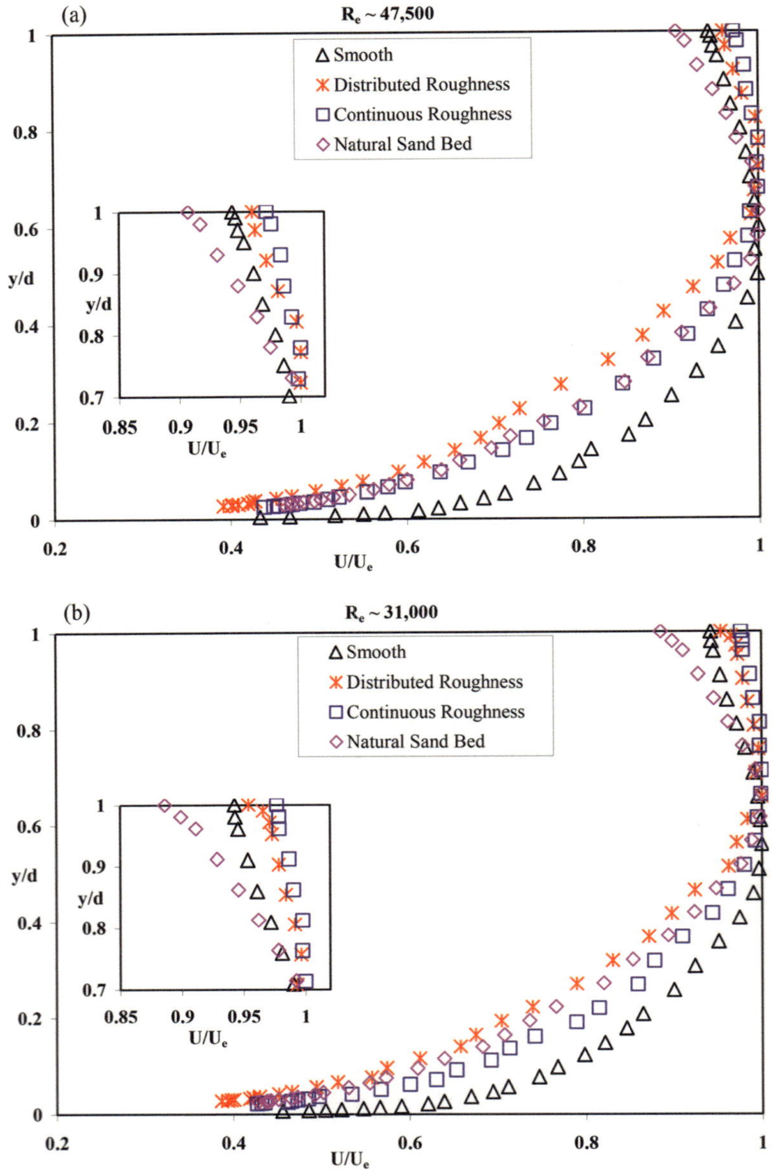

Figure 4.
Streamwise mean velocity profile for flow over different bed condition.

fitting the respective mean velocity profiles of different bed conditions with the classical log law, $U^+ = \kappa^{-1} \ln y^+ + B - \Delta U^+$. Log-law constants used here are $U^+ = U/U_\tau$, $y^+ = yU_\tau/\nu$, $\kappa = 0.41$, $B = 5$ and the downward shift of the velocity profile represented by the roughness function ΔU^+ with $\Delta U^+ = 0$ for the flow over the bed which is smooth. The present test data over the smooth bed has better agreement with the standard log-law represented by the solid line. For the flow over rough beds there are downward shift of the profile compared to the smooth bed which is fully expected and clearly visible. The effect of roughness can be measured by the downward shift of the profile and one can note from **Figure 5a** that the distributed roughness shows the highest deviation from the smooth bed with flow over natural sand bed shows the least deviation and flow over continuous roughness fall in-between. The variation of streamwise component of the velocity with respect to the

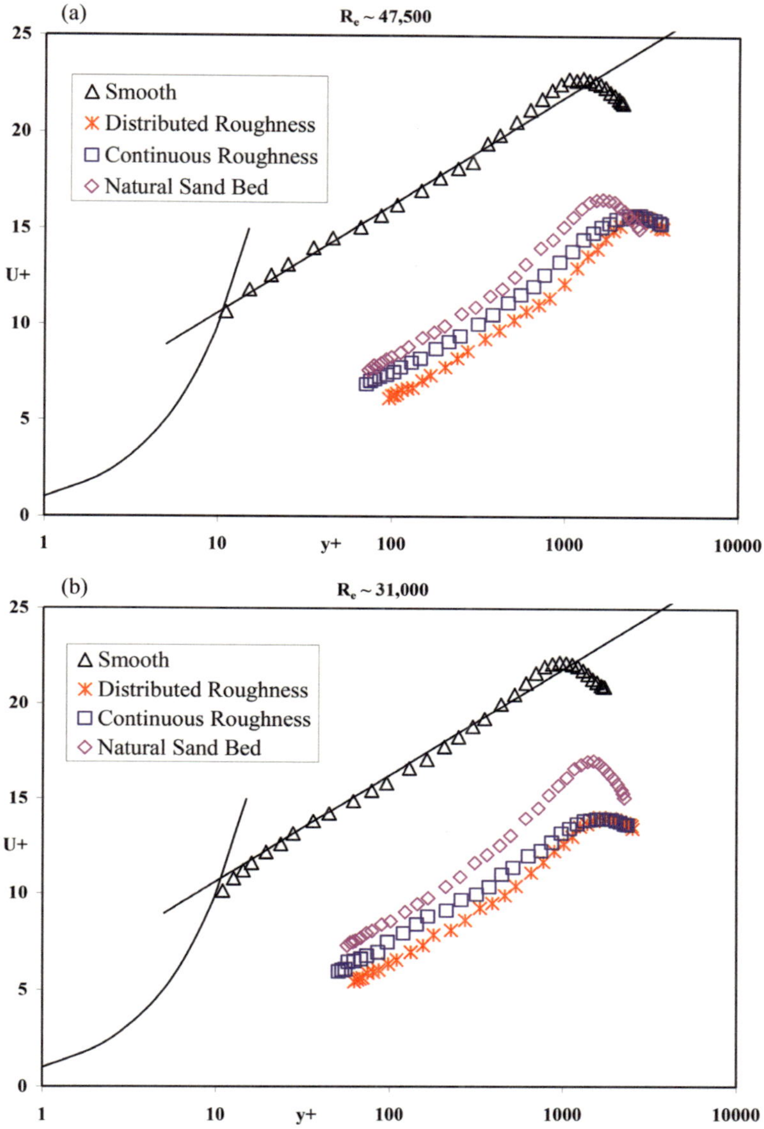

Figure 5.
Mean velocity profile in inner coordinates for flow over different bed condition.

depth of flow in inner coordinates with respect to the lower Reynolds number is shown in **Figure 5b**. The velocity profile characteristics are very similar for the lower Reynolds number flow compared to the flow for higher Reynolds number.

The magnitude of friction coefficient C_f ($C_f = 2(U_\tau/U_e)^2$) is found to be dependent on the type of bed roughness with distributed roughness has the highest value followed by the flow over the continuous roughness bed surface and the sand bed. The magnitude of friction coefficient is also found to be dependent on the Reynolds number with the reduction of the magnitude of friction coefficient with the increment of the Reynolds number. The magnitude of C_f is seen to be smaller for the flow over a permeable bed (natural sand bed) compared to the flow over an impermeable bed (distributed and continuous roughness bed). One can correlate this with the development of finite slip velocity across the permeable boundary layer causing the reduction of the magnitude of friction compared to the flow over impermeable layer. In contrary, [24] discovered that for the boundaries with similar rugosity the magnitude of friction resistance is seen to be higher for the flow over a permeable bed compared to the flow over an impermeable bed. Dissipation of energy happened in the transition zone of the porous permeable medium with added loss of energy due to interchange of fluid and momentum across the permeable boundary translated back into the main flow. They commented that the net effect of combined energy loss might be responsible for the higher resistance.

3.2 Turbulence intensity

3.2.1 Streamwise turbulence intensity

The distribution of the streamwise component of the turbulence intensity for flow over both smooth and rough beds is shown in **Figure 6**. Computed quantities

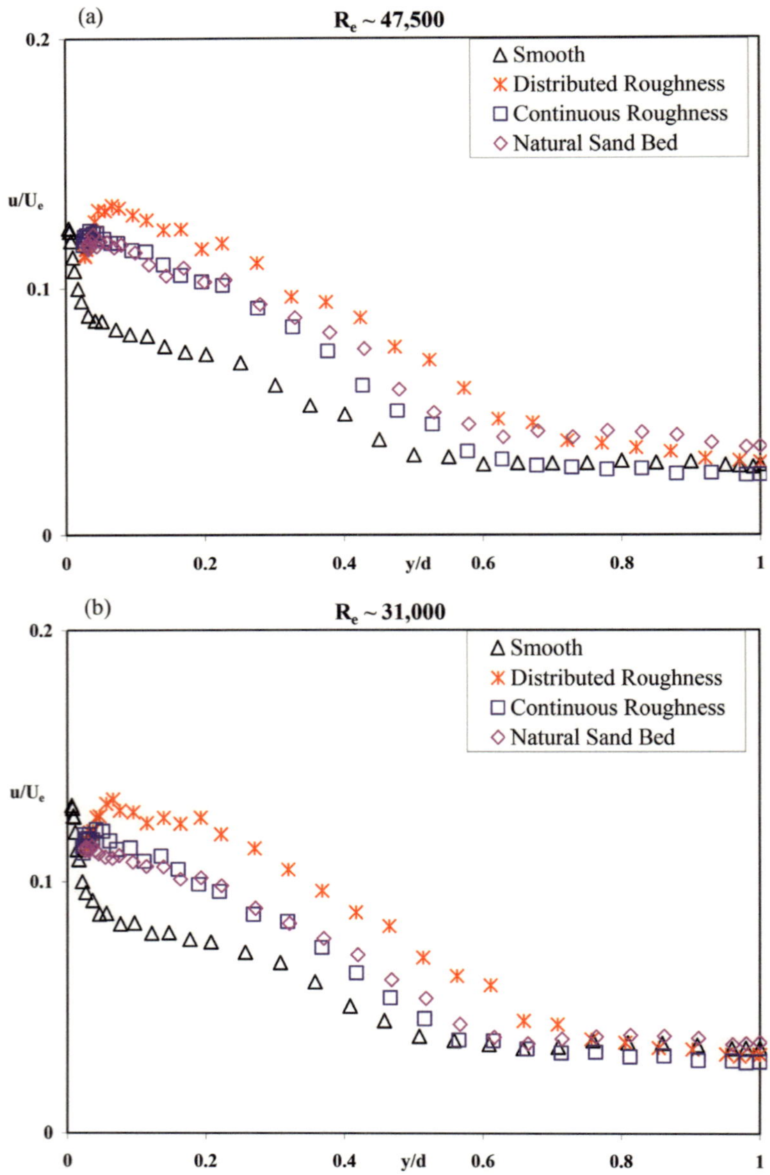

Figure 6.
Streamwise turbulence intensity for flow over different bed condition.

can bring additional uncertainties in relation to the scaling parameters and to avoid any additional uncertainties, the streamwise turbulent intensity (u) is non-dimensionalize by the maximum velocity (U_e) and the wall normal distance (y) is non-dimensionalize by the maximum flow depth (d). Magnitude of the streamwise component of the turbulence intensity reaches to the maximum at the location very close to the bed irrespective of the bed condition as one can note from **Figure 6a**. The location of maximum streamwise component of the turbulence intensity is different with different bed conditions. The location of the peak for the flow over smooth bed is very close to the bed at $y/d \sim 0$, whereas the peak for the flow over rough surfaces varies with the different type of roughness. As one can note from **Figure 6a** that the distributed roughness shows the highest peak compared to the flow over continuous roughness and flow over natural sand bed. The location of the peak for the flow over rough beds are also varied depending on the type of roughness. The location of the peak for the flow over distributed roughness is at around $y/d \sim 0.08$ whereas the location of the peak for the flow over continuous roughness and natural sand bed have occurred at the same location of $y/d \sim 0.04$ which is a distance closer to the bed compared to the flow over distributed roughness. Immediately after reaching the peak the streamwise component of the turbulence intensity for flow over both smooth and rough beds reduces but the trend of reduction is very different for the flow over smooth bed compared to the flow over rough surfaces. There is a sharp drop of the magnitude of the streamwise component of the turbulence intensity for the smooth bed before a more constant drop towards the free surface and reaching a near constant value at $y/d \sim 0.5$. For the flow over rough surfaces the drop of the value towards the free surface after the peak is linear and attains a near constant value but the location and magnitude of constant value is different for different rough surfaces (distributed roughness does not attain constant value but variation near free surface is minimal). The location of a near constant value for the flow over continuous roughness and natural sand bed is at the same level of $y/d \sim 0.62$. The streamwise component of the turbulence intensity near the free surface also shows the effect of roughness with natural sand bed shows the highest intensity followed by the distributed roughness with flow over continuous roughness is the lowest. The effect of roughness on the distribution of the streamwise component of the turbulence intensity is very evident throughout the flow depth with distributed roughness shows the highest deviation followed by natural sand bed and continuous roughness compared to the smooth surfaces with the exception at the location very close to the bed. Although the sand grain used to create all three bed roughness is of the same gradation characteristics but the geometry of the roughness formation is different causing the differences in the distribution of the streamwise component of the turbulence intensity.

The variation of streamwise component of the turbulence intensity with respect to the depth of flow for the flow conditions to the lower Reynolds number is shown in **Figure 6b**. The streamwise component of the turbulence intensity profile characteristics are very similar for the lower Reynolds number flow compared to the flow for higher Reynolds number with the exception of flow over distributed roughness bed, which shows much higher deviation than flow over the smooth bed at the lower Reynolds number. In lower Reynolds number flow, the differences in streamwise component of the turbulence intensity for continuous roughness and natural sand bed is negligible.

3.2.2 Vertical turbulence intensity

The distribution of the vertical component of the turbulence intensity for flow over both smooth and rough beds is shown in **Figure 7**. Significant effect of

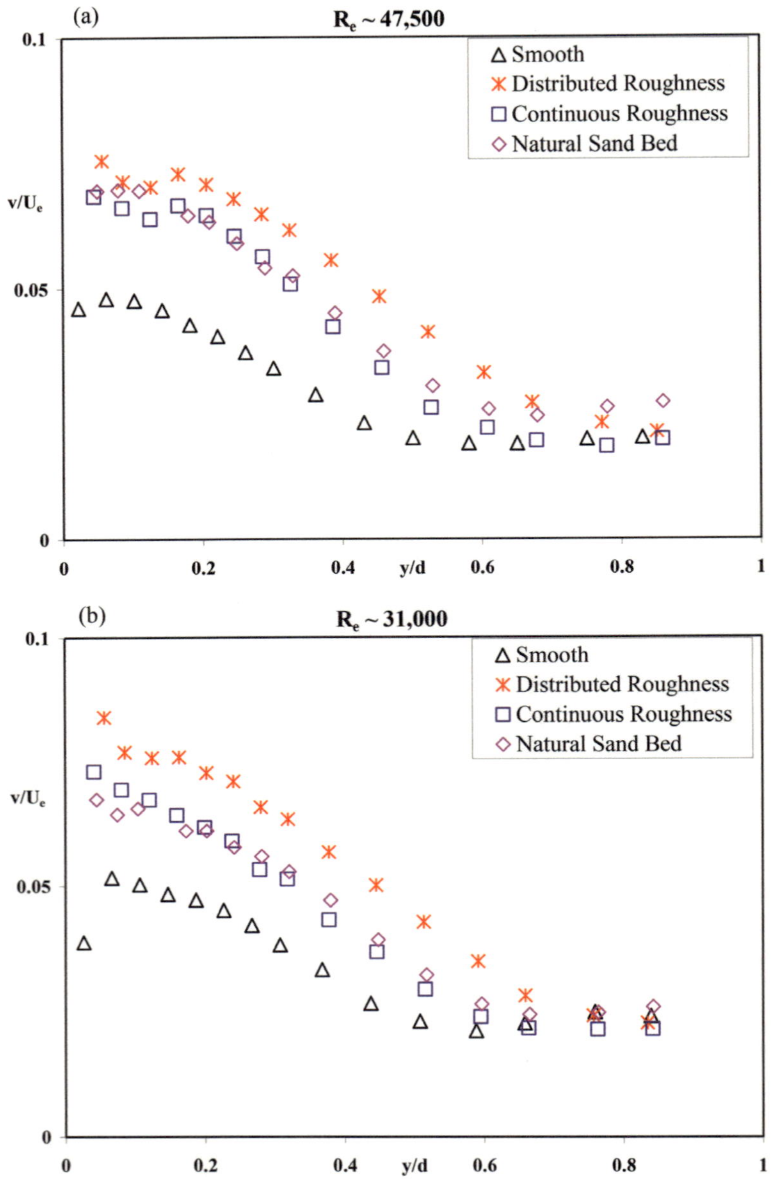

Figure 7.
Vertical turbulence intensity for flow over different bed condition.

roughness can be seen for lower two third of the depth of flow with the effect tapered off at the location closer to the surface. Comparing the effect of various type of bed roughness on the vertical component of the turbulence intensity as one can see from **Figure 7a** that distributed roughness profile has the biggest deviation from smooth bed profile with continuous roughness and natural sand bed shows identical deviation for most the depth of the flow. For the location closer to the surface, the flow over natural sand bed shows higher magnitude of the vertical component of the turbulence intensity compared to any other surfaces. The variation of the vertical component of the turbulence intensity for flow with respect to the lower Reynolds number is shown in **Figure 7b**. The profile characteristics are very similar for the lower Reynolds number flow compared to the flow for higher Reynolds number with the exception that there are almost no effect of roughness on the vertical component of the turbulence intensity for the location closer to the surface.

3.3 Reynolds shear stress

The distribution of the Reynolds shear stress in outer variables for flow over both smooth and rough beds is shown in **Figure 8**. Magnitude of the Reynolds shear stress reaches to the maximum at the location very close to the bed (y/d < 0.2) irrespective of the bed condition as one can note from **Figure 8a**. Effect of roughness on the Reynolds shear stress is very evident for lower two third of the depth of flow with the effect tapered off at the location closer to the surface. The peak for the flow over rough surfaces varies with the different type of roughness. As one can

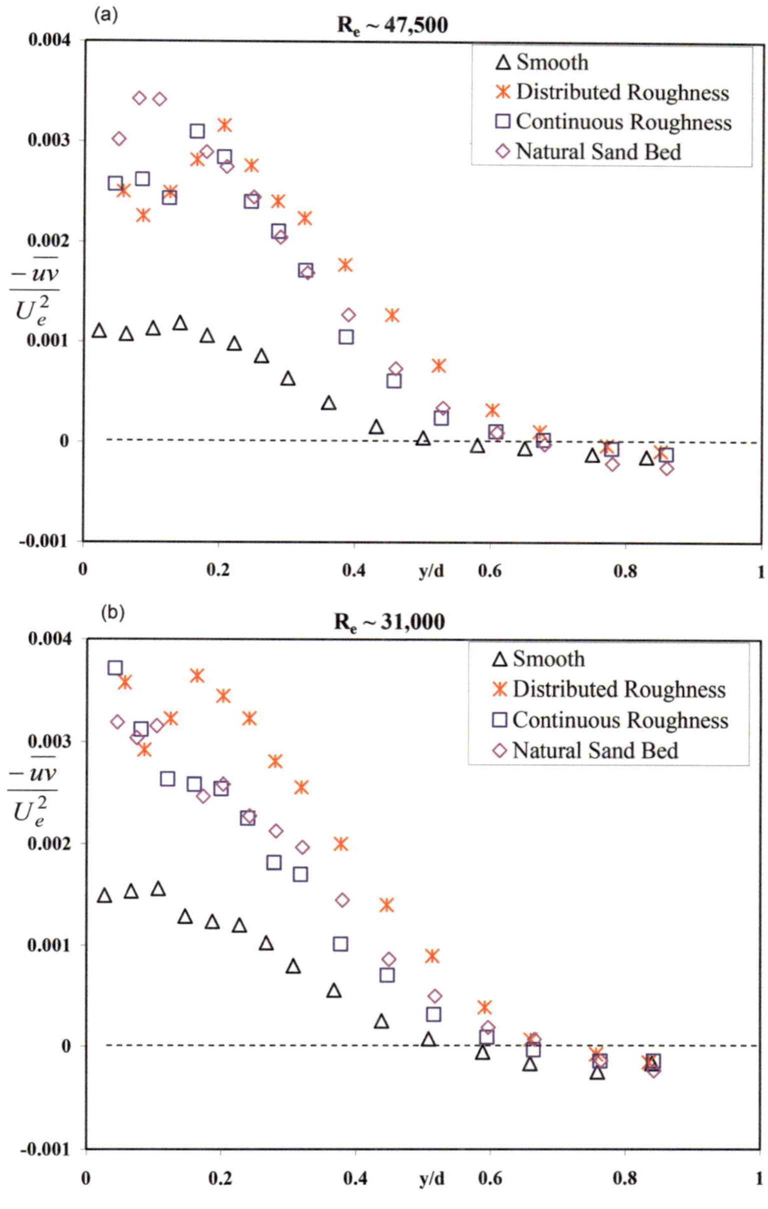

Figure 8.
Reynolds shear stress distribution for flow over different bed condition.

note from **Figure 8a** that the flow over natural sand bed shows the highest peak compared to the similar peak for flow over continuous roughness and flow over distributed roughness. Immediately after reaching the peak the Reynolds shear stress for flow over both smooth and rough beds reduces but the trend of reduction is very different for the flow over smooth bed compared to the flow over rough surfaces. There is a sharp drop of the magnitude of the Reynolds shear stress for the rough beds compared to the smooth bed before a more constant drop towards the free surface. For the region further away from the near bed (y/d > 0.2), flow over distributed roughness shows generation of higher Reynolds shear compared to the other two rough beds where the generation of the Reynolds shear stress is very similar. As one can see in **Figure 8a** that the Reynolds shear stress falls below zero and becomes negative in the location close to the free surface for flow over both smooth and rough beds. The location of zero Reynolds shear stress is different for flow over smooth bed (at y/d ~ 0.5) compared to the flow over rough beds (y/d ~ 0.7). The location of negative Reynolds shear stress for different bed conditions are on the same location where dU/d∂y is negative as one can see in **Figure 4**. Few other researchers [25, 6, 26] found the visible effect of roughness on Reynolds shear stress for the depth of flow y/d ≈ 0.2–0.3 but the distinct effect of roughness for the present study can be seen penetrating deep into the flow y/d ≈ 0.7. In case of the study by [3] where the researcher did not find any effect of roughness (2 mm sand and 9 mm pebbles) on Reynolds shear stress compared to the flow over smooth bed. The sample size used for the tests by [3] were rather very small rendered to the unexpected conclusion. The variation of the Reynolds shear stress for flow with respect to the lower Reynolds number is shown in **Figure 8b**. The profile characteristics are very similar for the lower Reynolds number flow compared to the flow for higher Reynolds number with one of the exception is that the flow over continuous roughness and flow over distributed roughness shows the similar highest peak compared to the flow over natural sand bed. Another exception can be seen as much higher generation of Reynolds shear stress for the flow over distributed roughness for the region further away from the near bed (y/d > 0.2) followed by flow over natural sand bed and continuous roughness.

3.4 Shear stress correlation coefficient

The distribution of the correlation coefficient of the Reynolds shear stress $\left(R = \frac{-\overline{uv}}{u \times v}\right)$ for flow over both smooth and rough beds is shown in **Figure 9**. One can state that R is the expression of a normalized covariance where degree of similarity between the streamwise component of the turbulence intensity and the vertical component of the turbulence intensity is established. The range of the R as $-1 \leq R \leq 1$ where the value of R = 1 is the indication that the linear relationship between the streamwise component and the vertical component of the turbulence intensity is increasing. The value of R = −1 is the indication that the linear relationship between the streamwise component and the vertical component of the turbulence intensity is decreasing. Local statistics of R at a particular location can be an indication of the presence or absence of any flow structures. The effect of roughness on the variation of R is mixed compared to the smooth bed flow. As one can see from **Figure 9a** that at the location close to the bed (y < 0.3d) the magnitude of R is very similar for flow over smooth bed compared to the flow over distributed roughness with much higher value of R for the flow over continuous roughness and natural sand bed. The effect of roughness for the outer layer (y > 0.3d) is very clear with value of R is consistently higher for the flow over all three rough beds compared to the flow over smooth bed. One can also see from **Figure 9a** that the value

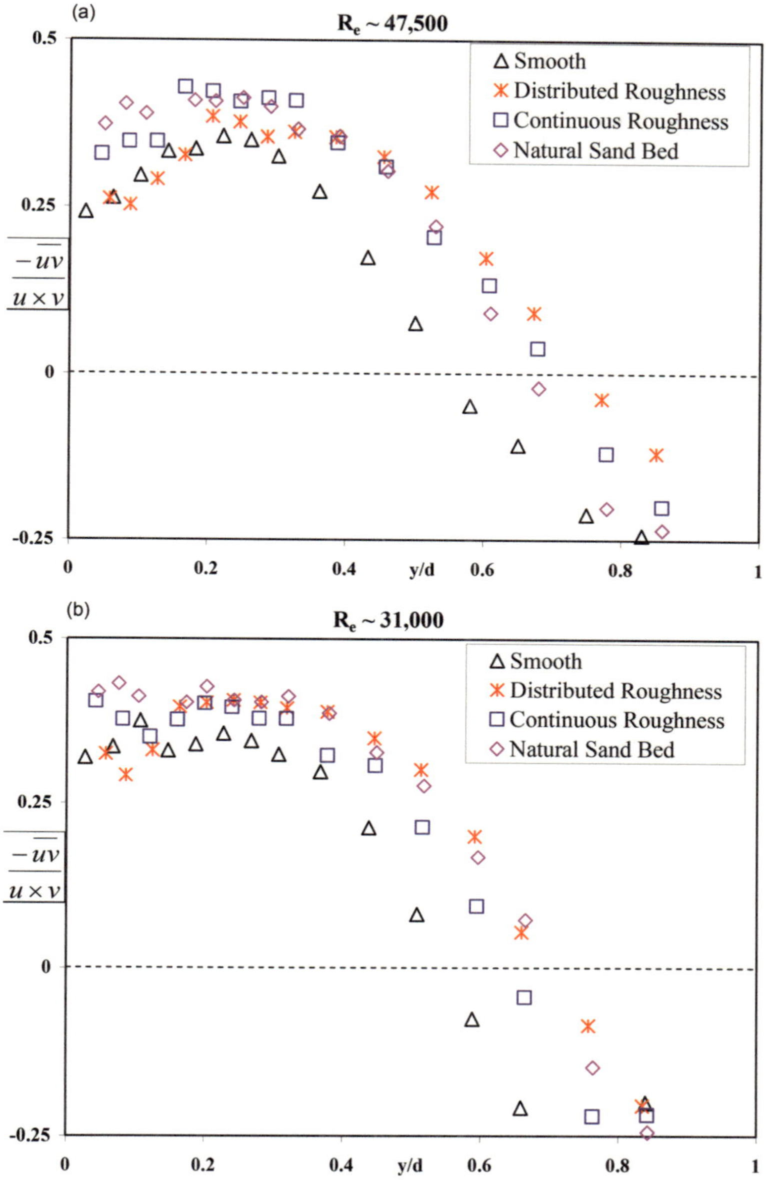

Figure 9.
Distribution of correlation coefficient for flow over different bed condition.

of R increases with the increasing distance from the bed and the trend reverses for the outer layer (y > 0.3d), indicating the changes of flow structure characteristics between the near-bed region and outer region. This observation is clearly different than the characteristics of R noted by [7, 5–6] where [7] called the distribution of R universal as they did not find any effect of roughness on the value of R. In Ref. [7] noted an existence of equilibrium region for $0.1 \leq y/d \leq 0.6$ with a value of R = 0.4–0.5 in open-channels, pipes, and boundary layers, irrespective of whether the wall bed is smooth or rough. In the inner region and for the flow over smooth bed, [6] found a much lower value of R and noted indifference of R value for the flow over rough and smooth bed for y > 0.15d with the peak values floating to 0.35 ± 0.02 range. Comparing the effect of various type of bed roughness on the correlation coefficient as one can see from **Figure 9a** that distributed roughness has the higher absolute value of R followed by continuous roughness and natural sand bed for the upper third of the flow. One can also note from **Figure 9a** that the value of R changes sign and become negative for flow over both smooth and rough bed surfaces at the locations where the Reynolds shear stress and dU/dy is negative. In the reference [6] also report similar observation. The value of R for the present study ranges from $-0.25 < R < 0.5$ and can be considered as small to medium correlation between the streamwise component and the vertical component of the turbulence intensity for all bed conditions and full depth of flow. The variation of the correlation coefficient of the Reynolds shear stress for flow over various bed surfaces with respect to the lower Reynolds number is shown in **Figure 9b**. The profile characteristics are very similar for the lower Reynolds number flow compared to the flow for higher Reynolds number with the exception that the profiles are more or less flatter for all bed conditions and bottom third of the flow. Another difference is that for the outer layer (y > 0.3d) flow the magnitude of R is higher for flow over natural sand bed compared to the flow over continuous roughness.

3.5 Higher-order moments

Velocity triple products $\overline{u^3}$, $\overline{u^2v}$, $\overline{v^3}$ and $\overline{v^2u}$ are very useful tools used by the researchers to extract valuable information of the flow structures and the distribution of different normalized velocity triple products are shown in **Figure 10**. To avoid any additional uncertainties by using computed quantities in relation to the scaling parameters, directly measured parameters of the maximum velocity (U_e) and the maximum flow depth (d) are used for normalizing all four velocity triple products. Streamwise flux of the turbulent kinetic energy u^2 and v^2 is defined by $\overline{u^3}$ and $\overline{v^2u}$ respectively whereas vertical transport/diffusion of the turbulent kinetic energy u^2 and v^2 is defined by $\overline{u^2v}$ and $\overline{v^3}$ respectively. Transportation in the direction normal to the bed for the Reynolds shear stress is also defined by $\overline{v^2u}$. Ejection-sweep cycle is considered to be the main turbulent motion contributing to the turbulent transport and velocity triple products are the tools used by the researchers to explain the ejection-sweep events effectively. Various bed conditions affect the variation of the different velocity triple products eventually provide insight about causing turbulent transport mechanisms change/modification.

For the flow condition over the smooth bed and very close to the bed, the magnitude of $\overline{u^3}$ is negative and $\overline{u^2v}$ is positive as one can note from **Figure 10a** and **b** indicating a fluid parcel slowly moving upward causing transportation of u momentum away from the bed representing an motion of ejection type. For the flow condition over the rough beds and very close to the bed, the magnitude of $\overline{u^3}$ is positive with very high comparable value and $\overline{u^2v}$ is negative as one can note from **Figure 10a** and **b** indicating a fast moving fluid parcel acting downwards causing transportation of u momentum towards the bed representing an motion of sweep type. Both triple products parameters of $\overline{u^3}$ and $\overline{u^2v}$ changes sign as one moves away from bed towards the free surface rendered changes of ejection-sweep cycle. The change of ejection-sweep cycle as one moving away from the bed is also observed by [27] and they had related this characteristic to the accompanying streaks of low-speed produced by the rough bed conditions and modification of the longitudinal vortices. The magnitude of $\overline{u^3}$ becomes more negative as one moves further away from the bed (y/d > 0.08) causing the sweeping events reduced substantially with the value of $\overline{u^3}$ fluctuates but stays negative for the depth throughout. The effect of roughness is also very evident for the value of $\overline{u^3}$ when compared with flow over smooth bed. The above mentioned differences between the flow over smooth bed

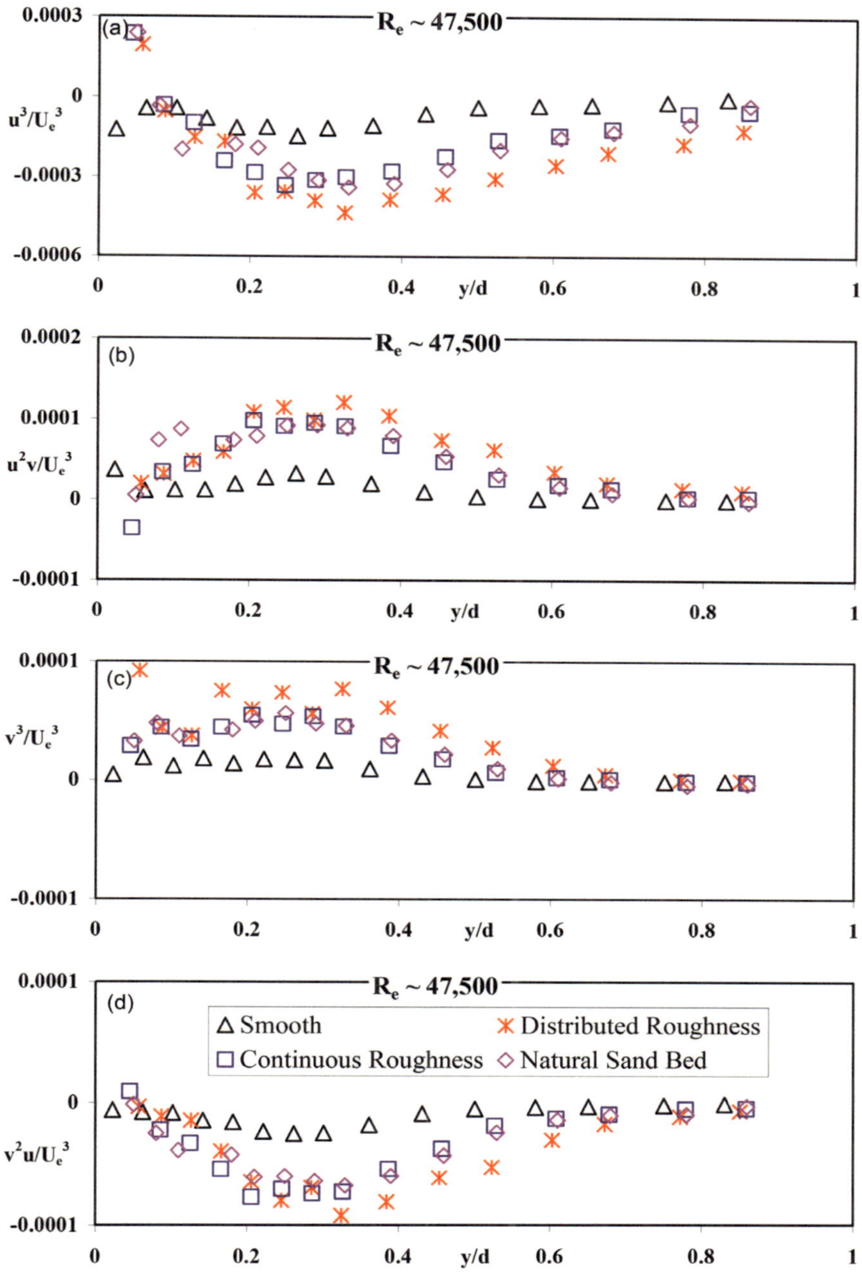

Figure 10.
Distribution of different velocity triple products for flow over different bed condition at $R_e \sim 47{,}500$.

and rough beds are in complete opposite to the observation of [27–28] who did not observe much variation at distance y/d > 0.2. For flows over transverse rod roughness, large differences in the variation of $\overline{u^3}$ were observed by [29] upto to the edge of the boundary layer. This difference as related by [3] is related to the lack of formation of long streamwise vortices near the rough wall. Comparing flow over rough bed conditions with flow over smooth bed, the mechanics of the entrainment of low momentum fluid at the wall differed as noted by [3]. The variation trend for of $\overline{v^3}$ (**Figure 10c**) and $\overline{u^2v}$ (**Figure 10b**) are very similar but there are exception in sign that throughout the depth the $\overline{v^3}$ is positive and much smaller magnitude (~60%) than $\overline{u^2v}$. The trend is qualitatively similar if one compare $\overline{v^2u}$ (**Figure 10d**) with $\overline{u^3}$ (**Figure 10a**) with the exception that the magnitude of $\overline{v^3}$ is about 60% less than that of $\overline{u^2v}$ but the magnitude of $\overline{v^2u}$ is much lower (about 20–25%) comparing magnitude of $\overline{u^3}$. In Refs. [6, 8] in their open channel flow experiments and [27–28] in their turbulent boundary layer experiments had also noted a similar reduction. The lower turbulent intensity in vertical direction is mainly the reason for the differences between $\overline{v^3}$ and $\overline{u^2v}$ and $\overline{v^2u}$ and $\overline{u^3}$. Comparing the magnitude of different velocity triple products for the open channel flow with the turbulent boundary layer flow, one can see the similarity as well as differences for magnitude and extent of the depth of the flow affected mainly in the outer layer by the roughness. As one can see from the **Figure 10** that the local peak (maxima/minima) for all normalized velocity triple products are in very similar location for the flow over smooth wall (≈0.26d). But the location of the peak (maxima/minima) for all normalized velocity triple products does not vary much with different type of roughness and occurs at a location of $y/d \approx 0.33$ for different rough beds. The magnitude of various velocity triple products changes in the range of 200–300% as one can note it from **Figure 10** when comparing the flow over the smooth bed to the flow over rough beds. The similar significant decrease/increase of the magnitude of various velocity triple products in the range of ~300% was also noticed by [8] when comparing the flow over the smooth bed to the flow over dunes. With the exception of the magnitude of $\overline{u^3}$ for the flow over distributed roughness, the magnitudes of the various velocity triple products approach zero for all smooth and rough surfaces as one moves from the location where the local maximum/minimum level achieved towards the free surface at $y/d > 0.85$.

The magnitude of various velocity triple product reaches near-zero at the location very close to the free surface irrespective of the bed surface conditions as one can note from **Figure 10** which is a clear indication of significant reduction of turbulent activity at near free surface. There is another significant finding one can note from the same figure that the type of bed roughness does not affect the location of maximum/minimum of various velocity triple product although there is clear effect of roughness on the magnitude of various velocity triple products. Flow over distributed roughness shows higher magnitude of various velocity triple product comparing with flow over other rough beds followed by very similar magnitude for flow over continuous roughness and natural sand bed. Turbulent activity at the near bed ($y/d < 0.1$) location also seen to be dependent on bed surface conditions. Flow over smooth bed shows the ejection type activity near bed location whereas the flow over rough beds show the sweep type activity at the location close to the bed. Interpolating this scenario to the real life stream or river flow, one can clearly note the influence of strong ejection/sweeping motion of the fluid parcels to the resuspension/transport of the bed particles. Ejection type events are very evident throughout the depth of flow with the exception of the location very close to the bed with flow over smooth bed only where one can observe some sweeping type of event. Bed surface conditions clearly affect the strength of the ejection like events with distributed roughness again shows the highest strength compared to similar strength from continuous bed roughness and natural sand bed. **Figure 11** shows the variation of same velocity triple products for the flow with respect to lower Reynolds number. The profile characteristics of all velocity triple products are very similar for flow with respect to lower Reynolds number compared to the flow with respect to higher Reynolds number. The differences in magnitude of various velocity triple products are seen to be reduced in the case of lower Reynolds number flow comparing the flow over smooth bed to the flow over continuous bed roughness and

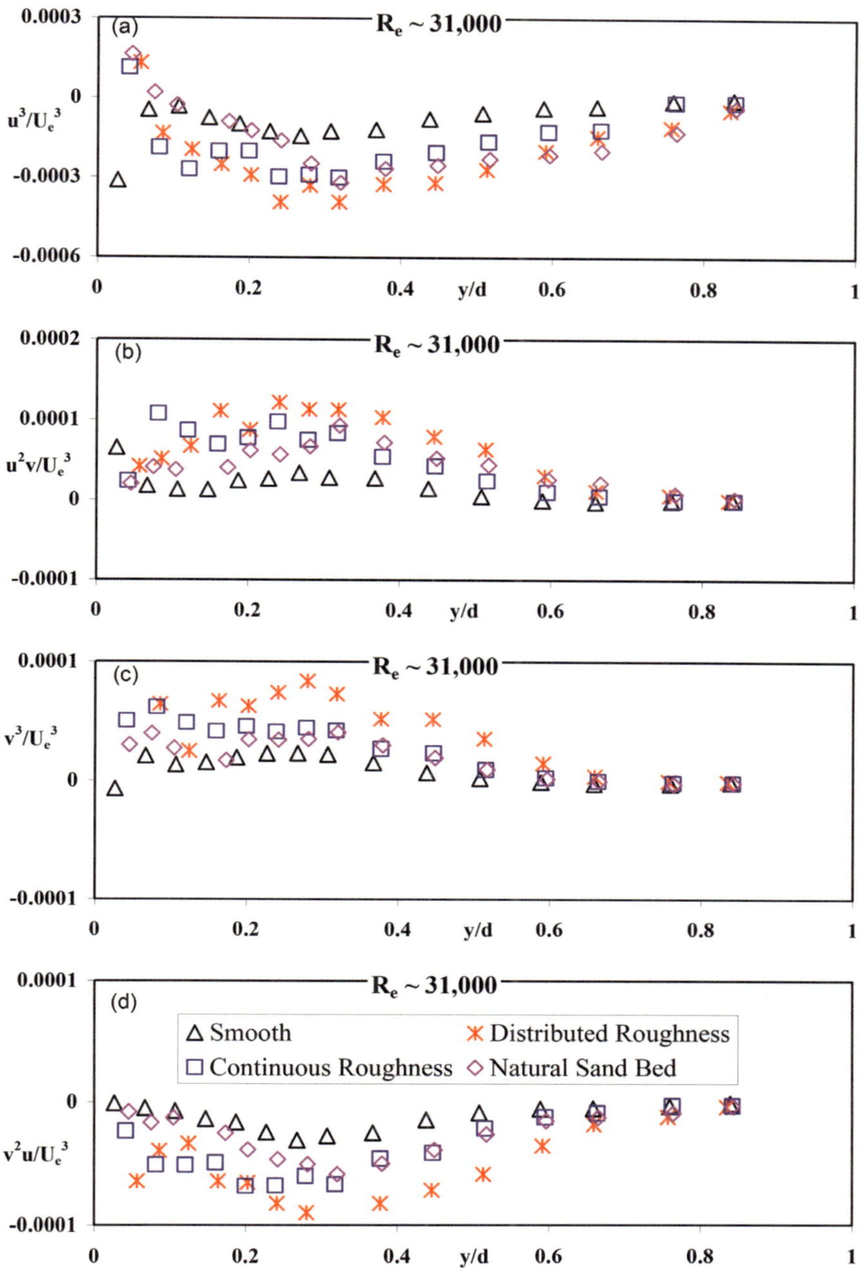

Figure 11.
Distribution of different velocity triple products for flow over different bed condition at $R_e \sim 31{,}000$.

natural sand bed. The value of $\overline{u^3}$ is also reaches near-zero close to the free surfaces irrespective of the bed conditions representing a vanishing turbulent activity at that location.

3.6 Vertical flux of the turbulent kinetic energy

The distribution of the vertical flux of the turbulent kinetic energy described as F_{kv} and which is normally measured as $0.5\left(\overline{v^3} + \overline{vu^2}\right)$ for a two-dimensional flow [8] is shown in **Figure 12** in outer variables for the flow over both smooth and rough beds. The LDA used to measure the velocity component is two-dimensional and not possible to measure the third component of turbulent intensity. An approximate method as proposed by [30] is used to overcome this shortcoming and the coefficient is changed from 0.75 to 0.5. The effect of roughness is very evident for the transport of the turbulent kinetic energy in the vertical direction as one can see from **Figure 12**. The effect of roughness is not only confined for near bed but can be seen throughout the depth of flow. This observation is in direct conflict with the observation of [8] who in their tests with large-bottomed roughness did not visualize notable differences in profile for the vertical flux of the turbulent kinetic energy when comparing open channel flow over smooth bed and rough bed conditions.

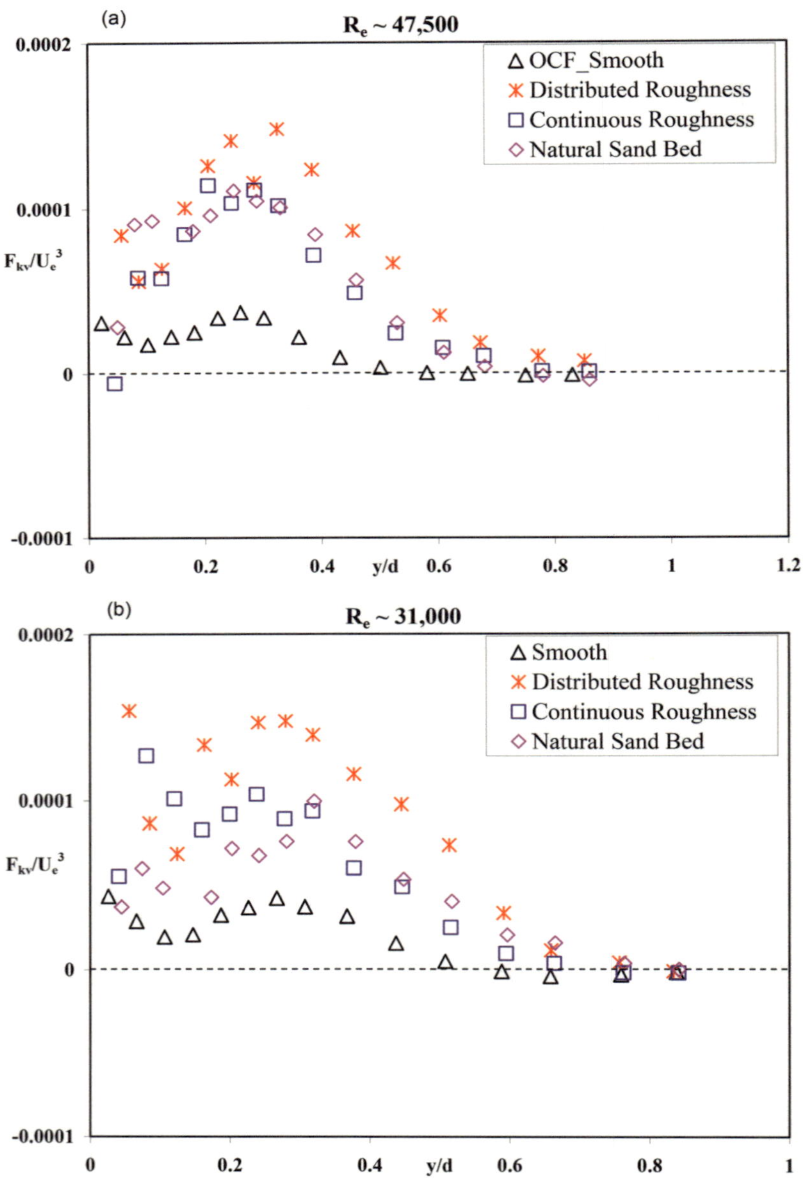

Figure 12.
Distribution of vertical flux of the turbulent kinetic energy for flow over different bed condition.

In Ref. [6] noted that the location of the outer (larger) peak of F_{kv} is closer to the wall (albeit slightly) as the roughness effect increases. The maximum value of F_{kv} is also noted in Ref. [8] where they found it occurred near the bed for the flow over rib roughness. As one can note from **Figure 12** that there are obvious effect of roughness on the variation of the vertical flux of the turbulent kinetic energy with the magnitude of the peak is very different for different type of rough surfaces but the location of the peak for all rough beds are more or less at around $y/d \sim 0.3$. The differences in magnitude of the vertical flux of the turbulent kinetic energy when comparing between smooth bed flow and flow over rough beds is a clear indication that the strength of the vertical flux of the turbulent kinetic energy is very different for flow over different surfaces. The slope of the variation of F_{kv} is different between smooth and rough beds representing difference in loss or gain of turbulent kinetic energy resulted from turbulent diffusion. Flow over distributed roughness shows the highest deviation compared to the flow over smooth bed. The vertical flux of the turbulent kinetic energy approaches near zero value after a peak value around $y/d = 1$ at the location near free surface for all bed conditions. Location of reaching zero value for the vertical flux of the turbulent kinetic energy also varies with the bed surface condition with flow over different rough beds show zero values closer to the free surface compared to the flow over smooth bed. **Figure 12b** shows the variation of the vertical flux of the turbulent kinetic energy for the flow with respect to lower Reynolds number. The profile characteristics are very similar for flow with respect to lower Reynolds number compared to the flow with respect to higher Reynolds number. The differences in magnitude of F_{kv} is seen to be reduced in the case of lower Reynolds number flow comparing the flow over smooth bed to the flow over continuous bed roughness and natural sand bed.

3.7 Quadrant analysis

In order to extract the magnitude of the Reynolds shear stress related to turbulent bursting events researchers often use quadrant decomposition as a convenient tool. A hydro dynamically unstable low-speed fluid particle lifted up from the surface because of the turbulent flow over a fixed bed can be swept away by comparatively high-speed fluid from the outer layer moving towards the bed surface. All different type of turbulent flow events that eventually contributed in the four different very important turbulent characteristics closer to the wall can be described by coupling streamwise and vertical fluctuating velocity components u and v based on their sign. Four different quadrants formed by using u and v with proper sign are related to four very important turbulent bursting events. Quadrant 1 represents the bursting effect called as outward interaction where the value of u is >0 and the value of v is >0. Quadrant 2 represents the bursting effect called as ejection where the value of u is <0 and the value of v is >0. Quadrant 3 represents the bursting effect called as inward interaction where the value of u is <0 and the value of v is <0. Quadrant 4 represents the bursting effect called as sweep where the value of u is >0 and the value of v is <0.

The contributions from Q2 and Q4 events for different threshold values to the Reynolds shear stress are shown in **Figure 13** with higher Reynolds number ($R_e = 47{,}500$). For the flow over rough walls and inclusive of all turbulent events, it was noted higher magnitude of Q2 and Q4 contributions as shown in **Figure 13a** and **b** compared to the flow over smooth wall for $H = 0$. The effect of roughness is not limited to the near-bed region but well progressed into the outer layer ($y/d \approx 0.7$). A local peak can be seen at $y/d = 0.1$–0.2 for the Q2 and Q4 contributions as one progresses from the bed towards the free surface for the flow over all rough beds. The peak magnitudes of both of the events eventually reduced

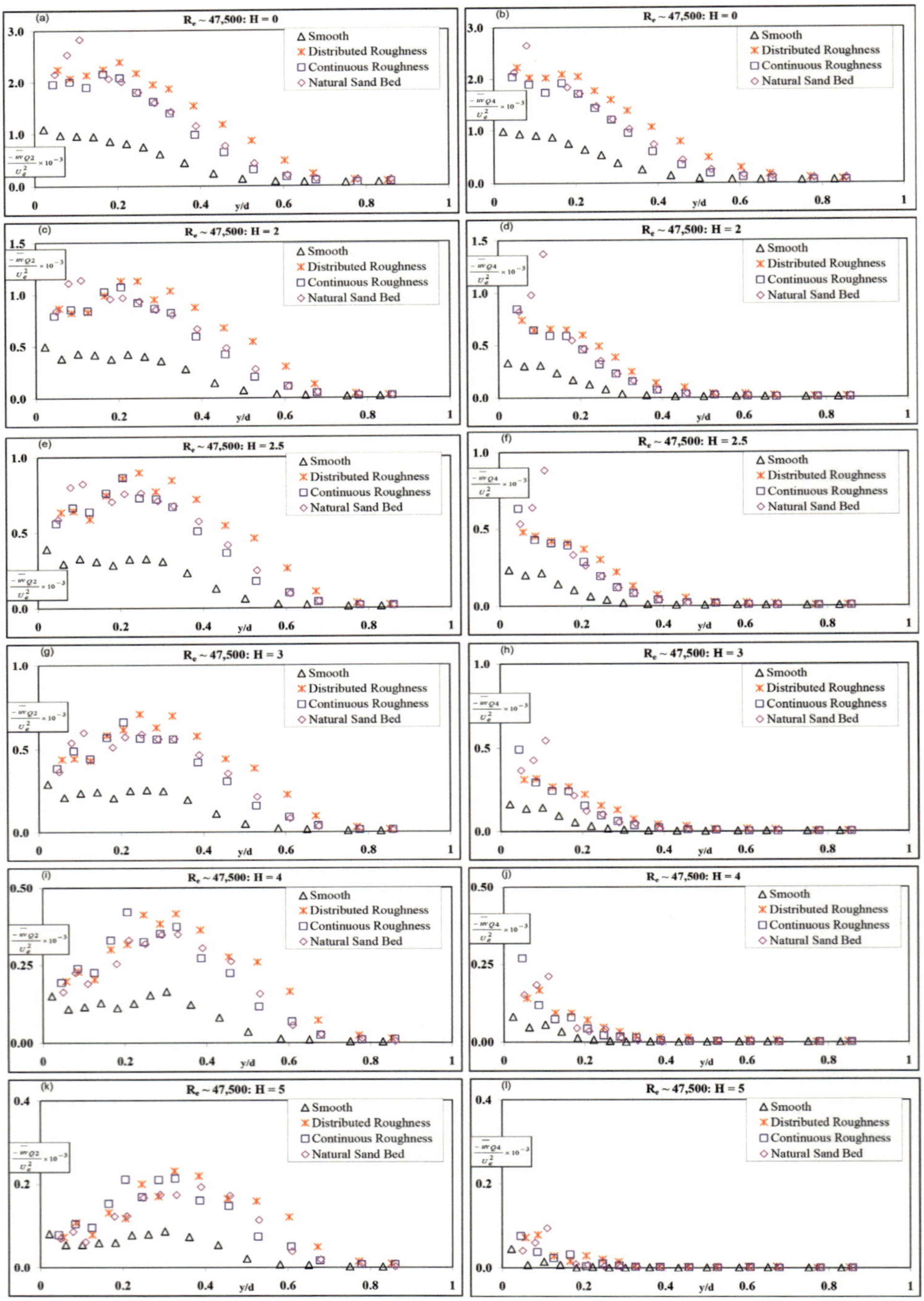

Figure 13.
Contribution of different quadrant events to the Reynolds shear stress for flow over different bed condition with higher Reynolds number.

to a near-zero constant value as flow moves towards the free surface. The location where the contributions from Q2 and Q4 events attains a near-zero constant value is not the same but varied with bed conditions. For the smooth bed condition the distance of the attainment of near-zero constant value is 0.5d from the bed, for the continuous roughness and sand bed condition the distance is 0.6d from the bed and for the distributed roughness the distance is 0.75d from the bed. Different rough bed conditions show different deviation from smooth wall with distributed roughness showing the highest deviation. The maximum deviation comparing the flow over smooth wall with the flow over rough bed occurs at a depth of around 0.2d from the bed with distributed roughness shows the highest deviation and continuous roughness and sand bed show almost equal deviation. In Ref. [31] found significantly higher magnitude of Q2 and Q4 events in the region very close to the bed but found very similar distribution for the flow over smooth bed and rough beds for the outer layer.

In order to investigate the contribution of the extreme turbulent events quadrant analysis at different threshold levels (H = 2–5) was also carried out. The respective approach was taken to take care of the contribution of the more energetic eddies and filtering out the small random turbulent fluctuations. The contributions from the extreme events whose amplitude exceeds the threshold value of H = 2 are shown in **Figure 13c** and **d**. Although due to the change of threshold value from 0 to 2, the number of events occurring corresponding to Q2 and Q4 reduce quite sharply but the events corresponding to H = 2 produced very large instantaneous Reynolds shear stress ($>5.5\,\overline{uv}$), which can potentially influence the sediment transport in the stream, causing resuspension of pollutant from the bed, bed formation/changes, downstream transportation of nutrients, entrainment and the exchange of energy and momentum in the flow. The trend of the data at H = 2 is very close to H = 0, however, the region of the flow depth affected for Q4 events reduces compared to H = 0. The contributions related to other threshold levels of H = 2.5–5 are shown in **Figure 13e–i** and one can observe that the region affected over the depth of flow for Q4 events reduces with respect to the increase of the threshold level of H but the affected region goes deep into the outer layer (y ~ 0.7d) for the Q2 events even for the value of H as high as 5. The incorporation of roughness is clearly visible in the increase in both Q2 and Q4 contribution to the Reynolds shear stress, irrespective of the affected region of the depth. Much stronger Q2 events were observed by [31–32] on a flow over a smooth wall when compared to the flow over a rough wall for the location close to the bed and they relate the phenomena for the smooth wall to the contributions of strongly favored Reynolds stress from ejection (Q2 events). The differences in observation between the turbulent boundary layer flow and open channel flow can confirm that turbulent bursting and eventual production of Reynolds shear stress due to ejection (Q2 events) and sweep (Q4 events) is different for the flow in open channel. Significant ejection and sweep components were noted by [8] with ejection events being dominant throughout the depth of flow and they also noted that different types of rib roughness result significant variations. One can notice in the present study that Q2 and Q4 events are dependent on the bed roughness accompanied by significant drop near the free surface for both events, signifies the important role the bed roughness type possesses on Q2 and Q4 events. At the location near bed, generation of turbulent activity varies with the type of bed roughness. Low momentum slow moving fluid from the near-bed is ejected and travels towards the outer layer/free surface and the same will happen for the fluid between the interstices of the roughness. In contrast, high momentum fast moving fluid from the outer layer travels towards bed, sweeping away the low momentum slow moving fluid parcels ejected earlier. The extent of depth of flow affected by the existence of universal intermittent sweep and ejection events is dependent on the type of bed and the flow condition. **Figure 14** shows the variation of the contributions from Q2 and Q4 events for different threshold values to the Reynolds shear stress with lower Reynolds number (R_e = 31,000). The profile characteristics are very similar for flow with respect to lower Reynolds number compared to the flow with respect to higher Reynolds number for the threshold values of H = 0–5.

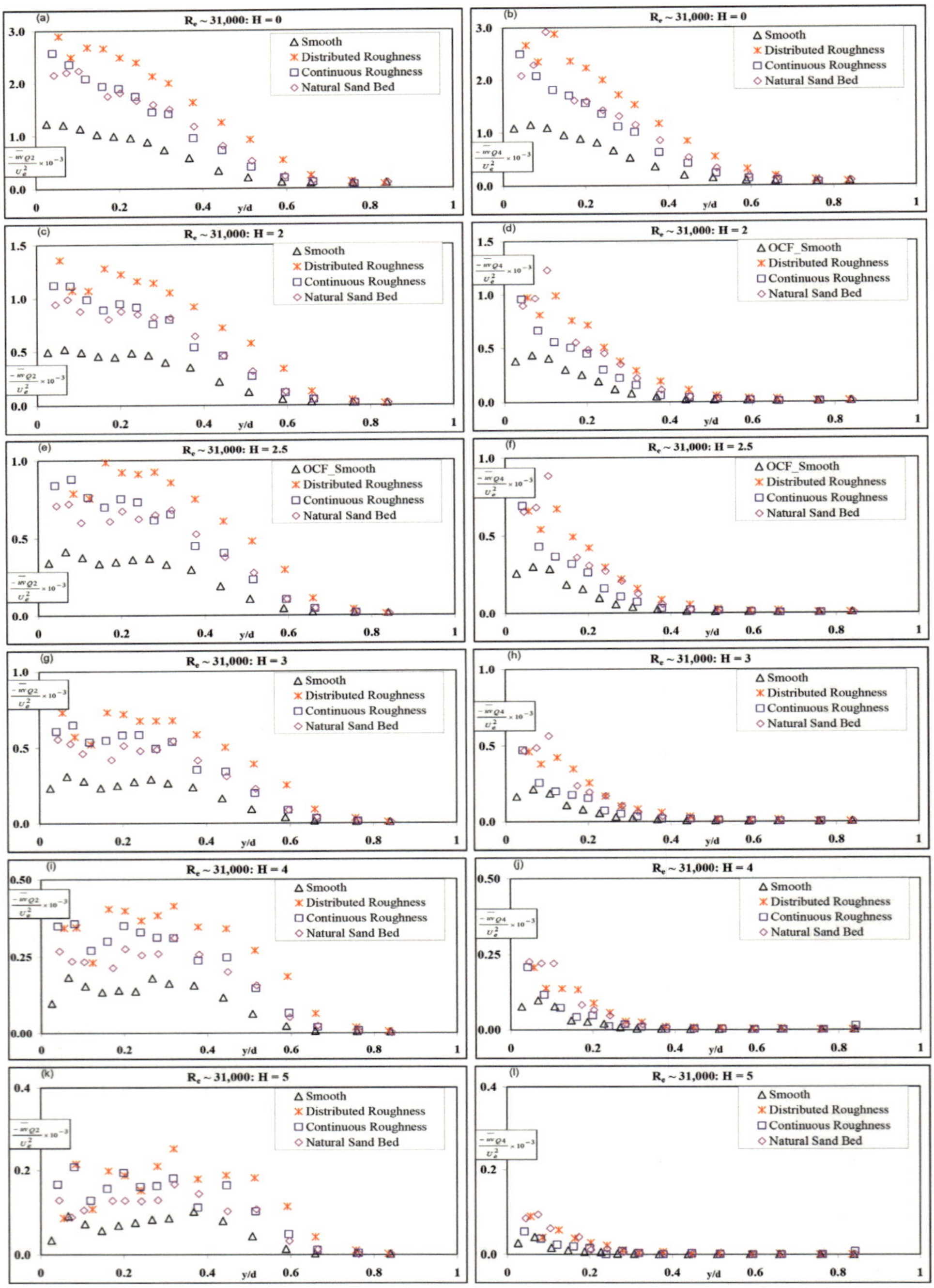

Figure 14.
Contribution of different quadrant events to the Reynolds shear stress for flow over different bed condition with lower Reynolds number.

Figure 15 shows the ratio to the Reynolds shear stress contributions of Q2/Q4 for H = 0–5 and for two different Reynolds numbers. The Q2/Q4 ratio is near unity at the location very close to the bed indicating identical strength of sweep and ejection event as one can note from **Figure 15**. The Q2/Q4 ratio increases from near unity to maximum at around mid-depth of the flow (y/d ~ 0.5) as one progress from the bed and towards the free surface which is an indication of relatively stronger

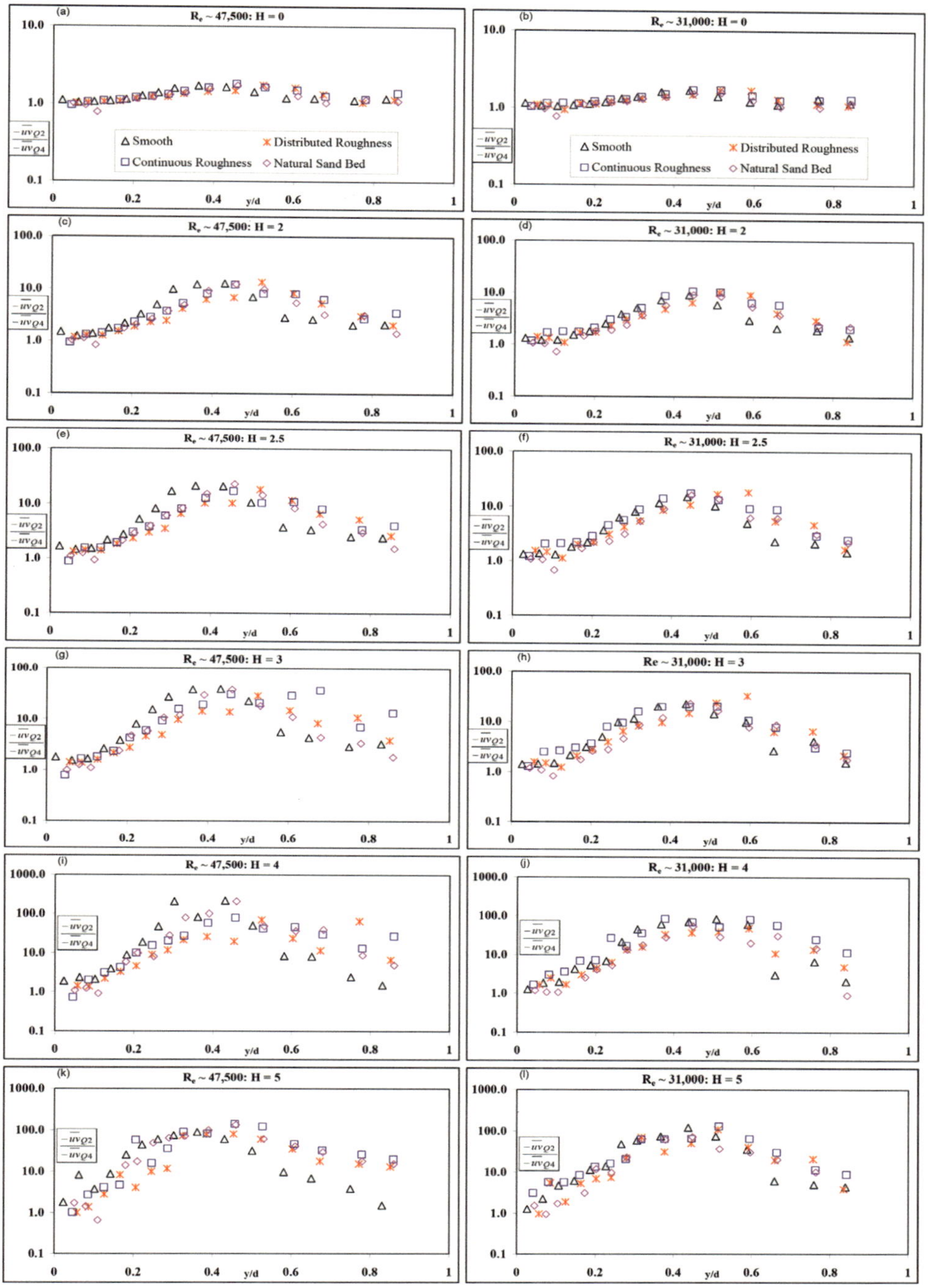

Figure 15.
Ratio of different quadrant events to the Reynolds shear stress for flow over different bed condition.

ejection events compared to the sweep events. The corresponding strength of the ejection events increases in comparison to sweep events with respect to increasing H and as one can note from the **Figure 15** that there is a 100 over fold increase for the threshold value of H = 5 compared to H = 0. As one can also note from **Figure 15** that there is little dependency on bed conditions of smooth and rough for H = 0 on the ratio of Reynolds shear stress in Q2 and Q4 but for the same value of H = 0 there are some effect of roughness for y > 0.5d.

Figure 16 shows the ratio to the number of events occurring/contributing in Q2 and Q4 for H = 0–3 and for two different Reynolds numbers. The ratio to the number of events occurring/contributing to Q2 and Q4 shows different trends for the threshold value of H = 0 (**Figure 16a** and **b**) compared to the threshold value of H = 2–3 (**Figure 16c–h**). This is very unlike to the ratio of the Reynolds shear stress contributions of Q2/Q4 as shown in **Figure 15**. The N_{Q2}/N_{Q4} ratio is near unity at the location very close to the bed indicating almost equal occurrence of ejection and sweep events as one can note from **Figure 16a** and **b**. The N_{Q2}/N_{Q4} ratio decreases from near unity to minimum at around mid-depth of the flow ($y/d \sim 0.5$) as one progress from the bed and towards the free surface which is an indication of relatively reduced ejection events compared to the sweep events. Moving farther away from bed ($y > 0.5d$) and towards the free surface, the ratio of N_{Q2}/N_{Q4} ratio is keep on increasing again and reaches to near unity indicating almost equal occurrence of ejection and sweep events. **Figure 16c–h** show a trend different from **Figure 16a** and **b**. As one progress from the bed towards the free surface, there is an increment of 30 over fold for the value of N_{Q2}/N_{Q4} at around $y \sim 0.5d$, indicating substantial increase of ejection events. As one can also note from **Figure 16** that there is little dependency on bed conditions of smooth and rough for H = 0 on the ratio of number of events in Q2 and Q4.

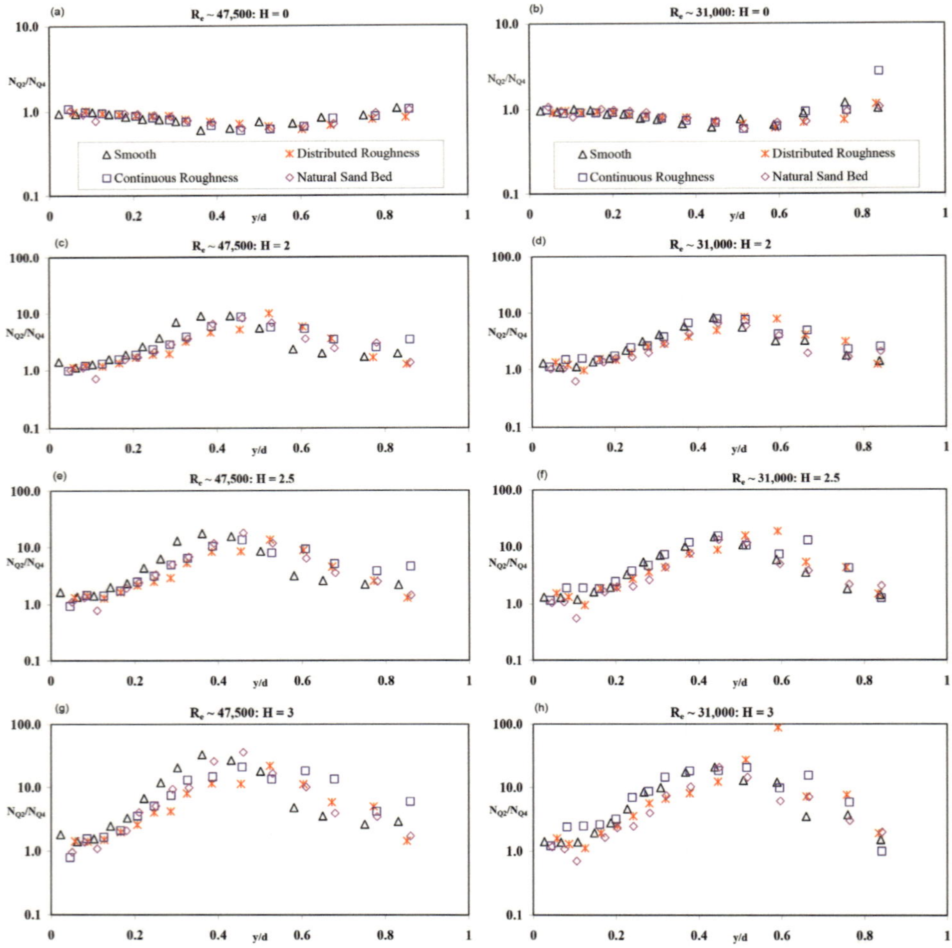

Figure 16.
Ratio of number of different quadrant events for flow over different bed condition.

4. Conclusions

The purpose of the present study [1] is to explain how the roughness and Reynolds number affect flow characteristics in an open channel flow (OCF). Tests were conducted with four different types of bed surface conditions and at two different Reynolds number for each and every bed surface. Instantaneous velocity components are used to analyze the streamwise mean velocity, turbulence intensity in both streamwise and vertical direction, Reynolds shear stress including shear stress correlation and higher-order moments including vertical flux of the turbulent kinetic energy. In order to extract the magnitude of the Reynolds shear stress related to turbulent bursting events quadrant decomposition was used. The main findings are summarized as follows:

1. Surface drug increases due to surface roughness making the mean streamwise velocity profile to be more fuller for the smooth bed compared to the rough beds. It is very much evident throughout the depth of the flow that the mean velocity profile is very much affected by the different type of bed roughness. Comparing the effect of various type of bed roughness on the streamwise velocity component and flow with higher flow Reynolds number, distributed roughness profile has the biggest deviation from smooth bed profile with continuous roughness and natural sand bed shows identical deviation. For the flow with lower flow Reynolds number, it was found that the flow over natural sand bed shows much higher deviation than flow over the bed of continuous roughness.

2. The magnitude of friction coefficient is found to be dependent on the type of bed roughness with distributed roughness has the highest value followed by the flow over the continuous roughness bed surface and the sand bed. The magnitude of friction coefficient is also found to be dependent on the Reynolds number with the reduction of the magnitude of friction coefficient with the increment of the Reynolds number. The magnitude of friction coefficient is seen to be smaller for the flow over a permeable bed (natural sand bed) compared to the flow over an impermeable bed (distributed and continuous roughness bed).

3. The effect of roughness on the distribution of the streamwise component of the turbulence intensity is very evident throughout the flow depth with distributed roughness shows the highest deviation followed by natural sand bed and continuous roughness compared to the smooth surfaces with the exception at the location very close to the bed. Comparing the effect of various type of bed conditions on the vertical component of the turbulence intensity, it was seen that distributed roughness profile has the biggest deviation from smooth bed profile with continuous roughness and natural sand bed shows identical deviation for most the depth of the flow. At locations very close to the bed and due to the introduction of roughness, streamwise turbulence intensity reduces but vertical turbulence intensity increases. Although the sand grain used to create all three bed roughness is of the same gradation characteristics but the specific geometry of the roughness formation is different causing the differences in the formation of turbulence structure.

4. Wall similarity hypothesis is disputed by the present experimental results where the researchers suggested that in the location of outside the roughness

layer, the turbulent mixing properties should be essentially the same for the flow over smooth and rough walls which was initially proposed by [33] and generalized by [34].

5. Effect of roughness on the Reynolds shear stress is very evident at the location close to the bed generating much higher Reynolds shear stress than the smooth bed. The distinct effect of roughness for the present study can be seen penetrating deep into the flow and distinctly visible at the location as high as $y/d \approx 0.7$.

6. The trend of the changes of the value of R (correlation coefficient) in the near-bed and outer layer indicating the changes of flow structure characteristics between the near-bed region and outer region. The present results clearly dispute the observation of [5] that the distribution of R is independent of the properties of the wall roughness, mean flow, and called the distribution of R is universal.

7. The magnitude of various velocity triple products changes in the range of 200–300% when comparing the flow over the smooth bed to the flow over rough beds. This is a clear indication that the transportation of turbulent kinetic energy and Reynolds shear stress is significantly affected by the bed roughness.

8. Turbulent activity at the near bed location also seen to be dependent on bed surface conditions. Flow over smooth bed shows the ejection type activity near bed location whereas the flow over rough beds show the sweep type activity at the location close to the bed. Interpolating this scenario to the real life stream or river flow, one can clearly note the influence of strong ejection/sweeping motion of the fluid parcels to the resuspension/transport of the bed particles.

9. Ejection type events are very evident throughout the depth of flow with the exception of the location very close to the bed with flow over smooth bed only where one can observe some sweeping type of event. Bed surface conditions clearly affect the strength of the ejection like events with distributed roughness again shows the highest strength compared to other form of bed roughness.

10. Effect of roughness is clearly visible well beyond the near-bed region and deep into the outer layer ($y \approx 0.7d$) from the analysis/result of turbulent bursting events (through quadrant decomposition). For the flow over rough walls and inclusive of all turbulent events, it was noted higher magnitude of Q2 and Q4 contributions compared to the flow over smooth wall for $H = 0$.

11. Analysis were also carried out to investigate the contribution of the extreme turbulent events at different threshold levels ($H = 2$–5). The region affected over the depth of flow for active sweep (Q4) events reduces with respect to the increase of the threshold level of H but the affected region goes deep into the outer layer ($y \sim 0.7d$) for the active ejection (Q2) events even for the value of H as high as 5. Although due to the change of threshold value from 0 to 2, the number of events occurring corresponding to Q2 and Q4 reduce quite sharply but the events corresponding to $H = 2$ produced very large instantaneous Reynolds shear stress ($>5.5\,\overline{uv}$), which can potentially influence the sediment transport in the stream, causing resuspension of pollutant from the bed, bed formation/changes, downstream transportation of nutrients, entrainment and the exchange of energy and momentum in the flow.

12. The ratio to the Reynolds shear stress contributions of Q2/Q4 is near unity at the location very close to the bed and location close to the free surface indicating identical strength of sweep and ejection. With the exception of near bed and near free surface, relatively stronger ejection events compared to the sweep events can be seen for throughout the flow depth and the strength of the ejection events increases many fold with increase of the threshold value of H.

13. The ratio to the number of events occurring/contributing in Q2 and Q4 is near unity at the location very close to the bed and location close to the free surface indicating almost equal occurrence of sweep and ejection events. With the exception of near bed and near free surface, relatively reduced ejection events compared to the sweep events can be seen for throughout the flow depth for $H = 0$ but shows substantial increase of ejection events compared to the sweep events for $H > 0$.

Author details

Abdullah Faruque
Civil Engineering Technology, Rochester Institute of Technology, Rochester, New York, USA

*Address all correspondence to: aafite@rit.edu

References

[1] Faruque MAA. Smooth and rough wall open channel flow including effects of seepage and ice cover [PhD thesis]. Windsor, Canada: University of Windsor; 2009

[2] Rashidi M, Hetsroni G, Banerjee S. Particle-turbulence interaction in a boundary layer. International Journal of Multiphase Flow. 1990;**16**(6):935-949

[3] Grass AJ. Structural features of turbulent flow over smooth and rough boundaries. Journal of Fluid Mechanics. 1971;**50**(2):233-255

[4] Nakagawa H, Nezu I. Prediction of the contributions to the Reynolds stress from bursting events in open-channel flows. Journal of Fluid Mechanics. 1977; **80**(1):99-128

[5] Nezu I, Nakagawa H. Turbulence in open-channel flows. IAHR Monograph. In: Balkema AA, editor. Rotterdam, Netherlands; 1993

[6] Tachie MF. Open-channel turbulent boundary layers and wall jets on rough surfaces [PhD thesis]. Saskatchewan, Canada: University of Saskatchewan; 2001

[7] Nezu I. Open-channel flow turbulence and its research prospect in the 21st century. Journal of Hydraulic Engineering. 2005;**131**(4):229-246

[8] Balachandar R, Bhuiyan F. Higher-order moments of velocity fluctuations in an open channel flow with large bottom roughness. Journal of Hydraulic Engineering. 2007;**133**(1):77-87

[9] Afzal B, Faruque MAA, Balachandar R. Effect of Reynolds number, near-wall perturbation and turbulence on smooth open channel flows. Journal of Hydraulic Research. 2009;**47**(1):66-81

[10] Patel VC. Perspective: Flow at high Reynolds number and over rough surfaces—Achilles heel of CFD. Journal of Fluids Engineering. 1998;**120**(3): 434-444

[11] Roussinova V, Biswas N, Balachandar R. Revisiting turbulence in smooth uniform open channel flow. Journal of Hydraulic Research. 2008;**46** (Suppl. 1):36-48

[12] Kirkgöz MS, Ardiçhoğlu M. Velocity profiles of developing and developed open channel flow. Journal of Hydraulic Engineering. 1997;**123**(2):1099-1105

[13] Tachie MF, Bergstrom DJ, Balachandar R. Roughness effects in low-Re_θ open-channel turbulent boundary layers. Experiments in Fluids. 2003;**35**:338-346

[14] Balachandar R, Patel VC. Rough wall boundary layer on plates in open channels. Journal of Hydraulic Engineering. 2002;**128**(10):947-951

[15] Tachie MF, Bergstrom DJ, Balachandar R. Rough wall turbulent boundary layers in shallow open channel flow. Journal of Fluids Engineering. 2000;**122**:533-541

[16] Kaftori D, Hetsroni G, Banerjee S. Particle behavior in the turbulent boundary layer. I. Motion, deposition, and entrainment. Physics of Fluids. 1995;7(5):1095-1106

[17] Dancey CL, Balakrishnan M, Diplas P, Papanicolaou AN. The spatial inhomogeneity of turbulence above a fully rough, packed bed in open channel flow. Experiments in Fluids. 2000; **29**(4):402-410

[18] Tachie MF, Bergstrom DJ, Balachandar R. Roughness effects on the mixing properties in open channel

turbulent boundary layers. Journal of Fluids Engineering. 2004;**126**:1025-1032

[19] Bigillon F, Niño Y, Garcia MH. Measurements of turbulence characteristics in an open-channel flow over a transitionally-rough bed using particle image velocimetry. Experiments in Fluids. 2006;**41**(6):857-867

[20] Faruque MAA, Sarathi P, Balachandar R. Clear water local scour by submerged three-dimensional wall jets: Effect of tailwater depth. Journal of Hydraulic Engineering. 2006;**132**(6): 575-580

[21] Sarathi P, Faruque MAA, Balachandar R. Scour by submerged square wall jets at low densimetric Froude numbers. Journal of Hydraulic Research. 2008;**46**(2):158-175

[22] Schlichting H. Boundary-Layer Theory. McGraw-Hill Classic Textbook Reissue Series. United States of America: McGraw-Hill, Inc; 1979

[23] Bey A, Faruque MAA, Balachandar R. Two dimensional scour hole problem: Role of fluid structures. Journal of Hydraulic Engineering. 2007;**133**(4): 414-430

[24] Zagni AFE, Smith KVH. Channel flow over permeable beds of graded spheres. Journal of Hydraulics Division: Proceedings of the ASCE. 1976;**102** (HY2):207-222

[25] Krogstad P-A, Andersson HI, Bakken OM, Ashrafian A. An experimental and numerical study of channel flow with rough walls. Journal of Fluid Mechanics. 2005;**530**:327-352

[26] Agelinchaab M, Tachie MF. Open channel turbulent flow over hemispherical ribs. International Journal of Heat and Fluid Flow. 2006;**27**(6): 1010-1027

[27] Schultz MP, Flack KA. The rough-wall turbulent boundary layer from the hydraulically smooth to the fully rough regime. Journal of Fluid Mechanics. 2007;**580**:381-405

[28] Flack KA, Schultz MP, Shapiro TA. Experimental support for Townsend's Reynolds number similarity hypothesis on rough walls. Physics of Fluids. 2005; **17**(3):35102-351-9

[29] Antonia RA, Krogstad P-A. Turbulence structure in boundary layers over different types of surface roughness. Fluid Dynamics Research. 2001;**28**(2):139-157

[30] Krogstad P-A, Antonia R. Surface roughness effects in turbulent boundary layers. Experiments in Fluids. 1999;**27**: 450-460

[31] Schultz MP, Flack KA. Outer layer similarity in fully rough turbulent boundary layers. Experiments in Fluids. 2005;**38**:328-340

[32] Krogstad P-A, Antonia R, Browne LWB. Comparison between rough and smooth-wall turbulent boundary layers. Journal of Fluid Mechanics. 1992;**245**: 599-617

[33] Townsend AA. The Structure of Turbulent Shear Flow. Cambridge, United Kingdom: Cambridge University Press; 1976

[34] Raupach MR, Antonia RA, Rajagopalan S. Rough wall turbulent boundary layers. Applied Mechanics Review. 1991;**44**(1):1-25

Chapter 4

Leading Edge Receptivity at Subsonic and Moderately Supersonic Mach Numbers

Marvin E. Goldstein and Pierre Ricco

Abstract

This chapter is a review of the receptivity and resulting global instability of boundary layers due to free-stream vortical and acoustic disturbances at subsonic and moderately supersonic Mach numbers. The vortical disturbances produce an unsteady boundary layer flow that develops into oblique instability waves with a viscous triple-deck structure in the downstream region. The acoustic disturbances (which have phase speeds that are small compared to the free stream velocity) produce boundary layer fluctuations that evolve into oblique normal modes downstream of the viscous triple-deck region. Asymptotic methods are used to show that both the vortically and acoustically-generated disturbances ultimately develop into modified Rayleigh modes that can exhibit spatial growth or decay depending on the nature of the receptivity process.

Keywords: boundary layer, boundary layer receptivity, compressible boundary layers, global instability

1. Introduction

This chapter is concerned with the effect of unsteady free-stream disturbances on laminar to turbulent transition in boundary layer flows. The exact mechanism depends on the nature and intensity of the disturbances. Transition at high disturbance levels (say >1%) usually begins with the excitation of low frequency streaks in the boundary layer flow that eventually break down into turbulent spots. This phenomena was initially studied by Dryden [1] and much later for compressible flows by Marensi et al. [2]. But the focus of this chapter is on low free steam disturbances levels (say less than 1%) where the transition usually results from a series of events beginning with the generation of spatially growing instability waves by acoustic and/or vortical disturbances in the free-stream. This so-called receptivity phenomenon results in a boundary value problem and therefore differs from classical instability theory which results in an eigenvalue problem for the Rayleigh or Orr-Sommerfeld equations that only apply when the mean flow can be treated as being nearly parallel (see, for example, Reshotko, [3]). The relevant boundary conditions cannot be imposed on the Orr-Sommerfeld or Rayleigh equations in the infinite Reynolds number limit being considered here but the free-stream disturbances can produce unsteady boundary layer perturbations in regions of rapidly changing mean flow that eventually produce unstable Rayleigh or Orr-Sommerfeld equation eigensolutions further downstream. These regions of nonparallel flow can result from surface roughness elements [4, 5], blowing or suction effects [6] or from the nonparallel mean flow that occurs near the boundary layer leading edge [7, 8].

The mechanism is similar in all cases but the simplest and arguably the most fundamental of these is the one resulting from the nonparallel leading edge flow and the focus here is, therefore, on that case. The initial studies were carried out for two dimensional incompressible flows. Ref. [7] used a low frequency parameter matched asymptotic expansion to show that there is an overlap domain where appropriate asymptotic solutions to the forced boundary layer equations (which apply near the edge) match onto the so-called Tollmien-Schlichting waves that satisfy the Orr-Sommerfeld equation in a region that lies somewhat further downstream. The coupling to the free-stream disturbances turns out to be fairly weak for the two dimensional incompressible flow considered in [7] due to the relatively large decay of boundary layer disturbances upstream of the Tollmien-Schlichting wave region where the Orr-Sommerfeld equation applies.

But there can be a much stronger coupling in supersonic flows which can support a number of different instabilities [9]. The coupling mechanism can be either viscous or inviscid and the instability can either be of the viscous Tollmien-Schlichting type or can be purely inviscid when the mean boundary layer flow has a generalized inflection point. The inviscid coupling, which was first analyzed in [10], tends to be dominant when the obliqueness angle θ of the disturbance differs from the critical angle, $\theta_c \equiv \cos^{-1}(1/M_\infty)$, where the M_∞ is the free-stream Mach number, by an $O(1)$ amount. **Figure 1** shows that the theoretical results of Ref. [10] are in good agreement with experimental data when $\Delta\theta \equiv \theta_c - \theta = O(1)$ but the agreement breaks down when $\theta \to \theta_c$ [12] and a new rescaled analysis was carried out in Ref. [11] to deal with this case.

Fedorov and Khokhlov [10] analyzed the generation of inviscid instabilities in a supersonic flat plate boundary layer by fast and slow acoustic disturbances in the free stream. They showed that the slow acoustic mode propagates downstream/upstream when the obliqueness angle θ of the acoustic disturbances is smaller/larger than the critical angle θ_c and that downstream propagating slow acoustic modes with $\Delta\theta>0$ generate unsteady boundary layer disturbances that match onto the inviscid 1st Mack mode instability without undergoing any significant decay. The

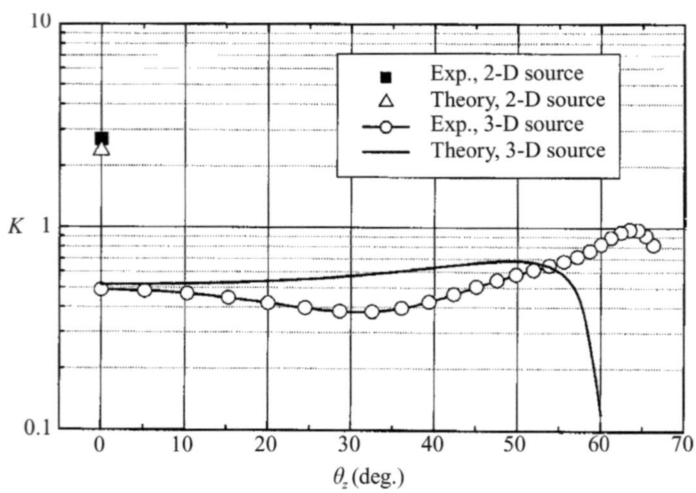

Figure 1.
Comparison of the Fedorov/Khokhlov solution with experiment [12].

Figure 2.
Low-sweep Aerion AS2 supersonic Bizjet. $M_\infty \leq 1.5$. Posted by Tim Brown on the Manufacturer Newsletter.

focus of that reference was on hypersonic flows while the interest here is in the moderately supersonic regime (Mach number less than 4), where the so called 1st Mack mode is the dominant instability, but (as shown in Section 6) emerges much too far downstream to be of practical interest when generated by the inviscid mechanism analyzed in [7]. The instability produced by the small $\Delta\theta$ analysis of Ref. [11] can, however, occur much further upstream when $\Delta\theta$ is sufficiently small. But there is a smallest value of $\Delta\theta$ for which the instability wave coupling can occur.

Smith [13] showed that viscous instabilities, which exhibit the same triple-deck structure as the subsonic Tollmien-Schlichting waves, can also occur at supersonic speeds when the obliqueness angles θ is greater than the critical angle θ_c. Their phase speeds are very small and they must therefore be produced by a viscous wall layer mechanism similar to the one identified in [7].

The analysis of Ref. [7] was extended to compressible subsonic and supersonic flat plate boundary layer flows by Ricco and Wu [14] who showed that highly oblique vortical disturbances can generate a limiting form of the Smith instability [13]. They found that the instability wave lower branch lies further upstream at supersonic speeds than the subsonic lower branch and much further upstream than the incompressible lower branch considered in [7], which means that the instability wave/free-stream disturbance coupling is much greater at supersonic speeds than it is in the incompressible flow considered in [7]. Goldstein and Ricco [11] show that the instability does not possess an upper branch in this case and matches onto a low frequency (short streamwise wavenumber) Rayleigh instability (that can be identified with the 1st Mack mode) when the downstream distance is slightly smaller than the downstream distance where acoustically generated instability corresponding to the smallest possible $\Delta\theta$ emerges. It therefore makes sense to consider both of these receptivity mechanisms simultaneously.

As noted above, the present chapter is concerned with the unsteady flow in a flat plate boundary layer generated by mildly oblique vortical disturbance and small $\Delta\theta$ acoustic disturbances in a moderately supersonic Mach number free stream. The results are expected to be relevant to transition in the straight wing boundary layers on supersonic aircraft such as the low-sweep Aerion AS2 Bizjet, shown in **Figure 2**.

2. Imposed free-stream disturbances

Since the boundary layer is believed to be convectively unstable, the receptivity phenomena are best illustrated by considering a small amplitude harmonic distortion with angular frequency ω^* superimposed on a subsonic or moderately low

Mach number supersonic flow of an ideal gas past an infinitely thin flat plate with uniform free-stream velocity U^*_∞, temperature T^*_∞, dynamic viscosity μ^*_∞ and density ρ^*_∞. The velocities, pressure fluctuations, temperature and dynamic viscosity are normalized by U^*_∞, $\rho^*_\infty(U^*_\infty)^2$, T^*_∞ and μ^*_∞, respectively. The time t is normalized by ω^* and the Cartesian coordinates, say $\{x, y, z\}$, are normalized by $L^* \equiv U^*_\infty/\omega^*$ with the coordinate y being normal to the plate.

As noted above the phenomenon is analyzed by requiring the Reynolds number $\mathrm{Re} = \rho^*_\infty U^*_\infty L^*/\mu^*_\infty$ to be large, or equivalently requiring the frequency parameter $\mathcal{F} \equiv 1/\mathrm{Re}$ to be small, and using asymptotic theory to explain how the imposed harmonic distortion generates oblique instabilities at large downstream distances in the viscous boundary layer that forms on the surface of the plate. The natural expansion parameter turns out to be

$$\varepsilon \equiv \mathcal{F}^{1/6}. \tag{1}$$

The free-steam disturbances will be inviscid at the lowest order of approximation and, as is well known [15], can be decomposed into an acoustic component that carries no vorticity, and vortical and entropic components that produce no pressure fluctuations. But only the first two will be considered here.

The vortical disturbance $\boldsymbol{u}_v$ is given

$$\boldsymbol{u}_v = \{u_v, v_v, w_v\} = \hat{\delta}\{u_\infty, v_\infty, w_\infty\} \exp\left[i(x - t + \gamma y + \beta z)\right], \tag{2}$$

where $\hat{\delta} \ll 1$ is a common scale factor and u_∞, v_∞, w_∞ satisfy the continuity condition

$$u_\infty + \gamma v_\infty + \beta w_\infty = 0 \tag{3}$$

but are otherwise arbitrary constants while the acoustic component is governed by the linear wave equation which has a fundamental plane wave solution

$$\{\boldsymbol{u}_a, p_a\} = \{u_a, v_a, w_a, p_a\} = \frac{\hat{\delta}}{1-\alpha}\{\alpha, \gamma, \beta, 1-\alpha\} e^{i(\alpha x + \gamma y + \beta z - t)}, \tag{4}$$

for the velocity and pressure perturbation $\{\boldsymbol{u}_a, p_a\}$ where

$$\gamma = \sqrt{(M^2_\infty - 1)(\alpha - \alpha_1)(\alpha - \alpha_2)}, \quad \alpha_{1,2} = \frac{M^2_\infty \pm \sqrt{M^2_\infty + \beta^2(M^2_\infty - 1)}}{M^2_\infty - 1} \tag{5}$$

and, as noted in Section 1, M_∞ denotes the free-stream Mach number.

The leading edge interaction will produce large scattered fields for $O(1)$ values of the incidence angles $\tan^{-1}(v_a/u_a) = \tan^{-1}(\gamma/\alpha)$ and $\tan^{-1}(v_v/u_v)$ of the acoustic and vortical disturbances, respectively. And, in order to focus on the fundamental mechanisms, we assume that the incidence angles of the vortical disturbances are small and that the incidence angles of the acoustic disturbances are zero, which requires that

$$v_\infty/u_\infty \ll 1 \tag{6}$$

for the former disturbances and that

$$\alpha = \alpha_\mp = M_\infty \cos\theta/(M_\infty \cos\theta \mp 1), \quad \theta \equiv \tan^{-1}(\beta/\alpha), \tag{7}$$

for the latter, where the subscripts $-/+$ refer to the slow/fast acoustic modes. Eq. (7) shows that the slow mode wavenumber becomes infinite when the obliqueness angle is equal to the critical angle referred to in the introduction.

3. Boundary layer disturbances

As indicated above our interest here is in explaining how the incident harmonic distortions generate oblique instabilities at large downstream distances in the viscous boundary layer that forms on the surface of the plate. We begin by considering the fluctuations imposed on this flow by the free-stream vortical disturbance (2).

3.1 Boundary layer disturbances generated by the free-stream vorticity

As noted in the introduction, these disturbances will generate oblique Tollmien-Schlichting instability waves which are known to exhibit a triple-deck structure in the vicinity of their lower branch which lies at an $O(\varepsilon^{-2})$ distance downstream [13] of the leading edge in the high Reynolds number flow being considered here. The Tollmien-Schlichting waves will have $O(\varepsilon^{-1})$ spanwise wavenumbers and we therefore require that

$$\overline{\beta} \equiv \varepsilon\beta = O(1) \tag{8}$$

since the spanwise wavenumber must remain constant as the disturbances propagate downstream.

The continuity condition (3) and the obliqueness restriction (6) will be satisfied if we put

$$\overline{w}_\infty \equiv w_\infty/\varepsilon = O(1), \quad \overline{v}_\infty \equiv v_\infty/\varepsilon = O(1), \overline{\gamma} \equiv \varepsilon\gamma = O(1). \tag{9}$$

The vortical velocity (2) will then interact with the plate to produce an inviscid velocity field [12] that generates a slip velocity at the surface of the plate which must be brought to zero in a thin viscous boundary layer whose temperature, density and streamwise velocity, say $T(\eta)$, $\rho(\eta)$, $U(\eta)$, respectively, are assumed to be functions of the Dorodnitsyn-Howarth variable

$$\eta \equiv \frac{1}{\varepsilon^3\sqrt{2x}} \int_0^y \rho(x,\tilde{y})d\tilde{y} \tag{10}$$

and are determined from the similarity equations given in Stewartson [16] and Ref. [14].

We begin by considering the flow in the vicinity of the leading edge where the streamwise length scale is $x = O(1)$. Since the inviscid velocity field can only depend on the streamwise coordinate through this relatively long streamwise length scale the solution for the velocity and temperature perturbation $\boldsymbol{u}' \equiv \{u', v', w', \vartheta'\}$ in this region is given by [14], [17]

$$\boldsymbol{u}' = \hat{\delta}\left[u_\infty\{\overline{u}, \overline{v}, 0, \overline{\vartheta}\} + \overline{\beta}(\overline{w}_\infty + i\overline{v}_\infty)\left\{\overline{u}^{(0)}, \overline{v}^{(0)}, \overline{w}^{(0)}, \overline{\vartheta}^{(0)}\right\}\right]e^{i\left(\overline{\beta}z/\varepsilon - t\right)}, \tag{11}$$

where $\left\{\overline{u}^{(0)}(x,\eta), \overline{v}^{(0)}(x,\eta), \overline{w}^{(0)}(x,\eta), \overline{\vartheta}^{(0)}(x,\eta)\right\}$ satisfies the three dimensional compressible linearized boundary layer equations (with unit spanwise wavenumber) subject to the boundary conditions [14]

$$\overline{u}^{(0)}, \overline{\vartheta}^{(0)} \rightarrow 0, \quad \overline{w}^{(0)} \rightarrow e^{ix}, \quad \text{as } \eta \rightarrow \infty, \tag{12}$$

while $\{\overline{u}(x,\eta), \overline{v}(x,\eta), 0, \overline{\vartheta}(x,\eta)\} \exp i(\overline{\beta}z/\varepsilon - t)$ is a quasi-two dimensional solution that satisfies the two dimensional linearized boundary layer equations subject to the boundary conditions

$$\overline{u} \rightarrow e^{ix}, \quad \overline{w}, \overline{\vartheta} \rightarrow 0 \text{ as } \eta \rightarrow \infty. \tag{13}$$

The lowest order triple-deck solution will match onto the quasi-two dimensional solution $\{\overline{u}, \overline{v}, 0, \overline{\vartheta}\} \exp i(\overline{\beta}z/\varepsilon - t)$ of the two dimensional boundary layer equations, where the spanwise dependence only enters parametrically through the exponential factor in(11) .

Prandtl [18], Glauert [19] and Lam and Rott [20] showed that

$$\overline{u}(x,\eta) = -\frac{B(x)U'(\eta)}{T\sqrt{2x}}, \overline{\vartheta}(x,y) = -\frac{B(x)T'(\eta)}{T(\eta)\sqrt{2x}}, \tag{14}$$

$$\overline{v}(x,\eta) = iB(x) + \frac{dB}{dx}U(\eta) - B(x)\frac{U'(\eta)\eta_c}{2x}, \tag{15}$$

where

$$\eta_c \equiv \frac{1}{T(\eta)} \int_0^{\eta} T(\tilde{\eta}) d\tilde{\eta} \tag{16}$$

is an exact eigensolution of the two-dimensional linearized unsteady boundary layer equations that satisfies the homogeneous boundary conditions $\overline{u}(x,\eta), \overline{w}(x,\eta), \overline{\vartheta}(x,\eta) \rightarrow 0$ as $\eta \rightarrow \infty$ for all $B(x)$, but does not necessarily satisfy the no-slip condition at the wall.

Lam and Rott [20], [21] analyzed the two dimensional flat plate boundary layer and showed that the linearized equations possess asymptotic eigensolutions that satisfy a no-slip condition at the wall when x becomes large. These solutions exhibit a two-layer structure consisting of an outer region that encompasses the main part of the boundary layer and a thin viscous region near the wall. The outer solution is given by (14) and (15) with the arbitrary function $B(x)$ determined by matching with the viscous wall layer flow.

Ref. [14] showed that the Lam and Rott [20, 21] analysis also applies to compressible flows when the full compressible solution (14) and (15) is used in the outer region and the viscous wall layer solution is slightly modified to account for the temperature and viscosity variations. The function $B(x)$ is then given by

$$B(x) = x^{3/2} B_n \exp\left[-\frac{2^{3/2} e^{i\pi/4}}{3\lambda \varsigma_n^{3/2}} \left(\frac{T_w}{\mu_w}\right)^{1/2} x^{3/2}\right] + \ldots . \tag{17}$$

where $T_w \equiv T(0)$, $\mu_w \equiv \mu(T(0))$, $\lambda \equiv U'(0)$ and ζ_n is a root of

$$Ai'(\varsigma_n) = 0, \quad \text{for } n = 0, 1, 2, 3.... \tag{18}$$

The only difference from the Lam-Rott result is the $(T_w/\mu_w)^{1/2}$ factor in the exponent. The asymptotic solution to the full inhomogeneous boundary value problem can now be expressed as the sum of a Stokes layer solution plus a number of these asymptotic eigensolutions. The first few B_n were determined from numerical solutions to the boundary layer problem in Ref. [8]. But we are primarily concerned with the lowest order $n = 0$ mode because that is the only one that matches onto a spatially growing oblique Tollmien-Schlicting wave further downstream [11]. The receptivity problem can then be solved by combining the numerical computations with appropriate matched asymptotic expansions to relate the instability wave amplitude to that of the free-stream disturbance. But we will analyze the boundary layer disturbances generated by the free-stream acoustic disturbances before considering these expansions.

3.2 Boundary layer disturbances generated by the Fedorov/Khokhlov mechanism for obliqueness angles close to critical angle

Fedorov and Khokhlov [10] used matched asymptotic expansions to analyze the generation of Mack mode instabilities by oblique acoustic waves of the form (4) where the wavenumbers α and β satisfy the dispersion relation (7) when the incidence angle γ is equal to zero, which, as noted, above is the case being considered here. Their focus was on hypersonic flows where the most rapidly growing disturbances are usually two dimensional 2nd Mack modes, while, as noted in the introduction, the focus of the present chapter is on the relatively low supersonic Mach number regime (say, less than about 4) where the most rapidly growing instability waves are highly oblique 1st Mack modes. Numerical results [9] show that the obliqueness angle of the most rapidly growing 1st mode lies between 50 and 70 degrees at Mach numbers between 2 and 6.

Ref. [10] shows that the boundary layer disturbance produced by diffraction of the slow acoustic wave by the nonparallel mean flow in the region where $x = o(\varepsilon^{-3})$ can be matched onto a 1st Mack mode instability in the downstream region where $x = O(\varepsilon^{-6})$ when the deviation

$$\Delta\theta \equiv \theta_c - \theta \tag{19}$$

of the obliqueness angle θ from the critical angle

$$\cos\theta_c \equiv 1/M_\infty \tag{20}$$

takes on $O(1)$ positive values. The diffraction region has a double layer structure which consists of a region that fills the mean boundary layer and an outer diffraction region of thickness $O(1/\varepsilon^{3/2})$. (The purely passive Stokes layer near the wall does not play a role in the diffraction process and can be ignored).

The instability emerges from the downstream limit of the solution in this region. But as noted in the introduction this occurs too far downstream to be of practical interest when scaled up to actual flight conditions if $\Delta\theta = O(1)$ [14] at the moderately supersonic Mach numbers being considered here. It will however emerge much further upstream when θ is close to the critical angle θ_c, i.e., when $\Delta\theta \ll 1$. But the solution in Ref. [10] does not apply when $\Delta\theta \ll 1$ and a new analysis was developed in Ref. [11] to extend their result into the small -$\Delta\theta$ regime.

It follows from (7) that

$$\alpha = \tilde{\alpha}/\Delta\theta + \tilde{\alpha}_1 + \dots, \quad \beta = \beta_1 = \tilde{\beta}/\Delta\theta \tag{21}$$

where

$$\tilde{\alpha} \equiv 1/\tan\theta_c, \tilde{\beta} \equiv 1, \tilde{\alpha}_1 \equiv 1/\sin^2\theta_c \tag{22}$$

when $\Delta\theta \ll 1$ since $\tan(\theta_c - \Delta\theta) = \tan\theta_c - \Delta\theta/\cos^2\theta_c + O(\Delta\theta)^2$ in that case.

This shows that α also becomes large when $\Delta\theta \ll 1$ and that α will expand in powers of $\Delta\theta$ as indicated in (21) if β is fixed at the indicated value to all orders in $\Delta\theta$ (which we now assume to be the case).

The spanwise wavenumber will equal the vortical spanwise wavenumber (8) when $\Delta\theta = O(\varepsilon)$ and as in that case the diffraction wave solution will eventually develop a triple-deck structure but the resulting solution will (as shown in [11]) not decay at large wall normal distances and is therefore invalid. This means that the diffraction region solution cannot be continued downstream for $\Delta\theta = O(\varepsilon)$.

Ref. [11] shows that the smallest value of $\Delta\theta$ is $\Delta\theta = O(\varepsilon^{2/3})$ and the diffraction region will then occur at an $O(\varepsilon^{-4/3})$ distance downstream. The relevant solution will have the triple-deck structure shown in **Figure 3**: a main boundary layer region that fills the mean boundary layer (region 1), an outer diffraction region of thickness $O(\varepsilon^{-1/3})$ (region 2) and an $O(\varepsilon^3)$ thick viscous wall layer in which the unsteady, convective and viscous terms all balance.

The pressure in region 2 is of the form

$$p = 1 + \hat{\delta} p_2(x_2, y_2) e^{i\left[\left(\tilde{\alpha}/\Delta\theta + \tilde{\alpha}_1\right)x + \tilde{\beta}z/\Delta\theta - t\right]}, \tag{23}$$

where

$$x_2 \equiv x\varepsilon^{4/3} = O(1), \quad y_2 \equiv y\varepsilon^{1/3} = O(1) \tag{24}$$

and the surface pressure $p_2(x_2, 0)$ is related to the up-wash velocity $v_1(x_2, \infty) \equiv \lim_{\eta\to\infty} v_1(x_2, \eta)$ at the outer edge of the boundary layer by

$$p_2(x_2, 0) = p_1(x_2) = 1 - \frac{x_2}{\sqrt{2\pi i\tilde{\alpha}\left(M_\infty^2 - 1\right)}} \int_0^1 \frac{\sqrt{\sigma}}{\sqrt{1-\sigma}} i\tilde{\alpha} \left[\frac{v_1(x_2\sigma, \infty)}{\sqrt{x_2\sigma}}\right] d\sigma, \tag{25}$$

where $p_1(x_2)$ denotes the pressure in the boundary layer region 1 (which is independent of the wall normal direction) and the wall normal velocity $v_1(x_2, \infty)$ is given in terms of

$$\xi_2 \equiv -i^{1/3}\left(\sqrt{2x_2}/\tilde{\alpha}\lambda\right)^{2/3} (T_w/\mu_w)^{1/3} \tag{26}$$

and the integral and the derivative of the Airy function $Ai(\xi)$ by

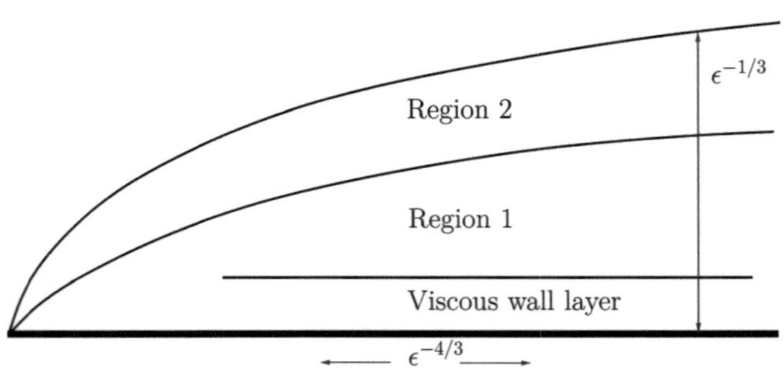

Figure 3.
Structure of diffraction region for $\Delta\theta = O(\varepsilon^{2/3})$.

$$\frac{v_1(x_2,\infty)}{\sqrt{2x_2}} = ip_1(x_2)\frac{\left(\tilde{\alpha}^2+\tilde{\beta}^2\right)T_w^2\xi_2}{\lambda Ai'(\xi_2)}\int_{\xi_2}^{\infty} Ai(\xi)d\xi, \tag{27}$$

which behaves like

$$v_1(x_2,\infty)/\sqrt{2x_2} \sim -ip_1(x_2)\left(\tilde{\alpha}^2+\tilde{\beta}^2\right)T_w^2/\lambda \tag{28}$$

as $x_2 \to \infty$ since ([22], pp. 446–447)

$$Ai'(\xi)/\int_{\xi}^{\infty} Ai(q)dq \to -\xi \text{ as } \xi \to \infty. \tag{29}$$

Inserting (28) and (27) into (25) shows that

$$p_1(x_2) = 1 - \gamma_0 x_2 \int_0^1 \frac{\sqrt{\sigma}}{\sqrt{1-\sigma}} p_1(\sigma x_2)d\sigma, \quad \text{as } x_2 \to \infty \tag{30}$$

where

$$\gamma_0 \equiv \frac{\left(\tilde{\alpha}^2+\tilde{\beta}^2\right)\tilde{\alpha}^{1/2}T_w^2}{\lambda\sqrt{2\pi i\left(M_\infty^2-1\right)}}, \tag{31}$$

which is formally the same as the equation considered in [10] who showed that the solution behaves like

$$p_1(x_2) \sim \exp\left[\gamma_0^2\pi(x_2)^2\right] \text{ as } x_2 \to \infty. \tag{32}$$

The acoustically and vortically generated boundary layer disturbances considered in this section will eventually evolve into propagating eigensolutions in regions that lie further downstream. The resulting flow will have a triple-deck structure of the type considered in [13], [23] and [14] in the former (i.e., vortically generated) case. But the acoustically generated disturbance will only develop an eigensolution structure much further downstream. The minimum distance occurs when $\Delta\theta = O\left(\varepsilon^{2/3}\right)$. We begin by considering the triple-deck region.

4. The viscous triple-deck region

Refs. [13, 14, 23] show that the linearized Navier-Stokes equations possess an eigensolution of the form

$$\{u,v,w,p\} = \hat{\delta}\Pi(y,\varepsilon)e^{i\left[\frac{1}{\varepsilon^3}\int_0^{x_1}\kappa(x_1,\varepsilon)dx_1+\bar{\beta}\bar{z}-t\right]} \tag{33}$$

in the triple-deck region where $\hat{\delta} \ll 1$ is the common scale factor introduced at the beginning of Section 2,

$$\Pi(y,\varepsilon)=\left\{\frac{\overline{A}(x_1)U'(\eta)}{T(\eta)},-i\kappa_0\overline{A}(x_1)U(\eta)\sqrt{2x_1},-\frac{\varepsilon^2\overline{\beta}\overline{P}T(\eta)}{\kappa_0U(\eta)},\varepsilon^2\overline{P}\right\} \tag{34}$$

in the main boundary layer where $\eta=O(1)$,

$$x_1\equiv\varepsilon^2x=O(1) \tag{35}$$

and

$$\overline{z}\equiv z/\varepsilon=z^*\omega^*/\varepsilon U^*_\infty \tag{36}$$

is a scaled transverse coordinate. The complex wavenumber κ has the expansion [11].

$$\kappa(x_1,\varepsilon)=\kappa_0(x_1)+\varepsilon\kappa_1(x_1)+\varepsilon^2\kappa_2(x_1)+\ldots., \tag{37}$$

where the lowest order term in this expansion satisfies the following dispersion relation ([13, 14, 23])

$$\kappa_0^2+\overline{\beta}^2=\frac{1}{(i\kappa_0)^{1/3}}\left(\frac{\lambda}{\sqrt{2x_1}}\right)^{5/3}\left(\frac{\mu_w}{T_w^7}\right)^{1/3}\frac{\left[\overline{\beta}^2-(M_\infty^2-1)\kappa_0^2\right]^{1/2}Ai'(\xi_0)}{\int\limits_{\xi_0}^{\infty}Ai(q)dq} \tag{38}$$

and

$$\xi_0=-i^{1/3}\left(\frac{\sqrt{2x_1}}{\kappa_0\lambda}\right)^{2/3}(T_w/\mu_w)^{1/3} \tag{39}$$

whose solution must satisfy the inequality

$$\mathrm{Re}\left[\overline{\beta}^2-(M_\infty^2-1)\kappa_0^2\right]^{1/2}\geq 0 \tag{40}$$

in order to insure that the eigensolution does not exhibit unphysical wall normal growth.

This requirement will be satisfied for all $M_\infty<1$ but will only be satisfied at supersonic Mach numbers when the obliqueness angle θ is greater than the critical angle $\theta_c\equiv\cos^{-1}(1/M_\infty)$ [11, 13]. The dispersion relation (38) and (39) reduces to the dispersion relation given by Eqs. (4.52), (5.2) and (5.3) of [7] when $\overline{\beta}$ and M_∞ are set to zero.

4.1 Matching with the Lam-Rott solution

The dispersion relation (38) and (39) will be satisfied at small values of x_1 if $\kappa_0\sim\sqrt{x_1}$ and $\xi_0\to\zeta_n$, for $n=0,1,2\ldots$ as $x_1\to 0$, where ς_n is the nth root of the Lam-Rott dispersion relation (18). Inserting this into (38) shows that

$$\kappa_0\to\frac{1}{\lambda\varsigma_n^{3/2}}\left(\frac{2T_wx_1}{i\mu_w}\right)^{1/2}\quad\text{as } x_1\to 0. \tag{41}$$

The cross flow velocity w drops out of (33) as $x_1\to 0$ and the flow in the main deck is therefore compatible with the quasi-two dimensional Lam-Rott solution (14)–(17).

4.2 Numerical results

The dispersion relation (38) is expected to have at least one root corresponding to each of the infinitely many roots of (18). But only the lowest order $n = 0$ root can produce the spatially growing modes of (38). The wall temperature T_w and viscosity μ_w can be scaled out of this equation by introducing the rescaled variables.

$$\kappa_0^\dagger = \kappa_0 T_w^{1/2}\mu_w^{1/6}, x_1^\dagger = x_1 T_w^2/\mu_w{}^{2/3}, \overline{\beta}^\dagger = \overline{\beta} T_w^{1/2}\mu_w^{1/6}. \tag{42}$$

The real and negative imaginary parts of $\kappa_0^\dagger$ calculated from (38) together with the $n = 0$ Lam-Rott initial condition (41) are plotted as a function of the scaled streamwise coordinate $\overline{x}_1^\dagger$ in **Figures 4** and 5 for three values of the frequency scaled transverse wavenumber $\overline{\beta}^\dagger \geq 2$. The insets are included to more clearly show the changes at small $\overline{x}_1^\dagger$. The dashed curves in the insets denote the real and imaginary parts of the small-$\overline{x}_1^\dagger$ asymptotic formula (41).

The triple-deck eigensolution (33) (which contains the Lam-Rott solution as an upstream limit) can undergo a significant amount of damping before it turns into a spatially growing instability wave at the lower branch of the neutral stability curve.

The exponential damping in Eq. (33) is proportional to $\mathcal{Im}\int_0^{x_{LB}} \kappa(x_1)dx =$ $\varepsilon^{-2}\mathcal{Im}\int_0^{(x_1)_{LB}} \kappa(x_1)dx_1$, where $(x_1)_{LB}$ and x_{LB} denote the scaled and unscaled streamwise location of the lower branch of the neutral stability, which implies that the total damping is proportional to the area under the growth rate curve between zero and the lower branch in **Figure 5**. The inset shows that the length $\Delta x_1^\dagger = 0.01$ of this upstream region is very short and therefore that the total amount of damping is relatively small.

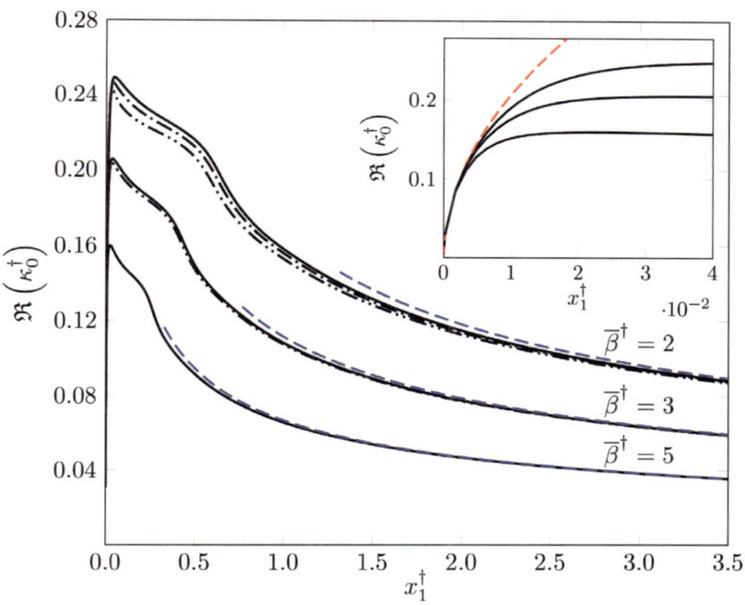

Figure 4.
$\mathcal{Re}(\kappa_0^\dagger)$ as a function of $x_1^\dagger$ calculated from (38) together with the initial condition (41) for $M_\infty = 2, 3, 4$ (double dot dashed, dot dashed, and solid lines, respectively) and three values of $\overline{\beta}^\dagger \geq 2$. The dashed curve in the main graph is the rescaled large-$x_1^\dagger$ asymptote (49).

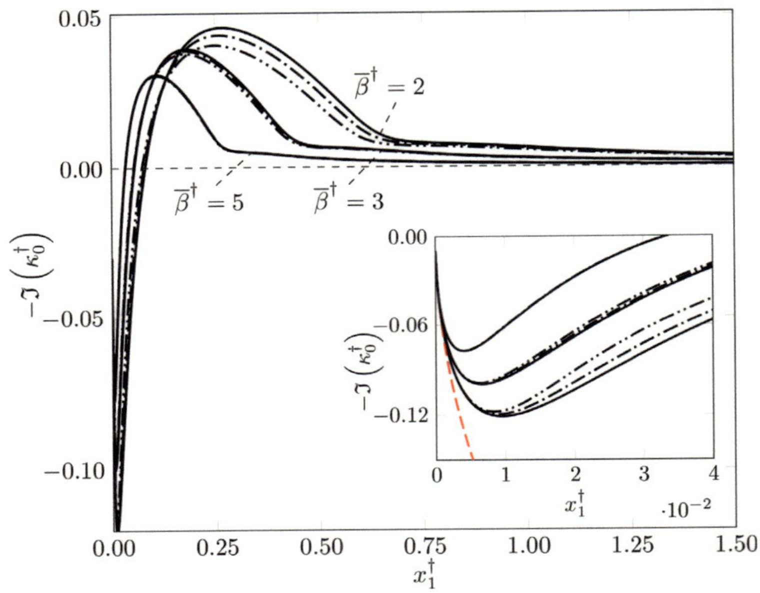

Figure 5.
$-\mathcal{I}m(\kappa_0^\dagger)$ as a function of $x_1^\dagger$ calculated from (38) together with the initial condition (41) for $M_\infty = 2, 3, 4$ (double dot dashed, dot dashed and solid lines, respectively) and three values of the frequency scaled transverse wavenumber.

5. The inviscid triple-deck region

As noted above the acoustically driven solution will only match onto an eigensolution in the downstream region when $O(\Delta\theta) \geq \varepsilon^{3/2}$. This region will lie downstream of the viscous triple-deck region considered above and will be closest to that region when $O(\Delta\theta) = \varepsilon^{3/2}$. It will have an inviscid triple- deck structure and the relevant dispersion relation can be obtained by putting $\varepsilon/\Delta\theta = O(\varepsilon^{1/3})$ in (21), inserting the rescaled variables

$$\overline{\overline{\beta}} = \overline{\beta}/\varepsilon^{1/3}, \overline{\kappa}_0 = \kappa_0/\varepsilon^{1/3}, \hat{x}_1 = x_1\varepsilon^{4/3} \tag{43}$$

into (38), using (29), and taking the limit as $\varepsilon \to 0$ with $\overline{\overline{\beta}}$, $\overline{\kappa}_0$ and $\hat{x}_1$ held fixed, to show that the rescaled wavenumber $\overline{\kappa}_0$ satisfies the inviscid dispersion relation

$$\overline{\kappa}_0^2 + \overline{\overline{\beta}}^2 = \frac{\lambda\left[\overline{\overline{\beta}}^2 - (M_\infty^2 - 1)\overline{\kappa}_0^2\right]^{1/2}}{\overline{\kappa}_0\sqrt{2\hat{x}_1}T_w^2} \tag{44}$$

when the square root $\left[\overline{\overline{\beta}}^2 - (M_\infty^2 - 1)\overline{\kappa}_0^2\right]^{1/2}$ is required to remain finite as $\varepsilon \to 0$.

5.1 Matching with the small $\Delta\theta$ Fedorov/Khokhlov solution

It can then be shown by direct substitution that the solution $\overline{\kappa}_0$ behaves like

$$\overline{\kappa}_0 \to \frac{\overline{\overline{\beta}}}{(M_\infty^2 - 1)^{1/2}} - \overline{\overline{\beta}}^5 \hat{\alpha}_0^2 \hat{x}_1 \quad \text{as } \hat{x}_1 \to 0, \tag{45}$$

where $\hat{\alpha}_0 \equiv M_\infty^2 T_w^2 / \left[\left(M_\infty^2 - 1\right)^{7/4} \lambda \right]$. The square root $\left[\bar{\bar{\beta}}^2 - \left(M_\infty^2 - 1\right)\bar{\kappa}_0^2 \right]^{1/2}$ still satisfies the inequality (40) when $\hat{x}_1 \to 0$ and (44) therefore remains valid in this limit.

The pressure component of the resulting solution will then match onto the downstream limit (32) and (30) of the acoustically generated diffraction region solution when $\bar{\bar{\beta}} = O\left(\varepsilon^{2/3}/\Delta\theta\right)$ and x_2 is given by (24) since it follows from (8),(35), (43) and (45) that

$$\begin{aligned}(1/\varepsilon^3)\int_0^{x_1}\kappa_0(x_1)dx_1 &= (1/\varepsilon^4)\int_0^{\hat{x}_1}\bar{\kappa}_0(\hat{x}_1)d\hat{x}_1 \to (\tilde{\alpha}/\Delta\theta)x - \varepsilon\bar{\beta}^5\hat{\alpha}_0{}^2x^2/2 \\ &= (\tilde{\alpha}/\Delta\theta)x - \beta^5\hat{\alpha}_0{}^2\left(\varepsilon^3 x\right)^2/2 = (\tilde{\alpha}/\Delta\theta)x - \hat{\alpha}_0{}^2x_2^2/2.\end{aligned} \quad (46)$$

5.2 Numerical results

Figure 6 is a plot of the scaled lowest order wavenumber $\bar{\kappa}_0/\bar{\bar{\beta}} = \kappa_0/\bar{\beta}$ as a function of the scaled streamwise coordinate $\left(\bar{\bar{\beta}}T_w\right)^4\hat{x}_1/\lambda^2 = \left(\bar{\beta}T_w\right)^4 x_1/\lambda^2$ for various values of the free-stream Mach number M_∞ calculated from the inviscid triple-deck dispersion relation(44) together with the asymptotic initial condition (45) which is shown by the dashed curves in the figure. The lowest order wave number $\bar{\kappa}_0$ is purely real which means that exponential growth (if it occurs) can only occur at higher order. This suggests that the acoustically generated instabilities will be less significant than the vortically-generated instabilities which appear upstream.

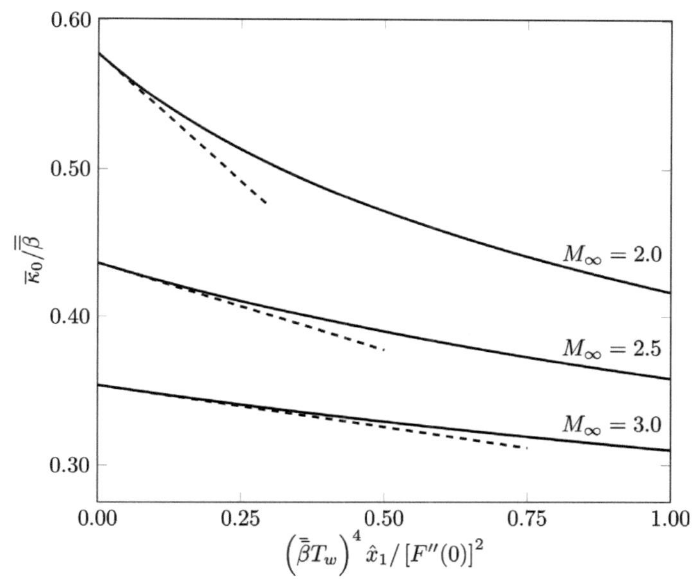

Figure 6.
Scaled wavenumber $\bar{\kappa}_0/\bar{\bar{\beta}} = \kappa_0/\bar{\beta}$ as a function of $\left(\bar{\bar{\beta}}T_w\right)^4\hat{x}_1/\lambda^2 = \left(\bar{\beta}T_w\right)^4 x_1/\lambda^2$ for various values of M_∞. The solid lines represent the numerical solution. Dashed lines are the asymptotic solution(45).

6. The next stage of evolution

6.1 Downstream behavior of the triple-deck solution

Eqs. (29), (38) and (39) show that

$$\overline{\beta} \to \frac{1}{\kappa_0^{1/3}}\left(\frac{\lambda}{\sqrt{2x_1}}\right)^{5/3}\left(\frac{1}{T_w^2}\right)\left(\frac{\sqrt{2x_1}}{\kappa_0\lambda}\right)^{2/3} = \frac{\lambda}{\kappa_0 T_w^2\sqrt{2x_1}} \tag{47}$$

when $x_1 \to \infty$ and, therefore, that

$$\kappa_0 \to \frac{\lambda}{\overline{\beta} T_w^2\sqrt{2x_1}}, \tag{48}$$

when κ_0 is allowed to approach zero as $x_1 \to \infty$.

The dashed curves in the main plot of **Figure 4** represent the re-scaled large-$\overline{x}_1^\dagger$ asymptote (48). It confirms that the numerical results are well approximated by the (appropriately rescaled) large-x_1 asymptote (48).

As noted in [11], the solution to the reduced dispersion relation (44) satisfies the rescaled version

$$\overline{\kappa}_0 \to \frac{\lambda}{\overline{\overline{\beta}} T_w^2\sqrt{2\hat{x}_1}} \quad \text{as } \hat{x}_1 \to \infty \tag{49}$$

of (48), which can be considered to be a special case of this result if we put

$$\overline{\overline{\beta}} = \overline{\beta}/\varepsilon^r, \overline{\kappa}_0 = \kappa_0/\varepsilon^r, \hat{x}_1 = x_1\varepsilon^{4r} \tag{50}$$

and allow r to be zero or 1/3.

The expansion (37) then generalizes to [11]

$$\overline{\kappa}(x_1,\varepsilon) = \overline{\kappa}_0(\hat{x}_1) + \varepsilon^{1-r}\overline{\kappa}_1(\hat{x}_1) + \varepsilon^{2(1-r)}\overline{\kappa}_2(\hat{x}_1) + ..., \tag{51}$$

where

$$\overline{\kappa}, \overline{\kappa}_1, \overline{\kappa}_2... \equiv \kappa/\varepsilon^r, \kappa_1, \kappa_2\varepsilon^r... \tag{52}$$

and $\hat{x}_1$ is defined in (43).

6.2 Derivation of the governing equations

Eq. (49) shows, among other things, that the lowest order wave number and streamwise growth rate approach zero but do not become negative as the disturbance propagates downstream. The boundary layer thickness which is $O(\varepsilon^3\sqrt{x})$ continues to increase and the triple-deck scaling breaks down at the streamwise location

$$\overline{x}_1 = x\varepsilon^{4+2r} = O(1), \tag{53}$$

where it becomes of the order of the spanwise length scale, which remains constant at $O(\varepsilon^{1-r})$. This region is located well upstream of the region where the unsteady flow is governed by the full Rayleigh equation considered in [9].

Eqs. (37), (43), (51) and (52) show that the Tollmien-Schlichting wave becomes more oblique and

$$\exp i\left[\frac{1}{\varepsilon^3}\int_0^{x_1}\kappa(x_1,\varepsilon)dx_1+\overline{\beta}\overline{z}-t\right]=\exp i\left[\frac{1}{\varepsilon^{3(1+r)}}\int_0^{\hat{x}_1}\overline{\kappa}_0(\hat{x}_1,\varepsilon)d\hat{x}_1+\frac{1}{\varepsilon^{2+4r}}\int_0^{\hat{x}_1}\overline{\kappa}_1(\hat{x}_1,\varepsilon)d\hat{x}_1\right.$$

$$\left.+\frac{1}{\varepsilon^{1+5r}}\int_0^{\hat{x}_1}\overline{\kappa}_2(\hat{x}_1,\varepsilon)d\hat{x}_1+O(\varepsilon^{-4r})+\varepsilon^r\overline{\overline{\beta}}\overline{z}-t\right]\to e^{i\left[\varepsilon^{-2(2+r)}\int_0^{\overline{x}_1}\overline{\alpha}(\overline{x}_1,\varepsilon)d\overline{x}_1+\overline{\overline{\beta}}\overline{\overline{z}}-t\right]} \tag{54}$$

as $\hat{x}_1\to\infty$, where $\overline{\alpha}(\overline{x}_1)$ is an $O(1)$ function of $\overline{x}_1$ (given by (53)) and

$$\overline{\overline{z}}\equiv\varepsilon^r\overline{z}=z/\varepsilon^{1-r}, \tag{55}$$

which means that the solution should be proportional to $\exp i\left[\varepsilon^{-(4+2r)}\int_0^{\overline{x}_1}\overline{\alpha}(\overline{x}_1,\varepsilon)d\overline{x}_1+\overline{\overline{\beta}}\overline{\overline{z}}-t\right]$, where $\overline{\alpha}(\overline{x}_1)$ is an $O(1)$ function of $\overline{x}_1$ that behaves like

$$\overline{\alpha}\to\frac{\lambda}{\overline{\overline{\beta}}T_w^2\sqrt{2\overline{x}_1}}+\ldots\ \text{as}\ \ \overline{x}_1\to 0 \tag{56}$$

in this stage of evolution. The solution should remain inviscid in the main boundary layer and the viscous wall layer (i.e., a Stokes layer) is expected to be completely passive.

The scaled variable

$$\overline{y}\equiv y/\varepsilon^{1-r} \tag{57}$$

will be $O(1)$ in the main boundary layer since its thickness is now of the order of the spanwise length scale, $O(\varepsilon^{1-r})$. It therefore follows from (53) and (57) that the transverse pressure gradients will come into play and the solution in this region should expand like

$$\{u,v,w,p\}=\{U,0,0,0\}+\hat{\delta}\mathcal{A}(\overline{x}_1)\{\overline{u}(\overline{y};\overline{x}_1),\varepsilon^{1-r}\overline{v}(\overline{y};\overline{x}_1),\varepsilon^{1-r}\overline{w}(\overline{y};\overline{x}_1),\varepsilon^{2(1-r)}\overline{p}(\overline{y};\overline{x}_1)\}$$

$$\exp i\left[\frac{1}{\varepsilon^{4+2r}}\int_0^{\overline{x}_1}\overline{\alpha}(\overline{x}_1,\varepsilon)d\overline{x}_1+\overline{\overline{\beta}}\overline{\overline{z}}-t\right]\ldots \tag{58}$$

where $\mathcal{A}(\overline{x}_1)$ is a function of the slow variable $\overline{x}_1$. Substituting (58) into the linearized Navier-Stokes equations shows that the wall normal velocity perturbation $\overline{v}$ is determined by the incompressible reduced Rayleigh equation

$$T\frac{d}{d\overline{y}}\frac{1}{T}\frac{d\overline{v}}{d\overline{y}}+\left[\frac{T\overline{\alpha}}{1-\overline{\alpha}U}\frac{d}{d\overline{y}}\left(\frac{dU/d\overline{y}}{T}\right)-\overline{\overline{\beta}}^2\right]\overline{v}=0 \tag{59}$$

whose solution must satisfy the following boundary conditions

$$\overline{v} \sim e^{-\overline{\overline{\beta}}\overline{y}} \quad \text{as } \overline{y} \to \infty, \quad \overline{v} = 0 \ \text{ at } \overline{y} = 0. \tag{60}$$

Matching with the upstream solution (33) and (37) requires that $\overline{\alpha}(\overline{x}_1)$ satisfy the matching condition (56) as $\overline{x}_1 \to 0$.

Inserting (10) and (57) into (59), using (60) and assuming the ideal gas law $\rho T = 1$ shows that

$$\frac{d}{d\eta}\frac{1}{T^2}\frac{d\overline{v}}{d\eta} + \left[\frac{\overline{\alpha}}{1-\overline{\alpha}U}\left(\frac{U'}{T^2}\right)' - \left(\overline{\overline{\beta}}\sqrt{2\overline{x}_1}\right)^2\right]\overline{v} = O\left(\varepsilon^{2(1-r)}\right), \tag{61}$$

$$\overline{v} = 0 \ \text{ at } \eta = 0, \tag{62}$$

which means that

$$\overline{\alpha} = f(\hat{\beta}), \tag{63}$$

where

$$\hat{\beta} \equiv \overline{\overline{\beta}}\sqrt{2\overline{x}_1}. \tag{64}$$

6.3 Matching with the triple-deck solution

Eq. (64) clearly approaches zero when $\overline{x}_1 \to 0$, which means that $\overline{\alpha}$ will be consistent with the matching condition (54) if we require that it behave like

$$\overline{\alpha} = \lambda/T_w^2\hat{\beta} + \alpha_1 + \alpha_2\hat{\beta} + \ldots \ \text{ as } \overline{x}_1 \to 0 \tag{65}$$

where $\alpha_1, \alpha_2 \ldots$ are (in general complex) constants such that

$$\alpha_1 = \lim_{\hat{x}_1 \to \infty} \overline{\kappa}_1(\hat{x}_1), \ \alpha_2 = \lim_{\hat{x}_1 \to \infty} \overline{\kappa}_2(\hat{x}_1)/\overline{\overline{\beta}}\sqrt{2\hat{x}_1}. \tag{66}$$

Ref. [11] proved that (60)–(64) possess an asymptotic solution of the form $\overline{v} = U(\eta) + \hat{\beta}v_1 + \hat{\beta}^2 v_2 + \ldots$ as $\hat{\beta} \to 0$ when $\overline{\alpha}$ satisfies (65) which implies that their solutions are able to match onto the lowest order triple-deck solution upstream and are consistent with the higher order solutions in this region.

6.4 Numerical results

The Rayleigh eigenvalues $\overline{\alpha}$ are determined by the boundary value problem (60), (61) and (62). We assume in the following that the Prandtl number is equal to unity and that the viscosity $\mu(T)$ satisfies the simple linear relation $\mu(T) = T(\eta)$.

Parts (a) and (b) of **Figure 7** are plots of the real and imaginary parts respectively of these eigenvalues as a function of $\hat{\beta}$. They show that the numerical solution for $\overline{\alpha}$ will be consistent with the matching conditions (65)and(66) if the higher order terms in the triple-deck expansion(51) satisfy $\mathcal{Im}\lim_{\hat{x}_1 \to \infty} \overline{\kappa}_1(\hat{x}_1) = 0$ and $\lim_{\hat{x}_1 \to \infty} \overline{\kappa}_2(\hat{x}_1)/\overline{\overline{\beta}}\sqrt{2\hat{x}_1} = \pm iC$, where the values of C are given in the caption of **Figure 7**. They also show that $\overline{\alpha}$ is initially real and eventually becomes complex. But these eigenvalues must occur in complex conjugate pairs since the coefficients in (61) are all real. The computations show that $\mathcal{Im}(\overline{\alpha})$ eventually goes to zero at

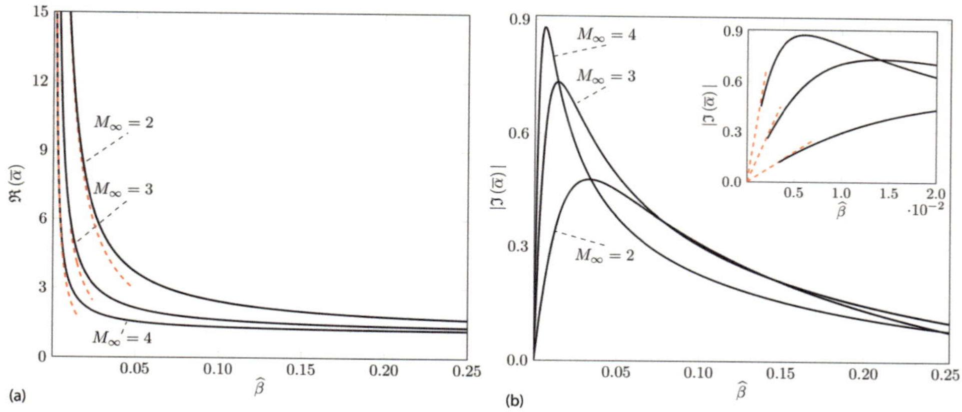

Figure 7.
(a) $\Re(\overline{\alpha})$ *and (b)* $|\Im(\overline{\alpha})|$ *vs.* $\hat{\beta}$ *calculated from the modified Rayleigh solution. The red dashed curves are calculated from the asymptotic formula(56). The red dashed lines in the inset are* $|\Im(\overline{\alpha})| = C\hat{\beta}$, *where* $C = 36$ *for* $M_\infty = 2C = 129.4$ *for* $M_\infty = 3$ *and* $C = 340.1$ *for* $M_\infty = 4$.

some finite value of $\hat{\beta}$ which is consistent with the fact that $\left(U'/T^2\right)'$ is equal to zero at some finite value of η and Eq. (61) therefore has a generalized inflection point there.

7. Conclusions

This chapter uses high Reynolds number asymptotics to study the nonlocal behavior of boundary layer instabilities generated by small amplitude free-stream disturbances at subsonic and moderate supersonic Mach numbers. The appropriate small expansion parameter turns out to be $\varepsilon = \mathcal{F}^{1/6}$, where $\mathcal{F}$ denotes the frequency parameter. The oblique 1st Mack mode instabilities generated by free-stream acoustic disturbances are compared with those generated by elongated vortical disturbances. The focus is on explaining the relevant physics and not on obtaining accurate numerical predictions.

The free-stream vortical disturbances generate unsteady flows in the leading edge region that produce short spanwise wavelength instabilities in a viscous triple-deck region which lies at an $O(\varepsilon^{-2})$ distance downstream from the leading edge. The mechanism was first considered for two dimensional incompressible flows in Ref. [7], but the instability onset occurs much further upstream in the supersonic case and is, therefore, much more likely to be important at the higher Mach numbers considered in this chapter. The lowest order triple-deck solution does not possess an upper branch and evolves into an inviscid 1st Mack mode instability with short spanwise wavelength at an $O(\varepsilon^{-4})$ distance downstream.

Fedorov and Khokhlov [10] used asymptotic methods to study the generation of inviscid instabilities in supersonic boundary layers by fast and slow acoustic disturbances in the free stream whose obliqueness angle θ deviated from its critical value by an $O(1)$ amount and showed that slow acoustic disturbances generate unsteady boundary layer disturbances that produce $O(1)$ spanwise wavelength inviscid 1st Mack mode instabilities a much larger $O\left(\varepsilon^{-6}\right)$ distance downstream. But the calculations in Ref. [11] show that the physical streamwise distance $x^* = \left(U_\infty^*\right)^3/(\omega^*)^2\nu_\infty^*$ corresponding to this scaled downstream location is at least equal to about 7 m for the typical supersonic flight conditions at

$M_\infty = 3$ $(U_\infty^* = 888\,m/s, \nu_\infty^* = 0.000264\,m^2/s)$ end an altitude of 20 km with an upper bound of 100 kHz for the characteristic frequency. This means that this instability occurs too far downstream to be of any practical interest at the moderately low supersonic Mach numbers considered in this chapter.

But, the inviscid instability, which first appears at an $O\left(\varepsilon^{-(4+2/3)}\right)$ distance downstream when $\Delta\theta$ is reduced to $O\left(\varepsilon^{2/3}\right)$ can be significant when scaled to flight conditions. It is therefore appropriate to compare the vortically-generated instabilities with the instabilities generated by oblique acoustic disturbances with obliqueness angles in this range as done in this chapter.

Acknowledgements

This research was sponsored by NASA's by the Transformative Aeronautics Concepts Program of the Aeronautics Research Mission Directorate under the Transformational Tools and Technologies (TTT) Project. PR was supported by the Air Force Office of Scientific Research award number AFOSR Grant FA9550-15-1-0248. We would also like to thank Dr. Meelan Choudhari for bringing the photograph in **Figure 2** to our attention.

Author details

Marvin E. Goldstein[1*] and Pierre Ricco[2]

1 National Aeronautics and Space Administration, Glenn Research Centre, Cleveland, OH, USA

2 Department of Mechanical Engineering, The University of Sheffield, Sheffield, UK

*Address all correspondence to: marvin.e.goldstein@nasa.gov

References

[1] Dryden HL. Air flow in the boundary layer near a plate. NACA Report no. 562; 1936. Available from: https://naca.central.cranfeld/ac.uk/reports/1937/naca-report-562.pdf

[2] Marensi E, Ricco P, Wu X. Nonlinear unsteady streaks engendered by the interaction of free-stream vorticity with a compressible boundary layer. Journal of Fluid Mechanics. 2017;**817**:80-121

[3] Reshotko E. Boundary layer stability and control. Annual Review of Fluid Mechanics. 1976;**8**:311-349

[4] Goldstein ME. Scattering of acoustic waves into Tollmien-Schlichting waves by small streamwise variations in surface geometry. Journal of Fluid Mechanics. 1985;**154**:509-531

[5] Ruban AI. On the generation of Tollmien-Schlicting waves by sound. Fluid Dynamics. 1985;**19**:709-716

[6] Choudhari M, Street CL. Theoretical prediction of boundary-layer receptivity. In: AIAA paper 94-2223, 25th Fluid Dynamics Conference, Colorado Springs, CO; 1994

[7] Goldstein ME. The evolution of Tollmein-Schlichting waves near a leading edge. Journal of Fluid Mechanics. 1983;**127**:59-81

[8] Goldstein ME, Sockol PM, Sanz J. The evolution of Tollmein-Schlichting waves near a leading edge. Part 2. Numerical determination of amplitudes. Journal of Fluid Mechanics. 1983;**129**: 443-453

[9] Mack LM. Boundary layer linear stability theory. AGARD Report 709; 1984

[10] Fedorov AV, Khokhlov AP. Excitation of unstable disturbances in a supersonic boundary layer by acoustic waves. Fluid Dynamics. 1991;**9**:457-467

[11] Goldstein ME, Ricco P. Non-localized boundary layer instabilities resulting from leading edge receptivity at moderate supersonic Mach numbers. Journal of Fluid Mechanics. 2018;**838**: 435-477. DOI: 10.101/jfm2017.889

[12] Fedorov AV. Receptivity of a high speed boundary layer to acoustic disturbances. Journal of Fluid Mechanics. 2003;**491**:101-129

[13] Smith FK. On the first-mode instability in subsonic, supersonic or hypersonic boundary layers. Journal of Fluid Mechanics. 1989;**198**:127-153

[14] Ricco P, Wu X. Response of a compressible laminar boundary layer to free-stream vortical disturbance. Journal of Fluid Mechanics. 2007;**587**:97-138. DOI: 10.1017/s0022112007007070

[15] Kovasznay LSG. Turbulence in supersonic flow. Journal of Aerosol Science. 1953;**20**(10):657-674

[16] Stewartson K. The Theory of Laminar Boundary Layers in Compressible Fluids. Clarendon Press; 1964

[17] Gulyaev AN, Kozlov VE, Kuznetson VR, Mineev BI, Sekundov AN. Interaction of a laminar boundary layer with external turbulence. Izvestiya Akademii Nauk SSSR, Mekhanika Zhidkosti i Gaza 1989;**6**:700-710

[18] Prandtl L. Zur Brechnung der Grenzschichten. Zeitchrift fuer Angewwandte Mathematik und Mechanik. 1938;**18**:77-82

[19] Glauert MB. The laminar boundary layer on oscillating plates and cylinders.

Journal of Fluid Mechanics. 1956;**1**(1): 97-110

[20] Lam SH, Rott. Theory of Linearized Time-Dependent Boundary Layers. Ithaca, NY: Graduate School of Aeronautical Engineering Report AFSOR TN-60-1100, Cornell University; 1960

[21] Lam SH, Rott. Eigen-functions of unsteady boundary layer equations. Journal of Fluid Engineering. 1995; **1995**:115

[22] Abramowitz M, Stegun IA. Handbook of Mathematical Functions. Washington: National Bureau of Standards; 1965

[23] Wu X. Generation of Tollmein-Schlichting waves by convecting gusts interacting with sound. Journal of Fluid Mechanics. 1999;**397**:285-316

Chapter 5

Transition Modeling for Low to High Speed Boundary Layer Flows with CFD Applications

Unver Kaynak, Onur Bas, Samet Caka Cakmakcioglu and Ismail Hakki Tuncer

Abstract

Transition modeling as applied to CFD methods has followed certain line of evolution starting from simple linear stability methods to almost or fully predictive methods such as LES and DNS. One pragmatic approach among these methods, such as the local correlation-based transition modeling approach, is gaining more popularity due to its straightforward incorporation into RANS solvers. Such models are based on blending the laminar and turbulent regions of the flow field by introducing intermittency equations into the turbulence equations. Menter et al. pioneered this approach by their two-equation γ-Reθ intermittency equation model that was incorporated into the k-ω SST turbulence model that results in a total of four equations. Later, a range of various three-equation models was developed for super-/hypersonic flow applications. However, striking the idea that the Reθ-equation was rather redundant, Menter produced a novel one-equation intermittency transport γ-equation model. In this report, yet another recently introduced transition model called as the Bas-Cakmakcioglu (B-C) algebraic model is elaborated. In this model, an algebraic γ-function, rather than the intermittency transport γ-equation, is incorporated into the one-equation Spalart-Allmaras turbulence model. Using the present B-C model, a number of two-dimensional test cases and three-dimensional test cases were simulated with quite successful results.

Keywords: transitional flow, correlation-based transition model, intermittency transport equation, boundary layer flow, turbulence modeling

1. Introduction

Industrial design aerodynamics heavily depends on development of new CFD methods that can be only as good as their experimental database. All these industrial design CFD codes, as they may be called, are constantly in search of better physical modeling starting with appropriate transition and turbulence modeling. To this end, although numerical representation of turbulence has reached the acceptable levels of accuracy for computational aerodynamics, transition modeling has yet to reach the level of turbulence modeling capability for routine calculations. Therefore, transition modeling as part of turbulence has always been standing as the crux of the matter with regard to turbulence modeling. Today, state of the art

Reynolds Averaged Navier-Stokes (RANS) solvers are widely available for numerically predicting fully turbulent part of flow fields by frequent use of, for instance, one- or two-equation turbulence closure models. However, none of these models are adequate to handle flows with significant transition effects due to the lack of practical transition modeling. Menter et al. [1] state that some of the main requirements for pragmatic transition modeling are the following: calibrated prediction of the onset and length of transition, allow inclusion of different mechanisms, allow local formulation, and allow a robust integration with background turbulence models.

Nevertheless, transition modeling as applied to CFD methods has followed certain line of evolution covering a range of methods starting from simple linear stability methods such as the e^N method [2, 3] to almost or fully predictive methods such as LES and DNS that are very costly for engineering applications [1]. The e^N method is the lowest level transition model based on linear stability theory. This method has found quite wide application in numerical boundary layer methods [4], but translating this into RANS methods has proven quite demanding as it requires a high-resolution boundary layer code that must work hand in hand with the RANS method. Also, this method is also dependent on the empirical factor-n that is not universal and depends on the type of flow.

Following the e^N method, a better level of complexity that is compatible with the CFD methods is the low Reynolds number turbulence models [5]. Yet, they do not reflect real flow physics and lack the true predictive capability. These methods take advantage of the fortuitous ability of the wall damping terms mimicking some of the effects of transition. Next in the line of increasing complexity comes the class of the so-called correlation-based transition models [1]. These models are based on the fundamental approach of blending the laminar and the turbulent regions of the flow field by introducing intermittency equations to the turbulence equations. In this line, based on the boundary layer methods, there are three similar examples of intermittency equation approach that was introduced by Dhawan and Narasimha [6], Steelant and Dick [7], and Cho and Chung [8]. First, Dhawan and Narasimha [6] used a generalized form of intermittency distribution function in order to combine the laminar and the turbulent flow regions. Second, Steelant and Dick [7] proposed an intermittency equation that behaves like an experimental correlation. Third, Cho and Chung [8] introduced the k-ε-γ model which was formulated by an additional transport equation-γ to the well-known k-ε turbulence model. Finally, Suzen and Huang [9] significantly improved intermittency equation approach for flow transition prediction by combining the last two methods with a model that simulates transition in both streamwise and cross-stream directions. However, these models all rely on nonlocal flow data, and it was difficult to embed these models into practical CFD codes. These models require calculating the momentum thickness Reynolds number-Re_θ, which is an integral parameter, and comparing it with a critical momentum thickness Reynolds number. For this reason, these early models are "nonlocal" methods that require exhausting search algorithms for flows with complex geometries.

After the success of the "nonlocal" transition models that use intermittency transport equations including experimental correlations, a range of new methods [10, 11] has been developed, called as the local correlation-based transition models (LCTM) by Menter et al. [1] that are compatible with the modern CFD codes. This compatibility has been achieved by the experimental observation that a locally calculated parameter called as the vorticity Reynolds number (Re_v) is proportional to the momentum thickness Reynolds number (Re_θ) in a Blasius boundary layer. This observation is also shown to be quite effective for a wide class of flow types with moderate pressure gradients. This is due to the fact that the relative error between the two parameters is less than 10% for such flows [1]. Therefore, the vorticity Reynolds number-Re_v would be used in order to avoid all the troublesome work that existed in the nonlocal models.

Following the success of the γ-Re_θ two-equation transition model of Menter et al. [1], some other two-, or three-equation models are proposed, such as the near/freestream intermittency model by Lodefier et al. [12], variations of the k-k_L-ω models of Walters and Leylek [13] and Walters and Cokljat [14], and the k-ω-γ model of Fu and Wang [15] with super/hypersonic flow applications. In addition, some researchers proposed extensions to local correlation-based transition models (LCTM) in order to take more physical phenomena into account. To this end, cross-flow instability effects by Seyfert and Krumbein [16], surface roughness effects by Dassler et al. [17], and compressibility effects by Kaynak [18] were included. Meanwhile, Bas et al. [19] proposed a very pragmatic approach by introducing an algebraic or a zero-equation model called later as the Bas-Cakmakcioglu (B-C) model [20]. Herein, it was shown that an equivalent level of prediction compared with the two- and three-equation models could be achieved with less equations provided that physics was correctly modeled. In parallel, Kubacki et al. [21] proposed yet another algebraic transition model with a good level of success vindicating this line of approach. Similarly, Menter et al. [22] proposed a new one-equation γ-model which is the simplification of their earlier two-equation γ-Re_θ model [11] without the Re_θ-equation that produced equal level of results as in the original model. Following this logical trend for reducing the total number of equations, the Wray-Agarwal (WA) wall-distance-free one-equation turbulence model [23] was complemented with the Menter et al. [22] one-equation intermittency transport-γ model to obtain the so-called two-equation Nagapetyan-Agarwal WA-γ transition model [24]. In the following, a brief review of the transition modeling is made that covers the practical applications of a range of models that are currently used in the industrial design aerodynamics. Based on the present authors' recent experiences, the Bas-Cakmakcioglu model [20] will be covered in some detail to display the viability of the algebraic intermittency equation approach vis-a-vis the one- and two-equation local correlation-based transition models (LCTM).

2. Review of transition models

2.1 e^N Method

The well-known e^N method is based on the linear stability theory [25], and it is developed by assuming that the flow is two-dimensional and steady, the boundary layer is thin and the level of disturbances in the flow region is initially very low. In this method, the Orr-Sommerfeld eigenvalue equations are solved by using the previously obtained velocity profiles over a surface in order to calculate the local instability amplification rates of the most unstable waves for each profile. By taking the integral of those rates after a certain point where the flow first becomes unstable along each streamline, an amplification factor is calculated. Transition is said to occur when the value of the amplification factor exceeds a threshold N value. Typical values of N vary between 7 and 9.

2.2 Low Reynolds number turbulence models

In the low Reynolds number turbulence models, the wall damping functions are modified in order to capture the transition effects [5]. To be able to predict the transition onset, these models depend on the diffusion of the turbulence from freestream into the boundary layer and its interaction with the source terms of the turbulence models. For this reason, these models are more suitable for bypass transition flows. Nonetheless, due to the similarities between a developing laminar boundary layer and a viscous sublayer, their success is thought to be coincidental, and thus these modes are mostly unreliable. These models also lack sensitivity to adverse pressure gradients and convergence problems arise for separation-induced transition cases.

2.3 Intermittency equation transition models

It has been known from experiment that turbulence has an intermittent character with large fluctuations in flow variables like velocity, pressure, etc. Based on this observation, transition to turbulence has been tried to be modeled using the so-called intermittency function. One-, two- or three-equation partial differential equations have been derived to include the intermittency equation as one of the equations of the complete equation set including relevant experimental calibrations that mimic the actual physical behavior. To this end, “nonlocal” [7–9] and “local” [1, 10, 11] correlation transition models have been proposed. In the following, a systematic line of progress is presented that reveals the evolution of such models.

2.3.1 Models depending on nonlocal flow variables

2.3.1.1 Dhawan and Narasimha model

Dhawan and Narasimha [6] proposed a scalar intermittency function-γ that would provide some sort of a measure of progression toward a fully turbulent boundary layer. Based on the experimentally measured streamwise intermittency distributions on flat plate boundary layers, for instance, Dhawan and Narasimha [6] introduced the following function for streamwise intermittency profile:

$$\gamma = \begin{cases} 0 & x < x_t \\ 1.0 - exp\left[-\frac{(x - x_t)^2 n\sigma}{U}\right] = 1.0 - exp\left(-0.41\xi^2\right) & x \leq x_t \end{cases} \tag{1}$$

In the above function, x_t is the known transition onset location, n is the turbulence spot formation rate per unit time per unit distance in the spanwise direction, σ is a turbulence spot propagation parameter, and U is the freestream velocity.

2.3.1.2 Cho and Chung model

Cho and Chung [8] developed the k-ε-γ turbulence model that is not designed for prediction of transitional flows but for free shear flows. In this model, the intermittency effect is incorporated into the conventional k-ε turbulence model with the addition of an intermittency transport equation for the intermittency factor γ. In this model, the turbulent viscosity is defined in terms of k, ε, and γ. The intermittency transport equation is given as:

$$u_j \frac{\partial \gamma}{\partial x_j} = D_\gamma + S_\gamma \tag{2}$$

where D_γ is the diffusion term and S_γ is the source term. This model is tested for a plane jet, a round jet, a plane far-wake, and a mixing layer case. As mentioned before, although the model was not designed for transition prediction, the γ intermittency profile for the turbulent-free shear layer flows was quite realistic.

2.3.1.3 Steelant and Dick model

Steelant and Dick [7] developed an intermittency transport model that can be used with the so-called conditioned Navier-Stokes equations. In this model, the intermittency function of Dhawan and Narasimha [6] is first differentiated along the streamline direction, s, and the following intermittency transport equation is obtained:

$$\frac{d\gamma}{\partial\tau} + \frac{\partial\rho u\gamma}{dx} + \frac{\partial\rho v\gamma}{\partial y} = (1 - \gamma)\rho\sqrt{u^2 + v^2}\,\beta(s) \quad (3)$$

In the above equation, $\beta(s)$ is a turbulent spot formation and propagation term, which is seen in the exponential function part of the Dhawan and Narasimha model. Steelant and Dick tested their model for zero, adverse and favorable pressure gradient flows by using two sets of the so-called conditioned averaged Navier-Stokes equations. Although their model reproduces the intermittency distribution of Dhawan and Narasimha for the streamwise direction, a uniform intermittency distribution in the cross-stream direction is assumed. Yet, this is inconsistent with the experimental observations of, for instance, Klebanoff [26] where a variation of the intermittency in the normal direction by means of an error function formula.

2.3.1.4 Suzen and Huang model

Suzen and Huang [9] proposed an intermittency transport equation model by mixing the production terms of the Cho and Chung [8] and Steelant and Dick [7] models by means of a new blending function. An extra diffusion-related production term due to Cho and Chung is also added to the resultant equation. This model successfully reproduces experimentally observed streamwise intermittency profiles and demonstrates a realistic profile for the cross-stream direction in the transition region. This model is coupled with the Menter's k-ω SST turbulence model [27] in which the intermittency factor calculated by the Suzen and Huang model is used to scale the eddy viscosity field computed by the turbulence model. This model is successfully tested against several flat plate and low-pressure turbine experiments. However, as mentioned before, this model is not a fully local formulation, and thus it cannot be implemented in straightforward fashion in the modern CFD codes.

2.3.2 Models depending on local flow variables

2.3.2.1 Langtry and Menter γ-Re_θ model

Langtry and Menter's formulation of the two-equation γ-Re_θ model [11] is one of the most widely used transition models as far as general CFD applications in aeronautics are concerned. This model is formulated in such a way that allows calibrated prediction of transition onset and length that are valid for both the 2-D and 3-D flows. It uses the so-called local variables and thus applicable to any type of grids generated around complex geometries with robust convergence characteristics. As mentioned in the introduction part, this model is based on an important experimental observation that a locally calculated parameter called as the vorticity Reynolds number (Re_v) and the momentum thickness Reynolds number (Re_θ) where

$$Re_\theta = \frac{Re_{vmax}}{2.193} \quad \text{and} \quad Re_v = \frac{\rho d_w^2}{\mu}\Omega \quad (4)$$

are proportional in a Blasius boundary layer. For most of the flow types, the relative error between the scaled vorticity Reynolds number and momentum thickness Reynolds number is reported [1] to be around 10%.

The model solves for two additional equations besides the underlying two-equation k-ω SST turbulence model, an intermittency equation (γ) that is used to trigger the turbulence production term of the k-ω SST turbulence model and a momentum thickness Reynolds number transport equation (Re_θ) that includes experimental correlations that relates important flow parameters such as turbulence intensity, freestream velocity, pressure gradients etc. and supplies it to the intermittency equation. The details of the model are available in the literature [1, 11].

2.3.2.2 Walters and Cokljat k-k_L-ω model

Walters and Cokljat's three-equation k-k_L-ω model [14] is proposed by the introduction of a transport equation for the laminar kinetic energy (k_L) into the conventional k-ω turbulence model and is used for natural and bypass transitional flows. This model is based on the understanding that the freestream turbulence is the cause of the high amplitude streamwise fluctuations in the pretransitional boundary layer, and these fluctuations are quite distinctive from the classic turbulence fluctuations. Also, growth of the laminar kinetic energy correlates with low frequency wall-normal fluctuations of the freestream turbulence. In this model, the total kinetic energy is assumed to be the sum of the large-scale energy which contributes to laminar kinetic energy and the small-scale energy which contributes to turbulence production. Thus, the transport equation for laminar kinetic energy (k_L) is solved in conjunction with the turbulent kinetic energy (k_T). Since the k-k_L-ω model uses a fully local formulation, it is suitable for the modern CFD codes and appears to be the first local model to specifically address pretransitional growth mechanism that is responsible for bypass transition [14].

2.3.2.3 Menter one-equation γ model

Menter's one-equation γ transition model [22] is a simplified version of the two-equation γ-Re_θ transition model [10, 11]. In the new model, the Re_θ equation is avoided, and the experimental correlations for transition onset is embedded into the γ equation in a simplified fashion. In effect, the simplified one-equation γ model still possesses the same level of predictive capabilities as the original model. Menter et al. [22] summarize the advantages and the key changes to the model as follows: the new model is still fully local with new correlations valid for nearly all types of transition mechanisms, solves for one less equation, which is computationally cheaper; it is Galilean invariant; it has less coefficients that makes the model easier to fine-tune for specific application areas; and the new model would be coupled to any turbulence model that has viscous sublayer formulation. Menter et al. tested their model against most of the test cases which they previously used for the two-equation model. The results show that the new one-equation model is quite successful, and it would be a viable replacement for the original model.

2.3.2.4 Nagapetyan and Agarwal two-equation WA-γ transition model

Following the trend for reducing the number of transition equations, a novel method was developed by integrating the recent Wray-Agarwal (WA) wall-distance-free one-equation turbulence model [23] based on the k-ω closure, with the one-equation intermittency transport γ-equation of Menter et al. [22] to construct the so-called two-equation Nagapetyan-Agarwal transition model WA-γ [24]. An important difference between the one-equation turbulence model derived earlier from k-ω models and the baseline turbulence model is the addition of a new cross diffusion term and a blending function between two destruction terms [23]. It was reported that the presence of destruction terms enables the Wray-Agarwal (WA) model to switch between a one-equation k-ω or one equation k-ε model. The new two-equation model was quite successfully validated for computing a number of two-dimensional benchmark experiments such as the transitional flows past flat plates in zero and slowly varying pressure gradients, flows past airfoils such as the S809, Aerospatiale-A, and NLR-7301 two-element airfoils.

2.3.2.5 Bas and Cakmakcioglu algebraic transition model

Bas and Cakmakcioglu (B-C) model [20] is an algebraic or zero-equation model that solves for an intermittency function rather than an intermittency transport (differential) equation. The main approach behind the B-C model follows the pragmatic idea of further reducing the total number of equations. Rather than deriving extra equations for intermittency convection and diffusion, already present convection and diffusion terms of the underlying turbulence model could be used. From a philosophical point of view, the transition, as such, is just a phase of a general turbulent flow. Addition of, in a sense, artificially manufactured transition equations appear to be rather redundant. Yet, for most of industrial flow types, the experimentally evidenced close relation between the scaled vorticity Reynolds number and the momentum thickness Reynolds number stood out as the primary reason for the success of so many intermittency transport equation models following the Langtry and Menter's original two-equation γ-Re_θ model [11].

In the application, the production term of the underlying turbulence model is damped until a considerable amount of turbulent viscosity is generated, and the damping effect of the transition model would be disabled after this point. The Spalart-Allmaras (S-A) turbulence model [28] is used as the baseline turbulence model, and rather than using an intermittency equation, just an intermittency function is proposed to control its production term. To this end, the B-C model is also a local correlation transition model that can be easily implemented for both 2-D and 3-D flows with reduced number of equations. For instance, for a 3-D problem, the B-C model solves for six equations (1 continuity + 3 momentum + 1 energy + 1 turbulence), whereas the two-equation γ-Re_θ model solves for nine equations (1 continuity + 3 momentum + 1 energy + 2 turbulence + 2 transition). In addition, in the B-C model formulation, the freestream turbulence intensity parameter is only present in the critical momentum thickness Reynolds number function that makes the calibration of the model quite easy for different problems. The details of the B-C model formulation are presented in the following.

The S-A one-equation turbulence model is used as the underlying turbulence model for the B-C model. The S-A model solves for a transport equation for a new working variable ν_T, which is related to the eddy viscosity. The B-C model's transition effects are included into the turbulence model is provided by multiplying the intermittency distribution function (γ_{BC}) with the production term of the S-A equation given as:

$$\frac{\partial \nu_T}{\partial t} + \frac{\partial}{\partial x_j}(\nu_T u_j) = \gamma_{BC} C_{b1} S \nu_T - C_{w1} f_w \left(\frac{\nu_T}{d}\right)^2 + \frac{1}{\sigma}\left\{\frac{\partial}{\partial x_j}\left[(\nu_L + \nu_T)\frac{\partial \nu}{\partial x_j}\right] + C_{b2}\frac{\partial \nu_T}{\partial x_j}\frac{\partial \nu_T}{\partial x_j}\right\} \quad (5)$$

The γ_{BC} function works in such a way that the turbulence production is damped ($\gamma_{BC} = 0$) until some transition onset criteria is fulfilled. After a point at which the onset criteria is ensured, the damping effect of the intermittency function γ_{BC} is checked, and the remaining part of the flow is taken to be fully turbulent ($\gamma_{BC} = 1$). For this purpose, an exponential function of the form (1-e^{-x}) is proposed for the γ_{BC} as follows:

$$\gamma_{BC} = 1 - exp\left(-\sqrt{Term_1} - \sqrt{Term_2}\right) \tag{6}$$

where $Term_1$ and $Term_2$ are defined as:

$$Term_1 = \frac{max(Re_\theta - Re_{\theta c}, 0.0)}{\chi_1 Re_{\theta c}}, Term_2 = \frac{max(\nu_{BC} - \chi_2, 0.0)}{\chi_2} \tag{7}$$

and,

$$Re_\theta = \frac{Re_v}{2.193} \quad \text{and} \quad Re_v = \frac{\rho d_w^{\ 2}}{\mu}\Omega, \ \nu_{BC} = \frac{\nu_t}{Ud_w} \tag{8}$$

In the above, ρ is the density, μ is the molecular viscosity, d_w is the distance from the nearest wall, ν_{BC} is a proposed turbulent viscosity-like nondimensional term where ν_t is the turbulent viscosity, U is the local velocity magnitude, d_w is the distance from the nearest wall, and χ_1 and χ_2 are calibration constants. $Re_{\theta c}$ is defined as the critical momentum thickness Reynolds number, which is a correlation that is based on a range of transition experiments. In effect, $Term_1$ checks for the transition onset point by comparing the locally calculated Re_θ with the experimentally obtained critical momentum thickness Reynolds number $Re_{\theta c}$. As soon as the vorticity Reynolds number Re_v exceeds a critical value, $Term_1$ becomes greater than zero and the intermittency function γ_{BC} begins to increase. However, the vorticity Reynolds number Re_v relation above is a function of the square of the wall distance d_w; therefore, it takes a very low value inside the boundary layer where the wall distance is quite low. Because of this, $Term_1$ alone is not enough for intermittency generation inside the boundary layer. To remedy this, $Term_2$ is introduced. Inspecting the $Term_2$ equation with the ν_{BC} relation shows that the regions close to wall is inversely related and the damping effect of the transition model would be disabled inside the boundary layer. In effect, $Term_2$ checks for the viscosity levels inside the boundary layer, and the turbulence production is activated wherever ν_{BC} exceeds a critical value χ_2. In order to determine the calibration constants' χ_1 and χ_2 values, the well-known zero pressure gradient flat plate test case of Schubauer and Klebanoff [29] is used. This test case represents a natural transition process due to the wind tunnel used in the experiment generates a freestream Tu around 0.2%. The model calibration is done by numerical experimentation; setting χ_1 and χ_2 such that the transition occurs at the same location as in the experiment. As a result, the χ_1 and χ_2 values are set to be 0.002 and 5.0, respectively.

Any experimental $Re_{\theta c}$ correlation could be used in the model. However, it should be noted that, since the S-A turbulence model does not solve for the local turbulent kinetic energy, local turbulence intensity values cannot be calculated. Due to this reason, the turbulence intensity Tu is assumed, for now, to be constant in the entire flow domain as Suluksna et al. [30] and Medida [31] have also suggested. For this lack of ability for calculating the local Tu values, the B-C model has some deficiency in this respect that it cannot handle some physical effects compared with the models that can dynamically calculate the local Tu levels. Whereas this deficiency makes the B-C model rather limited, there are quite a few aerodynamic flows for which the model is still viable. The transition onset correlation that was also used in the original two-equation γ-Re_θ model [1] is given by:

$$Re_{\theta c} = 803.73\left(Tu_\infty + 0.6067\right)^{-1.027} \tag{9}$$

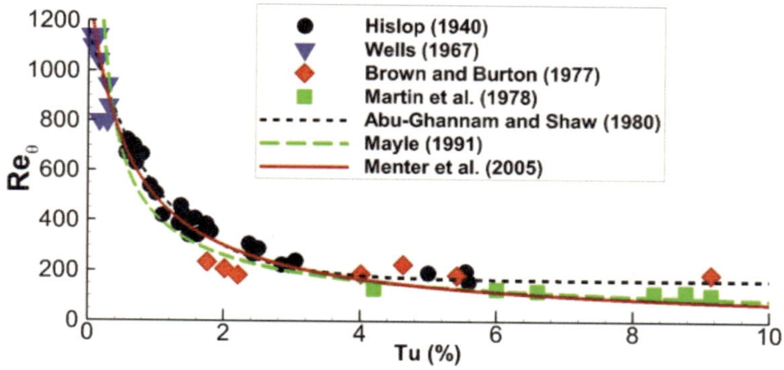

Figure 1.
Transition onset correlations compared with experiments.

As mentioned before, any transition onset correlation would be incorporated into the B-C model. For instance, a class of potential transition onset correlations along with the one preferred in the present B-C model is shown in **Figure 1**.

Currently, the B-C model is available in the SU2 (Stanford University Unstructured) v6.0, an open-source CFD solver by the ADL of Stanford University [32]. The SU2 can solve two- and three-dimensional incompressible/compressible Euler/RANS equations using linear system solver methods.

3. Two- and three-dimensional test cases for low to high speeds

Some outstanding test cases that make a good platform for measuring novel transition model performances are simulated by the foregoing transition models. These cases cover a wide range of flows from low speed two-dimensional flat plate and airfoil test cases to three-dimensional wind turbine blade and aircraft wing test cases from low to high speeds.

3.1 Low speed flat plate test cases

Well-known benchmark experiments such as the Schubauer and Klebanoff natural transition flat plate experiment [29] and the ERCOFTAC T3 series flat plate experiments by Savill [33] are used. The T3 series flat plate experiments consist of three zero pressure flat plate cases (T3A, T3B, and T3A-) and five variable pressure flat plate cases (T3C1, T3C2, T3C3, T3C4, and T3C5), in which the pressure gradients are generated using an adjustable upper tunnel wall. In all ERCOFTAC T3 test cases, the free stream turbulence intensities vary between 0.1 and 6%. **Table 1** summarizes the upstream conditions of the Schubauer and Klebanoff and the ERCOFTAC T3 flat plate experiments.

Figure 2 shows the numerical and experimental skin friction coefficients of the zero pressure gradient test cases of S&K, T3A, T3B and T3A-, respectively. The figures include numerical predictions of several researchers, including for instance Suzen and Huang [9], Langtry and Menter [11], Walters and Cokljat [14], Menter et al. [22], Nagapetyan and Agarwal [24], and Medida [31]. In the S&K calibration case, the B-C model displays a good agreement with the experiment for the transition onset point similar to other methods. For the T3A and T3B cases, the B-C model shows rather late transition onset, whereas the other models predict some early or late onset points. Specifically, Nagapetyan and Agarwal [24] show a very good agreement with the experiment as to the transition onset and rapid skin-friction

Case	U_{in}	Re_∞	*Tu%*
S&K	50.1	3.4E+6	0.18
T3A	5.4	3.6E+5	3.00
T3B	9.4	6.3E+5	6.00
T3A-	19.8	1.4E+6	0.90
T3C1	5.9	3.9E+5	6.60
T3C2	5.0	3.3E+5	3.00
T3C3	3.7	2.5E+5	3.00
T3C4	1.2	8.0E+4	3.00
T3C5	8.4	5.6E+5	3.00

Table 1.
Inlet conditions for the flat plate test cases.

rise characteristic. Finally, for the T3A- case, the B-C [20], Menter et al. [22], Walters and Cokljat [14], and Nagapetyan and Agarwal [24] display early transition onset points with rather rapid rise in skin-friction, whereas two-equation Langtry and Menter [11] and Medida [31] models show quite good onset point and a gradual rise in the skin friction.

Figure 3 depicts numerical and experimental skin friction coefficients for the T3C series variable pressure flat plate test cases. The T3C series flat plate test cases represent actual turbine characteristics by changing the pressure gradient by changing the upper wall profile of the wind tunnel over the flat plate. For the T3C1 case,

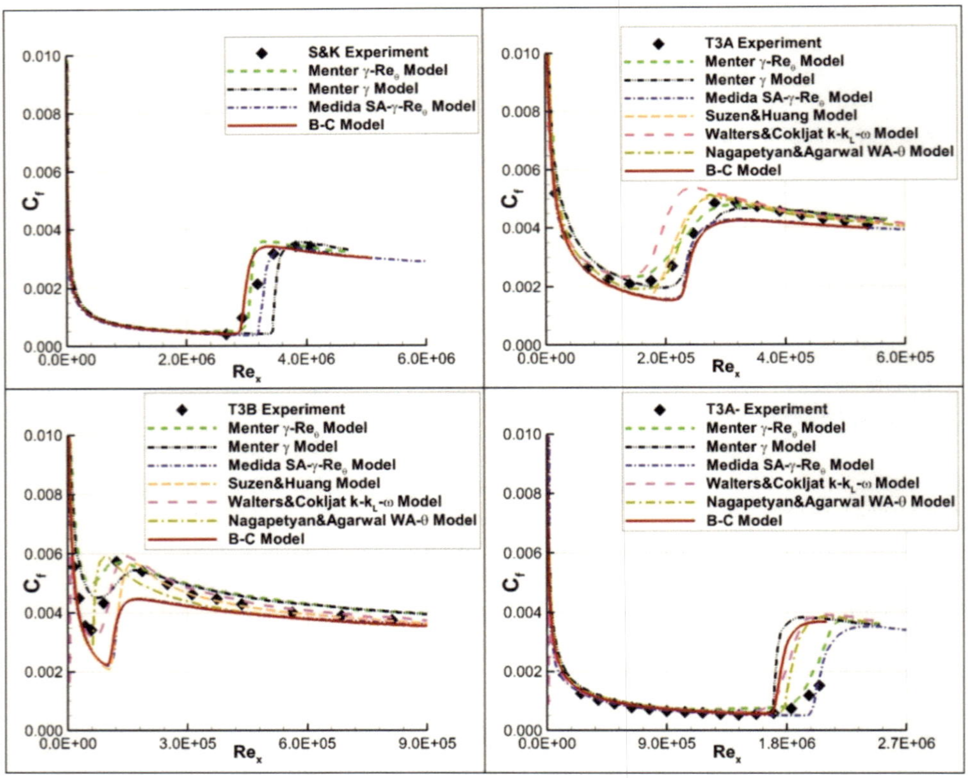

Figure 2.
Comparison of skin friction coefficients for the zero pressure gradient flat plate test cases.

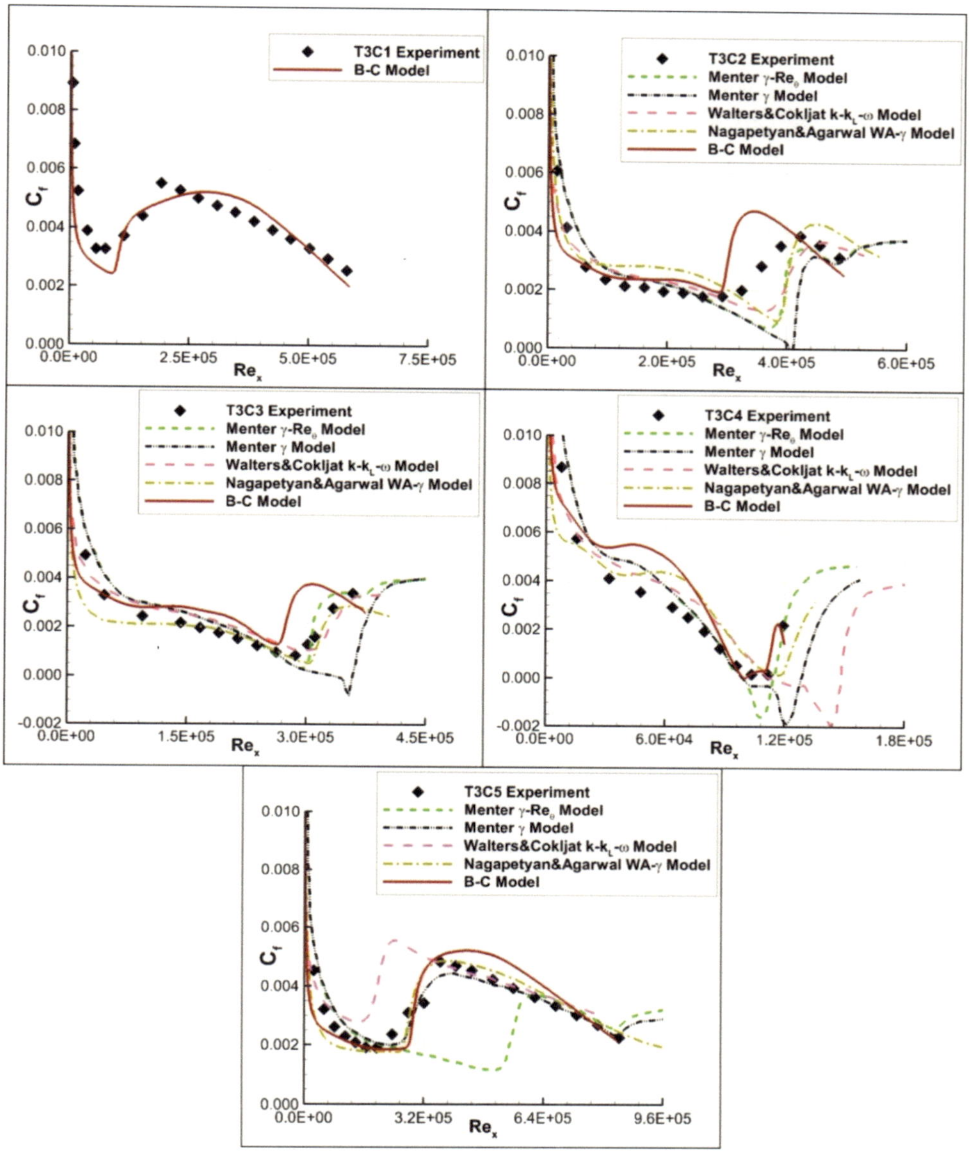

Figure 3.
Comparison of skin friction coefficients for the variable pressure gradient flat plate test cases.

which represents the highest turbulence intensity test case among the T3C series test cases, the B-C model results are quite in agreement with the experimental data as the transition onset location is predicted with decent accuracy. For the T3C2 case, it is observed that although the B-C model predicted a good transition onset point, the turbulent stress abruptly rises after the onset. All other models predicted the transition onset location rather late in general.

For the T3C3 case, it is observed that the γ-Re_θ model [11], k-k_L-ω model [14], and WA-γ model [24] outperform the other models as the B-C model prediction shows an early transition onset, whereas the one-equation γ model [22] predicts a rather late transition onset. For the T3C4 case, which represents the lowest Reynolds number case, all the models except for the B-C and WA-γ models show flow separation as their skin friction coefficients are below zero. Here, the B-C model obtained a quite good transition onset point that agreed with the experimental data although the laminar region was rather inaccurate. Finally, for the T3C5 case, solution of the zero-equation B-C model [20], Menter et al. one-equation γ model [22], and WA-γ model [24] well agree with the experiment in the laminar region, the onset of transition is also fairly good with some delay, and again quite good agreement in the subsequent variable pressure gradient region is obtained.

3.2 Airfoil and turbomachinery test cases

3.2.1 S809 airfoil

The S809 airfoil is a 21% thick profile, which specifically designed for horizontal-axis wind turbine applications. The S809 airfoil was tested in a low-turbulence wind tunnel (Tu = 0.2%) by Somers [34] at Re number of 2 million (based on chord length) and a Mach number of 0.15. Comparison of the numerical results by Langtry and Menter [11] γ-Re_θ, Walters and Cokljat [14] k-k_L-ω, and Medida [31] SA-γ-Re_θ and B-C models [20] with the experimental data is given in **Figures 4–6**. In general, all transition models agree well with the experimental data until the stall angle. Although the lift and drag coefficients (**Figure 4**) are rather inaccurate after the stall angle, it is observed that the experimental measurements of the transition locations are quite successfully predicted by all models (**Figure 5**). Also, comparing the experimental and numerical pressure coefficient distributions on the S809 airfoil at 1 angle of attack, it is observed that the separation bubble is predicted quite well by all the models (**Figure 6**).

3.2.2 T106 turbine cascade

T106 turbine cascade experiment was designed to investigate the interaction of a convected wake and a separation bubble on the suction surface of a highly loaded low-pressure turbine blade. In these experiments by Stieger et al. [35], five-blade cascade of T106 profile was placed downstream of a moving bar wake generator in order to simulate an unsteady wake passing environment of a turbomachine. In the experiment, the flow conditions correspond to a Reynolds number of nearly 91,000 based on the chord length of the T106 profile and the inlet velocity. The

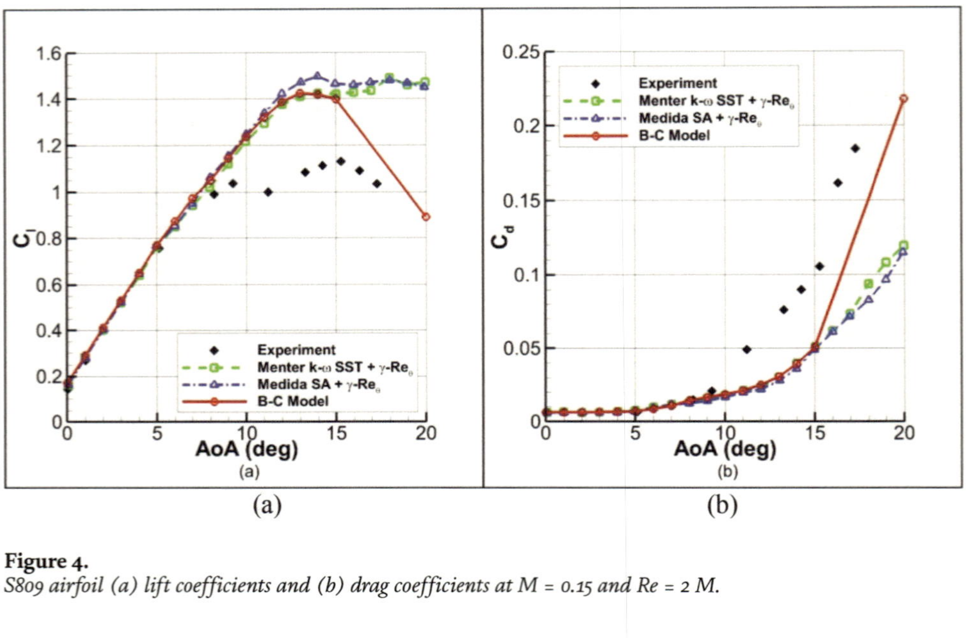

(a) (b)

Figure 4.
S809 airfoil (a) lift coefficients and (b) drag coefficients at M = 0.15 and Re = 2 M.

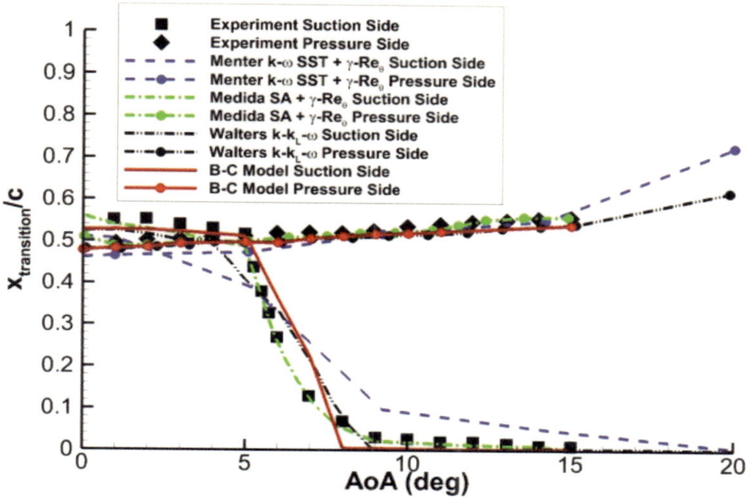

Figure 5.
S809 airfoil transition location comparison.

experimental turbulence intensity is specified to be 0.1%. Geometric details of the experimental cascade setup are given in **Table 2**. Comparison of the experimental and numerical pressure coefficient distributions for T106 cascade for the steady case is depicted in **Figure 7**. Looking at **Figure 7**, it is observed that the separation bubble on the blade predicted by the B-C model and the two-equation γ-Re_θ model is slightly smaller in size than the experimentally measured bubble.

3.3 3-D wing test cases from low to high speeds

3.3.1 Low speed rotating wind turbine blade

Two twisted and tapered 10-meter diameter turbine blades that use the S809 airfoil profile are tested in the NASA Ames Research Center wind tunnels [36, 37]. In the experiments, the NREL wind turbine rotation speed was set to 72 RPM for all cases, whereas the wind speeds varied from 7 to 25 m/s.

Figure 8 compares the pressure coefficient distributions over various spanwise locations on the turbine blades at the freestream velocity of 7 m/s. It is observed

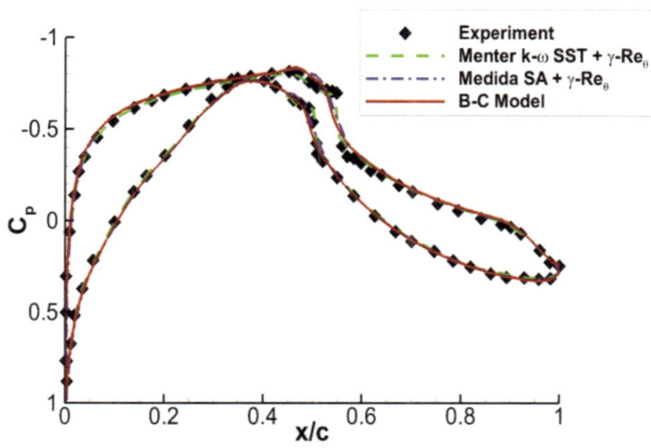

Figure 6.
Pressure coefficient distribution comparison for the S809 airfoil at 1̊.

Blade chord	198 mm
Blade stagger	59.3°
Cascade pitch	158 mm
Inlet flow angle	37.7°
Design exit flow angle	63.2°
Bar diameter	2.05 mm
Axial distance from bars to leading edge	70 mm

Table 2.
Geometric details of the T106 cascade experimental setup.

that both fully turbulent and the transitional solutions differ very slightly and both agree well with the experimental data. The skin friction contours and the surface streamlines obtained by Medida [31], Potsdam et al. [38], and Aranake et al. [39] for the same freestream velocity are compared to the B-C model and the S-A model solutions in **Figure 9**.

3.3.2 High subsonic flow over 3-D swept wing

DLR-F5 wing tested by Sobieczky [40] is a 0.65 m span wing with 20° sweep angle and an average chord length of 150 mm. The wing is mounted to the tunnel wall with a smooth blending region, and the angle of attack is set to be 2°. The square cross-section wind tunnel has dimensions of 1 × 1 × 4 meters. The experimental inlet Mach number and the turbulence intensity are specified as M = 0.82 and Tu <0.35%, respectively. The corresponding Re number based on the average chord is 1.5 million. In the experiment, the transition locations are determined by the sublimation technique, whereas measurements of pressure coefficients at different spanwise stations are available. In 1987, a workshop with several researchers were took place in Gottingen [41], where the results were compared against the experimental data.

Figure 10 shows the pressure coefficient distributions at different span locations. It is observed that the fully turbulent and the transitional solutions are very

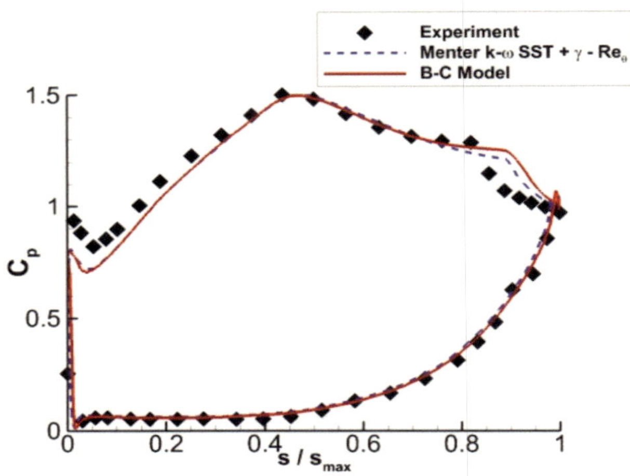

Figure 7.
Comparison of numerical and experimental pressure coefficient distributions on the T106 blade for Re = 91,000.

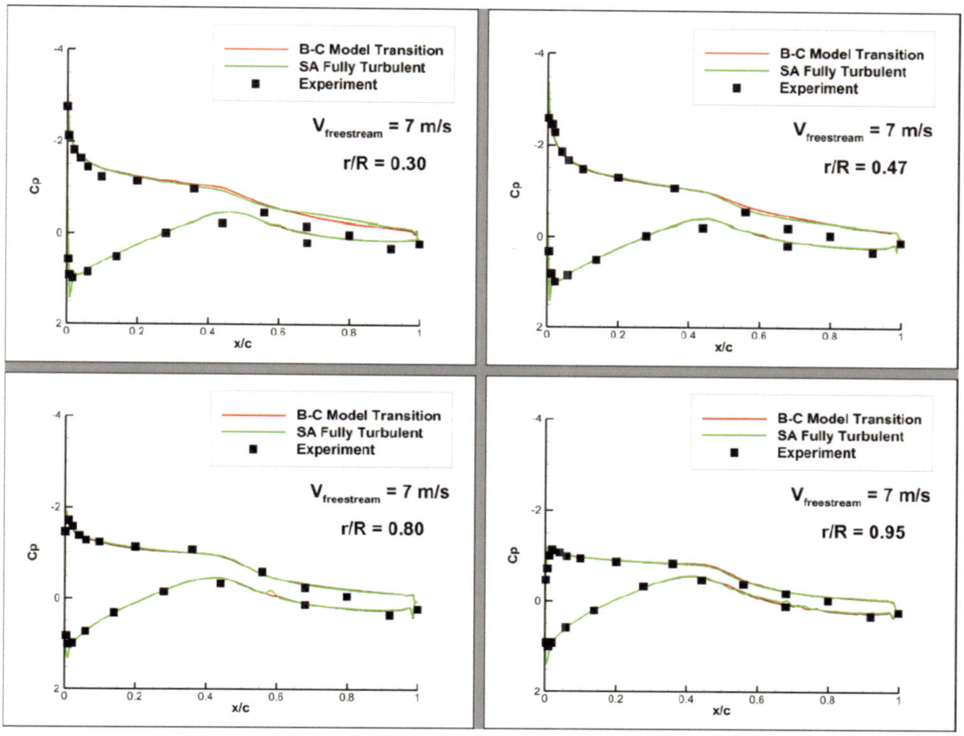

Figure 8.
Comparison of pressure coefficient distributions for the NREL phase IV blade for U = 7 m/s freestream velocity.

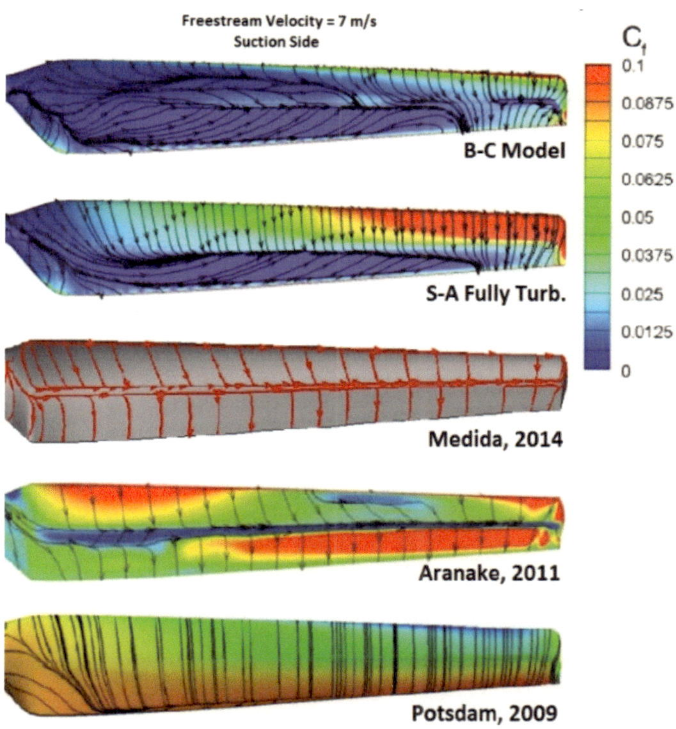

Figure 9.
Comparison of numerical skin friction contours obtained by several researchers.

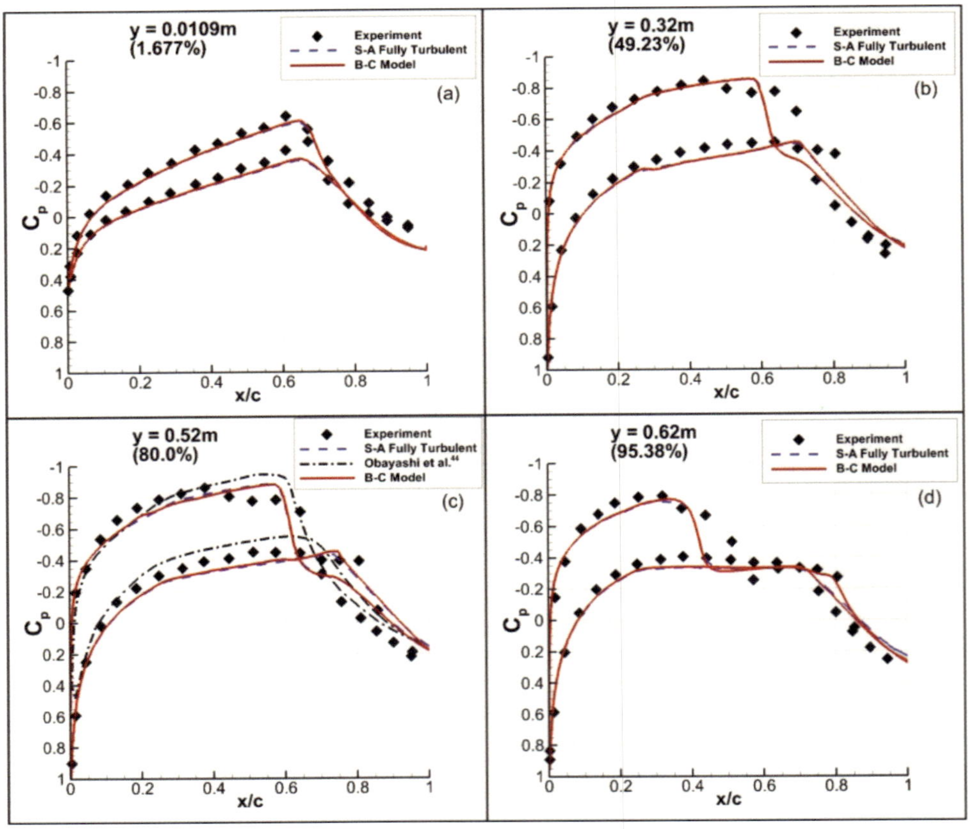

Figure 10.
Pressure coefficient distributions for the DLR-F5 wing at M = 0.82 and Re = 1.5 M.

similar to each other. **Figure 11** compares the skin friction contours of different numerical models with the experiment [40]. As seen, the B-C model predicts a somewhat similar transition and separation region with the experiment obtained by the sublimation and pressure measurement techniques.

Finally, in order to emphasize the difference between the fully turbulent and the transitional solutions, comparison of the skin friction coefficients at 80% span on

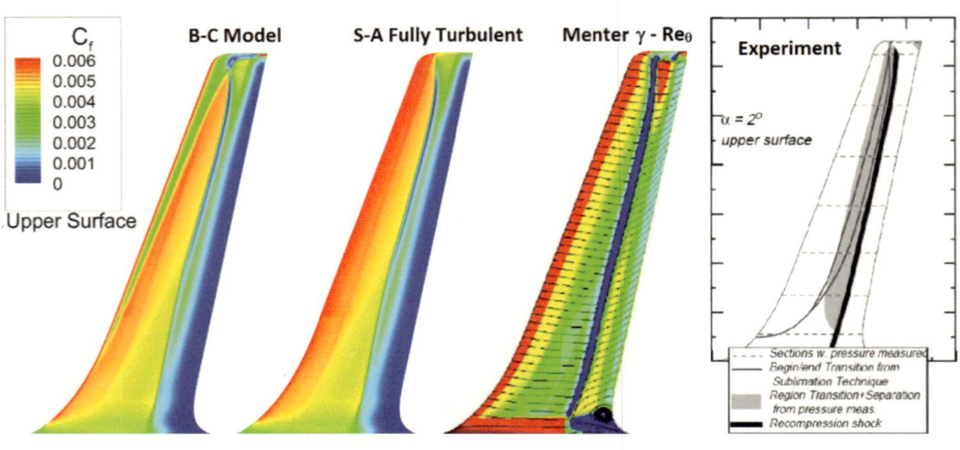

Figure 11.
Skin-friction coefficient comparisons for the DLR-F5 wing.

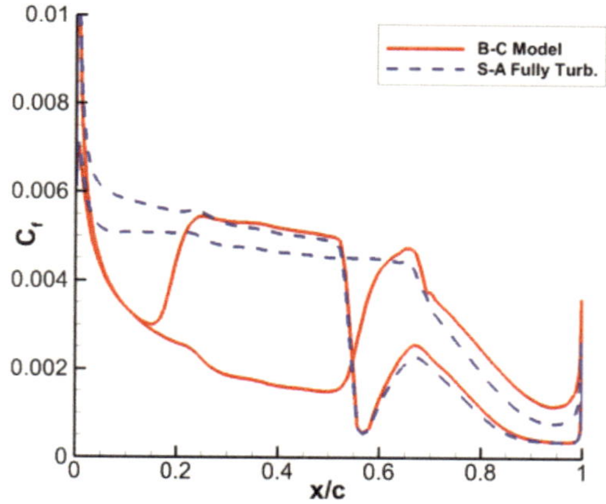

Figure 12.
Comparison of the skin friction coefficients predicted by the S-A turbulence model and the B-C transition model at 80% span on the DLR-F5 wing.

the DLR-F5 wing is depicted in **Figure 12**. It can be clearly observed that the B-C model predicts marked extent of laminar regions for both the suction and pressure sides of the wing, which is in agreement with the contours shown in **Figure 11**.

4. Conclusions

Local correlation-based transition models in the sense of empirical correlations incorporated into Reynolds-averaged Navier-Stokes methods have been discussed. A logical path for the development of such models is highlighted such that a variety of combinations of turbulence and transition equations lead to different modeling alternatives. For instance, the pioneering work by Menter et al. [1] two-equation γ-Re_θ transition model sums up to a total of four-equation model by the incorporation of the two-equation k-ω SST turbulence model of Menter et al. [27]. In the same line of development but in a leaner approach, Walters and Cokljat [14] developed a three-equation k-k_L-ω model. Similarly, Medida [31] developed a three-equation S-A-γ-Re_θ transition model that is a sum of the Menter et al. [1] two-equation γ-Re_θ transition model and the one-equation S-A turbulence model [28].

In fact, in a recent work, Menter [22] reached to the conclusion that the Re_θ equation was rather redundant. Without any loss of accuracy, Menter produced a leaner three-equation k-ω SST-γ transition model by incorporating a novel one-equation intermittency transport γ-model [22] with the two-equation k-ω SST turbulence model of Menter et al. [27]. In the same line of thought, Nagapetyan-Agarwal constructed the so-called two-equation transition model of WA-γ [24] by incorporating the Wray-Agarwal (WA) wall-distance-free one-equation turbulence model [23] based on the k-ω closure with the one-equation intermittency transport γ-equation of Menter et al. [22]. These two models paved the way for developing yet another leaner transition model by Bas et al. [19] with the introduction of the algebraic Bas-Cakmakcioglu (B-C) model by incorporating an algebraic γ-function with the one-equation S-A turbulence model [28].

The Bas-Cakmakcioglu (B-C) [19] model qualifies as a zero-equation model that solves for an intermittency function rather than an intermittency transport (differential) equation. The main approach behind the B-C model follows again the pragmatic idea of further reducing the total number of equations. Thus, rather than deriving extra equations for intermittency convection and diffusion, already present convection and diffusion terms of the underlying turbulence model could have been used. From a philosophical point of view, the transition, as such, is just a phase of a general turbulent flow. In a sense, addition of artificially manufactured transition equations may appear to be rather redundant. Yet, for most of industrial flow types, there is experimental evidence that a close relation between the scaled vorticity Reynolds number and the momentum thickness Reynolds number exists. This fact stands out as the primary reason for the success of the class of so many intermittency transport equation models following the Menter's pioneering two-equation γ-Re_θ model [1]. Using the present B-C model, a number of two-dimensional test cases including flat plates, airfoils, turbomachinery blades, and three-dimensional low speed wind turbine and high-speed transport plane wing were simulated with quite successful results. These results may be regarded to vindicate this leaner approach of using even lesser equations for industrial design aerodynamics problems.

Author details

Unver Kaynak[1*], Onur Bas[2], Samet Caka Cakmakcioglu[3] and Ismail Hakki Tuncer[4]

1 Eskisehir Technical University, Eskisehir, Turkey

2 TED University, Ankara, Turkey

3 ASELSAN Inc., Ankara, Turkey

4 Middle East Technical University, Ankara, Turkey

*Address all correspondence to: unkaynak@gmail.com

References

[1] Menter FR, Langtry RB, Völker S. Transition modeling for general purpose CFD codes. Flow, Turbulence and Combustion. 2002;**77**:277-303

[2] Smith AMO, Gamberoni N. Pressure Gradient and Stability Theory. Report No. ES 26388. Long Beach, CA: Douglas Aircraft Company; 1956

[3] Van Ingen JL. A Suggested Semi-Empirical Method for the Calculation of the Boundary Layer Transition Region. Report VTH-74. The Netherlands: Delft University of Technology, Department of Aerospace Engineering; 1956

[4] Drela M, Giles MB. Viscous-inviscid analysis of transonic and low Reynolds number airfoils. AIAA Journal. 1987;**25**(10):1347-1355

[5] Jones WP, Launder BE. The calculation of low Reynolds number phenomena with a two-equation model of turbulence. International Journal of Heat and Mass Transfer. 1973;**15**(2):301-314

[6] Dhawan S, Narasimha R. Some properties of boundary layer during transition from laminar to turbulent flow motion. Journal of Fluid Mechanics. 1958;**3**(04):418-436

[7] Steelant J, Dick E. Modeling of laminar-turbulent transition for high freestream turbulence. ASME Journal of Fluids Engineering. 2000;**123**(1):22-30

[8] Cho JR, Chung MK. A k—ε—γ equation turbulence model. Journal of Fluid Mechanics. 1992;**237**:301-322

[9] Suzen YB, Huang PG. Modeling of flow transition using an intermittency transport equation. ASME Journal of Fluids Engineering. 2000;**122**(2):273-284

[10] Menter FR, Langtry RB, Likki SR, Suzen YB, Huang PG, Völker S. A correlation based transition model using local variables: Part I—Model Formulation. In: ASME Turbo Expo 2004, Power for Land, Sea, and Air, Vienna, Austria; 14-17 June 2004; Paper No. GT2004-53452

[11] Langtry RB, Menter FR. Transition modeling for general cfd applications in aeronautics. In: 43rd AIAA Aerospace Sciences Meeting and Exhibit, Reno, NV, USA; 10-13 January 2005; Paper No. 2005-522

[12] Lodefier K, Merci B, De Langhe C, Dick E. Transition Modeling with the SST Turbulence Model and Intermittency Transport Equation. In: ASME Turbo Expo 2003, Power for Land, Sea, and Air, Atlanta, GA, USA; 16-19 June 2003; Paper No. GT2003-38282

[13] Walters DK, Leylek JH. A new model for boundary-layer transition using a single point RANS approach. ASME Journal of Turbomachinery. 2004;**126**(1):193-202

[14] Walters DK, Cokljat D. A three-equation Eddy-viscosity model for Reynolds-averaged Navier-stokes simulations of transitional flow. ASME Journal of Fluids Engineering. 2008;**130**(12):121401

[15] Fu S, Wang L. A transport intermittency model for supersonic/hypersonic boundary layer transition. In: 5th European Congress on Computational Methods in Applied Sciences and Engineering (ECCOMAS), Venice, Italy; 30 June-5 July 2008

[16] Seyfert C, Krumbein A. Correlation-based transition transport modeling for three-dimensional aerodynamic configurations. In: 50th AIAA Aerospace Sciences Meeting Including the New Horizons Forum and Aerospace Exposition, Nashville, Tennessee,

USA; 09-12 January 2012; Paper No. 2012-0448

[17] Dassler P, Kozulovic D, Fiala A. An approach for modelling the roughness-induced boundary layer transition using transport equations. In: 6th European Congress on Computational Methods in Applied Sciences and Engineering (ECCOMAS), Vienna, Austria; 10-14 September 2014

[18] Kaynak U. Supersonic boundary-layer transition prediction under the effect of compressibility using a correlation based model. Proceedings of the Institution of Mechanical Engineers Part G Journal of Aerospace Engineering. 2011;**226**(7):722-739

[19] Bas O, Cakmakcioglu SC, Kaynak U. A novel intermittency distribution based transition model for low-re number airfoils. In: 31st AIAA Applied Aerodynamics Conference, San Diego, CA, USA; 24-27 June 2013; Paper No. 2013-2531

[20] Cakmakcioglu SC, Bas O, Kaynak U. A correlation-based algebraic transition model. Proceedings of the Institution of Mechanical Engineers Part C Journal of Mechanical Engineering Science. 2018;**232**(21):3915-3929

[21] Kubacki S, Dick E. An algebraic model for bypass transition in turbomachinery boundary layer flows. International Journal of Heat and Fluid Flow. 2016;**58**:68-83

[22] Menter FR, Smirnov PE, Liu T, Avancha R. A one-equation local correlation based transition model. Flow, Turbulence and Combustion. 2015;**95**(4):583-619

[23] Han X, Rahman MM, Agarwal RK. Development and application of a wall distance free Wray-Agarwal turbulence model. In: AIAA SciTech Forum, Kissimmee, FL; 8-12 January 2018; Paper No. 2018-0593

[24] Nagapetyan HJ, Agarwal RK. Development of a new transitional flow model integrating the Wray-Agarwal turbulence model with an intermittency transport equation. In: AIAA Aviation Forum, Atlanta, GA, USA; 25-29 June 2018; Paper No. 2018-3384

[25] White FM. Viscous Fluid Flow. New York: McGraw-Hill Book Company; 1974

[26] Klebanoff PS. Characteristics of Turbulence in a Boundary-Layer with Zero Pressure Gradient. NACA Report No. 1247

[27] Menter FR. Two-equation Eddy-viscosity turbulence model for engineering applications. AIAA Journal. 1994;**32**(8):1598-1605

[28] Spalart PR, Allmaras SR. A one-equation turbulence model for aerodynamic flows. In: 30th Aerospace Sciences Meeting and Exhibit, Reno, NV, USA; 6-9 January 1992; Paper No. 92-0439

[29] Schubauer GB, Klebanoff PS. Contribution on the Mechanics of Boundary Layer Transition. NACA Technical Note No. TN-3489. 1955

[30] Suluksna K, Juntasaro E. Assessment of intermittency transport equations for modeling transition in boundary layers subjected to freestream turbulence. International Journal of Heat and Fluid Flow. 2008;**29**(1):48-61

[31] Medida S. Correlation-Based Transition Modeling for External Aerodynamic Flows [PhD Thesis]. USA: University of Maryland; 2014

[32] Economon TD, Palacios F, Copeland SR, et al. SU2: An open-source suite for multi-physics simulation and design. AIAA Journal. 2016;**54**:828-846

[33] Savill AM. Some recent progress in the turbulence modelling of by-pass transition. In: So RMC, Speziale CG,

Launder BE, editors. Near-Wall Turbulent Flows. New York: Elsevier; 1993. p. 829

[34] Somers DM. Design and Experimental Results for the S809 Airfoil. State College, PA: Airfoils, Inc.; 1989

[35] Stieger R, Hollis D, Hodson H. Unsteady surface pressures due to wake induced transition in a laminar separation bubble on a LP turbine cascade. In: ASME Turbo Expo 2003, Power for Land, Sea, and Air, Atlanta, GA, USA; 16-19 June 2003; Paper No. GT2003-38303

[36] Hand MM, Simms DA, Fingersh LJ, Jager DW, Cotrell JR, Schreck S, Lawood SM. Unsteady Aerodynamics Experiment Phase VI: Wind Tunnel Test Configurations and Available Data Campaigns. NREL/TP-500-29955. 2001

[37] Simms D, Schreck S, Hand M, Fingersh LJ. NREL Unsteady Aerodynamics Experiment in the NASA-Ames Wind Tunnel: A Comparison of Predictions to Measurements. NREL/TP-500-194494. 2001

[38] Potsdam MA, Mavriplis DJ. Unstructured mesh CFD aerodynamic analysis of the NREL Phase VI Rotor. In: 47th AIAA Aerospace Sciences Meeting, Orlando, FL, USA; 5-8 January 2009; Paper No. 2009-1221

[39] Aranake AC, Lakshminarayan VK, Duraisamy K. Assessment of transition model and CFD methodology for wind turbine flows. In: 42nd AIAA Fluid Dynamics Conference and Exhibit, New Orleans, LA, USA; 25-28 June 2012; Paper No. 2012-2720

[40] Sobieczky H. DLR-F5: Test Wing for CFD and Applied Aerodynamics, Test Case B-5. AGARD FDP Advisory Report AR 303: Test Cases for CFD Validation. 1994

[41] Kordulla W, Sobieczky H. Summary and evaluation of the workshop "numerical simulation of compressible viscous-flow aerodynamics". In: Zierep J, Oertel H, editors. Symposium Transsonicum III. International Union of Theoretical and Applied Mechanics. Berlin: Springer; 1989. pp. 3-18

Section 3

Fundamentals of Boundary Layer Theory

Chapter 6

Physical Models of Atmospheric Boundary Layer Flows: Some Developments and Recent Applications

Adrián R. Wittwer, Acir M. Loredo-Souza, Mario E. De Bortoli and Jorge O. Marighetti

Abstract

Experimental studies on wind engineering require the use of different types of physical models of boundary layer flows. Small-scale models obtained in a wind tunnel, for example, attempt to reproduce real atmosphere phenomena like wind loads on structures and pollutant dispersion by the mean flow and turbulent mixing. The quality of the scale model depends on the similarity between the laboratory-generated flow and the atmospheric flow. Different types of neutral atmospheric boundary layer (ABL) including full-depth and part-depth simulations are experimentally evaluated. The Prof. Jacek Gorecki wind tunnel of the UNNE, Argentina, and the Prof. Joaquim Blessmann closed-return wind tunnel of the UFRGS, Brazil, were used to obtain the experimental data. Finally, some recent wind engineering applications of this type of physical wind models are shown.

Keywords: wind tunnel, turbulence, aerodynamic loads, atmospheric dispersion, physical simulation

1. Introduction

Wind tunnels are designed to realize similarity in model studies, with the confidence that actual operational conditions will be reproduced. The first step is the evaluation of the flow characteristics with the empty wind tunnel, and then, different flow characteristics are achieved or reproduced at the test section to be applied in the experimental tests. To perform aerodynamic studies of structural models, the distribution of the incident flow must be such that the atmospheric boundary layer (ABL) at the actual location of the structure is reproduced. This is obtained by surface roughness elements and vortex generators, so that natural wind simulations are performed.

The atmospheric boundary layer (ABL) is the lowest part of the atmosphere where the effects of the surface roughness, temperature, and others properties are transmitted by turbulent flows. Turbulent exchanges are very weak when there are conditions of weak winds and very stable stratification [1]. On the other hand, the atmospheric boundary layer over nonhomogeneous terrain is not well defined, and topographical features could cause highly complex flows. The depth of the atmospheric boundary layer is typically 100 m during the nighttime stable conditions, and this could reach 1 km in daytime unstable conditions [2]. The Prandtl logarithmic law (Eq. (1)), proposed from similarity theories, can be used near the surface in the case of a neutral boundary layer.

$$\frac{U(z)}{u*} = \frac{1}{0.4}\ln\frac{z - z_d}{z_0} \qquad (1)$$

where U is the mean velocity, u^* is the friction velocity, z_0 is known as the roughness height, and z_d is defined as the zero-plane displacement for very rough surface.

The potential law (Eq. (2)) is also widely used in wind engineering to characterize the vertical velocity distribution. The values for the exponent α vary between 0.10 and 0.43 and the boundary layer thickness z_g between 250 and 500 m, according to the terrain type [2]. This law is verifiable in the case of strong winds and neutral stability conditions that must be considered for structural analysis.

$$\frac{U(z)}{U(z_g)} = \left(\frac{z}{z_g}\right)^{\alpha} \qquad (2)$$

Similarity requirements corresponding to studies of atmospheric flow in the laboratory can be obtained by the dimensional analysis. The equations are expressed in dimensionless form by means of reference parameters that lead to the following set of non-dimensional groups or numbers: Reynolds number, Prandtl number, Rossby number, and Richardson number. These dimensionless parameters must be in the same value with the model and prototype to obtain the exact similarity, and, in addition, there must be geometric similarity and similarity of the boundary conditions, including incident flow, surface temperature, heat flow, and longitudinal pressure gradient [3].

Geometric scales defined between the simulated laboratory boundary layer and the atmospheric boundary layer are generally <1:200, velocities in the model and prototype have values of the same order, and the viscosity is the same for both cases. This results in the impossibility of reproducing the Reynolds number in low-speed wind tunnels; however, the effects of Reynolds number variation can be taken into account according to the type of wind tunnel test. On the other hand, the equality of the Prandtl number is obtained simply by using the same fluid in model and prototype, as in this case. The equality of the Rossby and Richardson numbers may not be considered for simulation of neutral ABL since Coriolis forces and thermal effects are negligible.

In most laboratories it is more common to simulate the neutrally stratified boundary layer. This implies modeling the distribution of mean velocities, turbulence scales, and atmospheric spectrum [4]. The quality of these approximate models is simply evaluated by comparing the results expressed in dimensionless form with design values. Turbulence intensity distribution is commonly compared with values obtained by other authors [5] and by using Harris-Davenport formula for atmospheric boundary layer [2].

Atmospheric velocity fluctuations with frequencies upper than 0.0015 Hz define the micrometeorological spectral region. Interest of wind engineering is concentrated on this spectral turbulence region. von Kármán suggested an expression for the turbulence spectrum in 1948, and today this spectral formula is still used for wind engineering applications. According to Reference [2], the expression for the dimensionless spectrum of the longitudinal component of atmospheric turbulence is given by Eq. (3):

$$\frac{fS_u}{(u_{RMS})^2} = \frac{1.6\left(fz_{ref}/U\right)}{\left[1 + 11.325\left(fz_{ref}/U\right)^2\right]^{5/6}} \tag{3}$$

where S_u is the spectral density function of the longitudinal component, f is the frequency in Hertz, and ${u_{RMS}}^2$ is the quadratic mean, or the variance, of the longitudinal velocity fluctuations in m^2/s^2. The dimensionless frequency fz_{ref}/U is defined using an appropriate z_{ref}, generally gradient height, and the mean velocity U.

Different boundary layer flows are experimentally analyzed in this work. First, three types of boundary layer flows developed at the UNNE wind tunnel: one corresponding to a naturally developed boundary layer with the empty wind tunnel and the other two generated by different ABL simulation methods. Then, simulated ABL flows obtained with different velocities at the UFRGS wind tunnel are analyzed, and the results are compared with each other. Finally, some recent applications of ABL simulations are described, among them wind effects in high-rise buildings considering the urban environment and the surrounding topography, low buildings, aerodynamics of cable-stayed bridges, pollutant atmospheric dispersion, and flow in the wake of wind turbines.

2. Boundary layer flows at the UNNE wind tunnel

Next, measurement results obtained at the Prof. Jacek Gorecki wind tunnel of the UNNE (**Figure 1**) in three different boundary layer flows are analyzed. The UNNE wind tunnel is a 39.56-m-long channel where the air enters through a contraction to reach the test section. This is connected to the velocity regulator and to the blower, and then, the air passes through a diffuser before leaving the wind tunnel. The contraction has a honeycomb and a screen to uniform the airflow. The test section is a 22.8-m-long rectangular channel (2.40 m width, 1.80 m height) where two rotating tables are located to place test models. Conditions of zero-pressure-gradient boundary layers can be obtained by the vertical displacement of the upper wall. The blower has a 2.25 m diameter and is driven by a 92 kW electric motor at 720 rpm.

The first of these flows correspond to a boundary layer developed on the smooth floor of the wind tunnel test section. Then, the results obtained for two ABL simulations are analyzed. The first model corresponds to a full-depth simulation of the neutrally stable ABL and the second to a part-depth model.

Figure 1.
The Prof. Jacek Gorecki boundary layer wind tunnel of the UNNE.

Velocity and longitudinal velocity fluctuations were measured by a Dantec 56C constant temperature hot-wire anemometer connected to a Stanford amplifier with low- and high-pass analogic filters. Hot-wire signals were digitalized by a DAS-1600 A/D converter board controlled by a computer which was also used for the analysis of the results. Voltage output from hot wires was converted in mean velocity and velocity fluctuations [6, 7] by the probe calibration curves previously determined. Full spectra from longitudinal velocity fluctuations were obtained by juxtaposing three different partial spectra from three different sampling series, registered in the same location, each with a specific sampling frequency, designed as low, mean, and high frequencies. Then, the fast Fourier transform (FFT) algorithm was applied to each numerical series, and the corresponding longitudinal turbulence spectra were obtained.

2.1 The boundary layer obtained with empty tunnel

The uniformity of the flow corresponding to the empty boundary layer wind tunnel is evaluated previously to implement physical models of turbulent flows. Deviations of mean velocity and turbulence intensity are measured to determine uniform flow zones and boundary layer thickness in the test section. Finally, longitudinal component of the turbulence spectrum is obtained from the boundary layer flow and from the uniform flow.

Dimensionless velocity profiles measured with the empty tunnel along a vertical line on the center of the rotating table of the test section (see reference [8]) and at positions 0.6 m to the right and left of this line are presented in **Figure 2**. The vertical coordinate z is measured from the floor, and H is the test section height equal to 1.80 m. Measurements are presented only for the lower half of the test section. The depth of the boundary layer is of about 0.3 m, and a good uniformity can be observed from the vertical velocity distributions. A maximal deviation of the mean velocity of about 3% is verified outside the boundary layer, by taking the velocity at the center of the channel as reference.

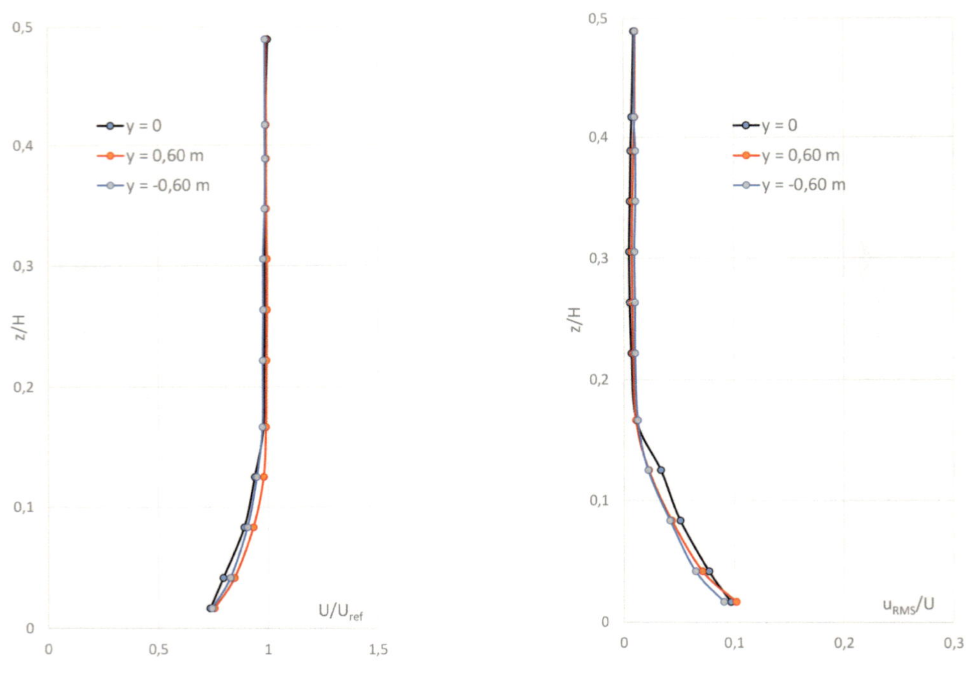

Figure 2.
Vertical profiles of mean velocity and turbulence intensity with the empty wind tunnel.

Turbulence intensity distribution at the same locations shows values around 1% outside the boundary layer increasing, as expected, inside the boundary layer. Reference velocity for these tests was the velocity at the center of the channel, 27 m/s. The value of Reynolds number calculated with the hydraulic diameter of the test section was 3.67×10^6.

Turbulence spectra obtained inside and outside the boundary layer with the empty tunnel are presented in **Figure 3**. Inside the boundary layer, it is possible to observe higher values of fluctuation energy and a clear definition of the 5/3 declivity, characterizing Kolmogorov's inertial subrange. Outside the boundary layer, low turbulence levels produce a spectral definition only for frequencies minor than 70 Hz.

2.2 Full-depth simulation of the atmospheric boundary layer

The complete boundary layer thickness of the ABL is simulated when a full-depth simulation is developed. The Counihan method [9] was applied at the UNNE wind tunnel, and four 1.42-m-high elliptic vortex generators and a 0.23 m (b) barrier were used, together with prismatic roughness elements placed on the test section floor along 17 m (see **Figure 4**).

Velocity and longitudinal velocity fluctuations were measured by the same hot-wire anemometer system described above. Measurements of the mean velocity distribution were made along a vertical line on the center of rotating table and along lines 0.30 m to the right and left of this line. **Figure 5** shows the vertical velocity distribution, and at center, the same measured values are presented in a log-graph to verify the low part of the profile where the distribution of mean speeds is logarithmic. There is a good similarity among the three measured velocity profiles, and the value of the exponent α obtained by fitting to Eq. (2) is 0.24.

Turbulence intensity distribution at the same locations is also shown in **Figure 5** on the right. The values are lower than those obtained by using Harris-Davenport formula for atmospheric boundary layer [2]. This behavior was verified by other authors [5] mainly in the points located above ($z/H > 0.5$).

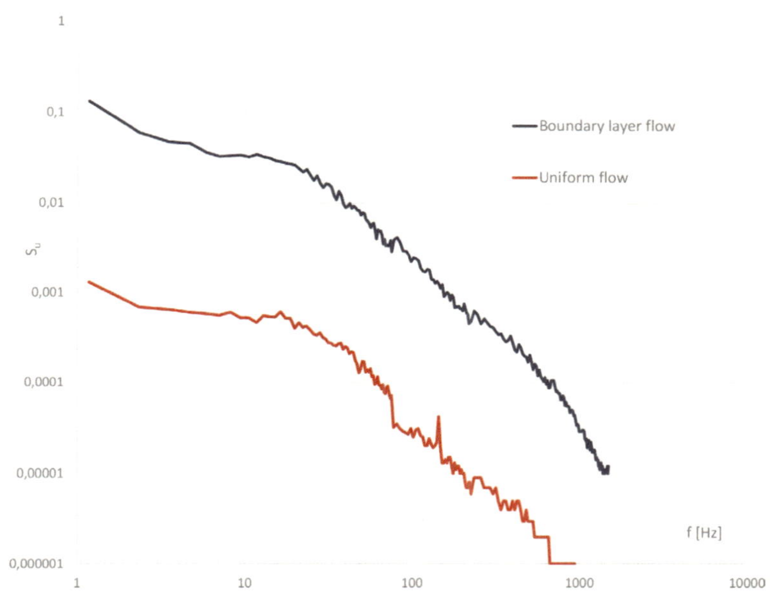

Figure 3.
Spectral density function with the empty wind tunnel.

Figure 4.
Counihan vortex generators, barrier, and roughness elements of the full-depth boundary layer simulation.

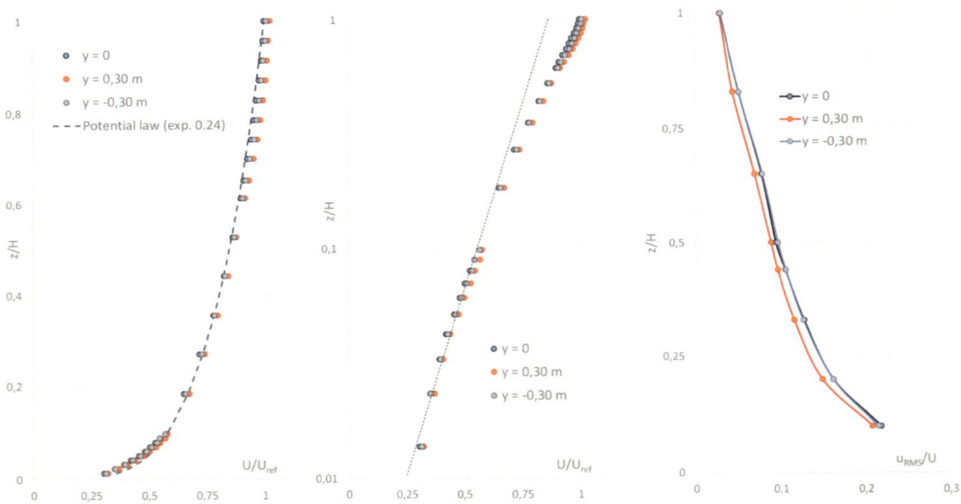

Figure 5.
Vertical mean velocity and turbulence intensity profiles measured for the full-depth boundary layer simulation.

Three spectra obtained at positions z = 0.23, 0.58, and 0.97 m are presented in **Figure 6**. An important characteristic of the spectra is the presence of a clear region of the Kolmogorov's inertial subrange. The comparison of the results obtained through the simulations with the atmospheric boundary layer is made by means of dimensionless variables of the auto-spectral density and of the frequency using the von Kármán spectrum (Eq. (3)). A good agreement is observed at z = 0.23 m, but this agreement diminishes at positions z = 0.58 and 0.97 m, and this behavior is coincident with the behavior observed for the turbulence intensities.

These measurements were realized at velocity $U_{ref} \approx 27$ m/s, being U_{ref} measured at gradient height z_g = 1.16 and the corresponding Reynolds number value of $Re \approx 2.10 \times 10^6$. A scale factor of 250 was calculated through the Cook's procedure [5], using the roughness length z_0 and the integral scale L_u as key parameters. The value of the roughness length was obtained by fitting mean velocity values to the logarithmic law, and the integral scale values were determined through the fitting of the measured spectrum to the design spectrum.

2.3 Part-depth simulation of the atmospheric boundary layer

Two Irwin-type generators separated 1.5 m were used to simulate the part-depth boundary layer by means of the Standen method [10]. The windward plate of the

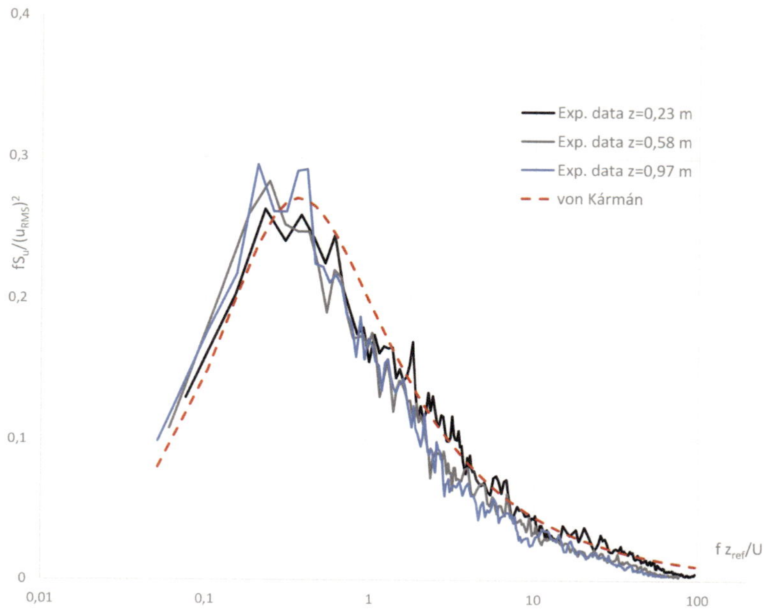

Figure 6.
Dimensionless spectra obtained at different heights for the full-depth boundary layer simulation and the von Kármán spectrum.

Figure 7.
Irwin vortex generators and roughness elements of the part-depth boundary layer simulation.

simulator has a trapezoidal shape of 1.50 m height, 0.53 and 0.32 m sides. The roughness elements distributed on the test section floor is the same that was used for the Counihan method (**Figure 7**).

Measurements of mean velocity and longitudinal velocity fluctuations were made along a vertical line on the center of rotating table and along lines 0.60 m to the right and left of this line. Vertical velocity distribution and the corresponding log-graph representation to verify the extension of the logarithmic behavior are shown in **Figure 8**. The three measured velocity profiles are quite similar, and the fit to Eq. (2) determines a value of the exponent α of 0.23.

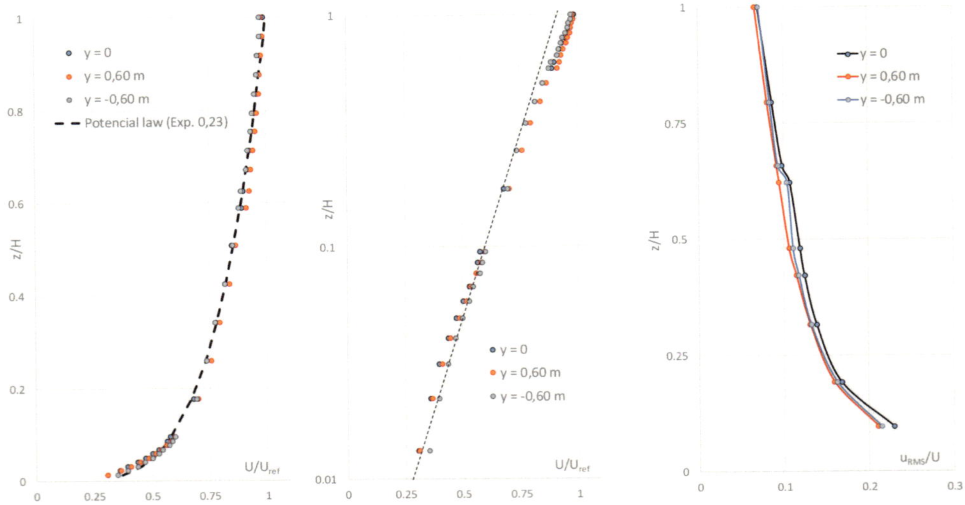

Figure 8.
Vertical mean velocity and turbulence intensity profiles measured for the part-depth boundary layer simulation.

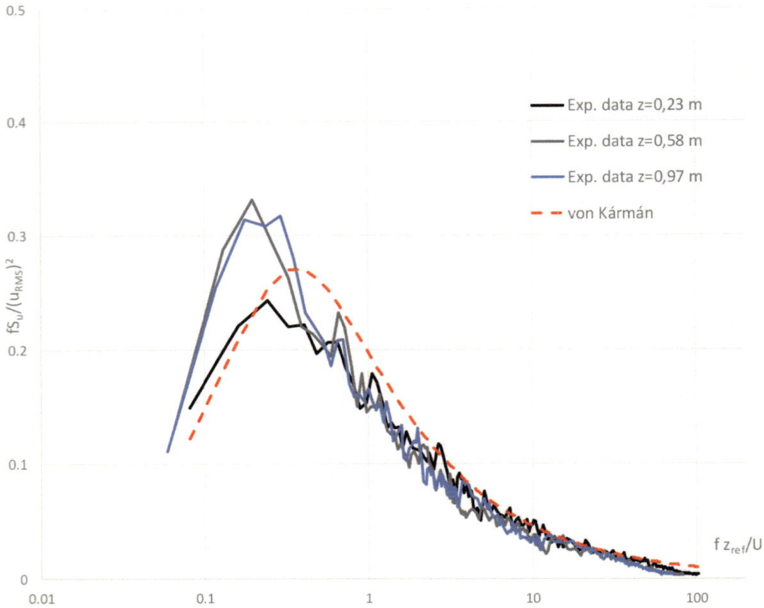

Figure 9.
Dimensionless spectra obtained at different heights for the part-depth boundary layer simulation and the von Kármán spectrum.

Figure 8, to the right, also shows turbulence intensity distribution at the same locations. The values are higher than those obtained by full-depth boundary layer, mainly in the positions located above, but lower than those obtained using Harris-Davenport formula if the condition of part-depth is considered. However, higher turbulence levels in these positions indicate a coherent behavior when a part-depth ABL is simulated.

Figure 9 has shown spectra obtained at positions z = 0.23, 0.58, and 0.97 m. Dimensionless spectral comparison indicates a shift of the experimental peak toward low frequencies with respect to the von Kármán spectrum. Higher differences of energy contents are also observed between the spectrum obtained at z = 0.23 m and spectra measured at upper positions.

The reference velocity for these tests was $U_{ref} \approx 25.5$ m/s and the gradient height $z_g = 1.21$ m. The corresponding Reynolds number value of Re $\approx 2.01 \times 10^6$. Finally, the Cook's procedure [5] was applied, and a scale factor of 150 was calculated.

3. Boundary layer flows at the UFRGS wind tunnel

Next, tests made at the wind tunnel of the UFRGS (**Figure 10**) are analyzed. The Prof. Joaquim Blessmann boundary layer wind tunnel at the Laboratório de Aerodinâmica das Construções of UFRGS, Brazil, is a closed-return circuit, and it has a cross-section of 1.30 m × 0.90 m at downstream end of the main working section that is 9.32 m long (**Figure 10**). A detailed description of the characteristics of the tunnel is indicated in Blessmann's previous work [11].

3.1 Simulation of atmospheric boundary layers with different velocities

Four perforated spires, a barrier, and surface roughness elements were used to simulate a full-depth boundary layer. The arrangement of the simulation hardware is shown/illustrated in **Figure 11**. Velocity and longitudinal velocity fluctuations were measured by means of a TSI hot-wire anemometer along a vertical line on the center of rotating table located downstream of the working section.

Figure 10.
The Prof. Joaquim Blessmann boundary layer wind tunnel of the UFRGS.

Figure 12 shows the non-dimensional profiles obtained with low velocities U_{ref} = 1 and 3.5 m/s, respectively. These profiles are compared with the values obtained with the highest mean velocity achievable in the wind tunnel ($U_{ref} \approx$ 35 m/s). The mean velocity profile given by the power law expression (Eq. (2)) is also included in this graph, being the power law exponent α equal to 0.23 and the boundary layer thickness H = 0.60 m.

Also, turbulence intensities measured in correspondence to U_{ref} = 1, 3.5, and 35 m/s are shown in **Figure 12**. Turbulence intensity values corresponding to 3.5 m/s are slightly higher than those obtained at high velocity, which is a behavior commonly observed at low velocities. For measurements at velocity U_{ref} = 1 m/s, it is possible to observe even larger deviations in comparison with 3.5 and 35 m/s cases that can be attributed to extremely low velocity. It is worth noting that with these velocity magnitudes, the relative errors affecting the hot-wire anemometer technique are larger than for measurements at high velocities. This kind of measurement deviation was also observed in similar wind tunnel tests using three-dimensional laser Doppler velocimetry [12].

Power spectra of the velocity fluctuations obtained at two different positions, z = 0.15 and 0.35 m with low velocities U_{ref} = 1 and 3.5 m/s, respectively, are presented in **Figure 13**. Sampling series used for the spectral analysis were obtained with an acquisition frequency of 1024 Hz. A poor definition of the Kolmogorov's inertial subrange is observed for the spectra measured at velocity U_{ref} = 1 m/s.

3.2 Analysis of the model scale factor

The evaluation of the model scale factor was only realized with high velocity U_{ref} = 35 m/s. The Cook's procedure [5] was applied using the form proposed by Blessmann [11] and a value of the scale factor at each measurement position by means of the roughness length z_0 and the integral scale L_u. Finally, a mean value of the model scale factor of 400 was calculated, and it is considered the same in the case of low velocities based on the maintenance of the mean statistical parameters.

Figure 11.
Perforated spires, barrier, and roughness elements to simulate a full-depth atmospheric boundary layer.

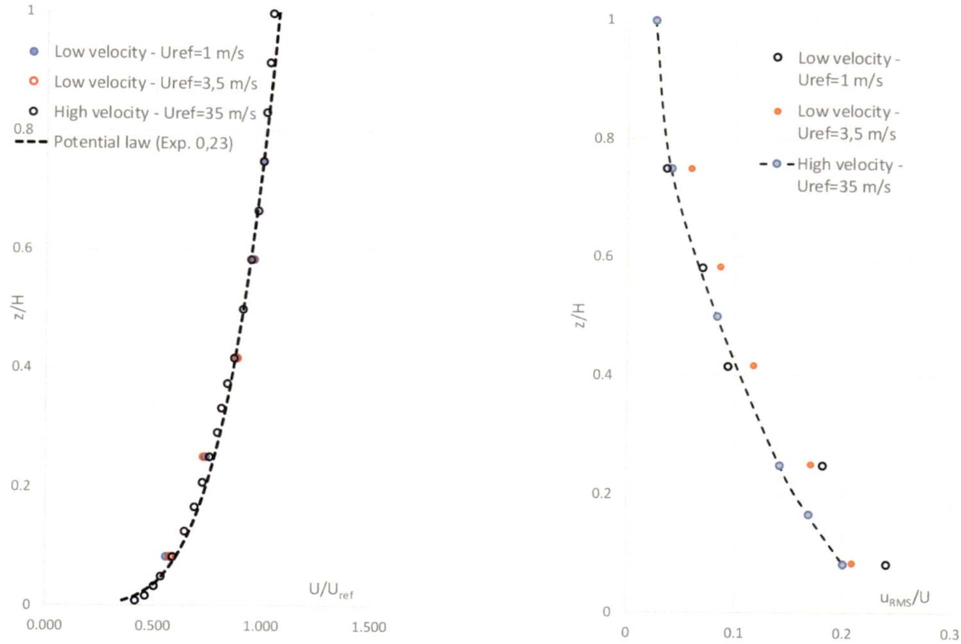

Figure 12.
Vertical mean velocity and turbulence intensity profiles measured with different reference velocities for a full-depth boundary layer simulation.

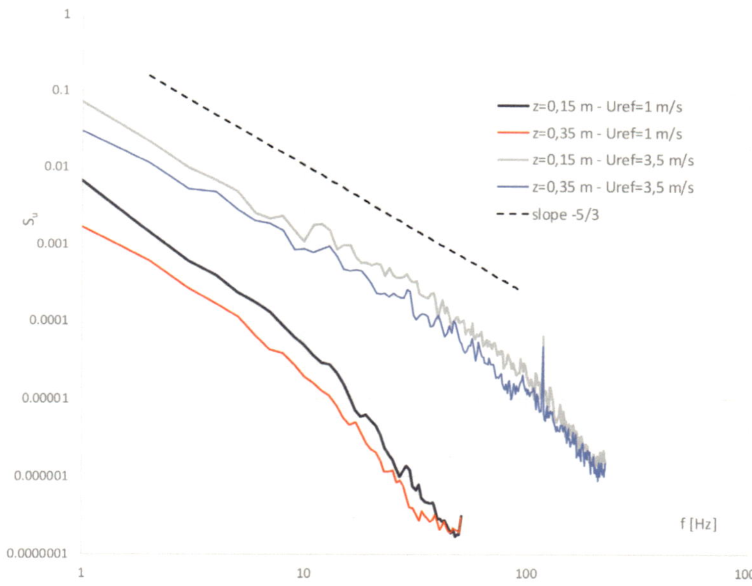

Figure 13.
Spectral density function measured with low velocity at two positions z = 0.15 m and 0.35 m for a full-depth boundary layer simulation.

4. Recent applications of simulated boundary layer flows

Some recent wind engineering applications of the ABL simulations are presented. In general, this type of applications was referred to wind action on civil structures, but new studies related with ambient evaluation, urban design, and wind energy are being developed. In this work, experimental studies related to high-rise and low building, atmospheric pollutant dispersion, rain-wind action on structures, and the turbulent wake of wind turbines will be shown.

4.1 Wind tunnel study of the local aerodynamic loads on a high-rise building

A study of the wind loads on a high-rise building was realized in the Prof. Jacek Gorecki wind tunnel using a 1/400 scale model [13]. The building is 240 m high; it is named Infinity Tower and it is built in Camboriu, RS, Brazil. Local mean, maximum, minimum, and *rms* pressure coefficients were measured by means of a Scanivalve pressure system.

Atmospheric boundary layer simulations similar to the full-depth simulation described in Section 2.2 were used. Some characteristics of the incident wind were modified according to the terrain features upwind model. Thus, two different mean velocity profiles were used according to the incident wind direction. Aerodynamic details were reproduced in the building model (**Figure 14**), and fluctuating local pressures in 511 tower points for 24 wind directions were measured. The effects of urban vicinity and topographic surrounding were considered by means of a detailed modeling.

Some considerations referred to the extreme values approximation were realized in this work. The graph in **Figure 14** illustrates a fluctuating pressure registered at a measurement point. Mean values associated with different duration times of wind gusts (1, 4, and 16 s full-scale) can be obtained by means of this technique, and statistical extreme value analysis can be applied to improve the calculation of local wind loads.

4.2 Wind tunnel study of the aerodynamic loads on a low structure

There exist structures that due to its size, complexity, or importance justify turning to wind tunnel tests in order to optimize the structural design. A wind tunnel study of the Ezeiza Airport located in that village of Buenos Aires was realized in 2010 [14]. The study comprised the determination of both local and global wind actions.

It was carried out at the Prof. Jacek Gorecki wind tunnel of the UNNE, using different 1/200 scale models (**Figure 15**) that were compatible with the scale factor of the wind simulated in the wind tunnel. The real neighbor conditions were taken into account as well as the turbulent features of the atmospheric wind in agreement with the type of terrain. In this case, an ABL flow similar to the part-depth simulation described in Section 2.3 was used.

In addition to the mean load coefficients, peak coefficients were obtained by extreme value analysis using the Cook and Mayne method [15]. It is shown how

Figure 14.
High-rise building scale model in the test section of the UNNE wind tunnel and detail of the fluctuating local pressure.

the application of this kind of analysis is influenced in this particular case and, in general terms, how the design of structures could be optimized by means of this kind of studies.

4.3 Aerodynamic analysis of cable-stayed bridges

The prediction of the aerodynamic performance of concrete cable-stayed bridges can be realized by means of wind tunnel testing. The analyses of the structural stability under the aerodynamic actions must be included into the design verifications. The structural characteristics of cable-stayed bridges and the dynamic aspects of the aerodynamic actions implicate the application of special analyses of aerodynamic stability including flutter and vortex shedding. The determination of the critical velocity is very important in the design of cable-stayed bridges.

First, static forces are obtained through force balance measurements for simple models of the bridge deck and towers. Aerodynamic coefficients varying with wind incidence for the deck may be easily measured with pressure systems or force balance.

A sectional model is used for the dynamic modeling of the deck. The sectional model must be ideally rigid to avoid the influence of the own model vibration in the experimental results. Details of the bridge deck must be represented. **Figure 16**, left, shows a picture of a 1:60 dynamic sectional model mounted in the test section of the wind tunnel of Prof. Joaquim Blessmann of the UFRGS. The deck corresponds to the Octávio Frias de Oliveira cable-stayed bridge, and the obtained results permitted to observe the differences in the deck vertical and torsional responses [16].

The relevant parameters in aeroelastic modeling are length, specific mass (density), and acceleration. The design of a full-aeroelastic model must reproduce the aerodynamic and dynamic characteristics of the structure of interest. The flow and geometric similarities must be respected and the Reynolds number considered for aerodynamic similarity. The most relevant frequencies and mode shapes must be reproduced to obtain dynamic similarity. The design of full-aeroelastic model includes bridge deck, cables, masts, and end supports. The complexity of this type of model can be observed in **Figure 16**, right, showing the full-aeroelastic model, the Octávio Frias de Oliveira cable-stayed bridge tested at the wind tunnel Prof. Joaquim Blessmann of the UFRGS [17].

The incident wind used to test sectional models is normally a turbulent uniform flow similar to the flow obtained with empty tunnel out the wall boundary layer and described in Section 2.1. Meanwhile, the full-aeroelastic model of the Octávio Frias de Oliveira cable-stayed bridge was tested using a full-depth boundary layer simulated

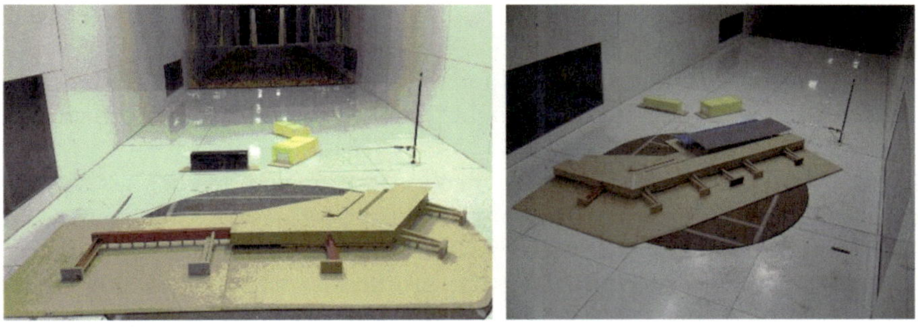

Figure 15.
Ezeiza Airport 1/200 models (partial) in the test section of the UNNE wind tunnel.

Figure 16.
Sectional model and full-aeroelastic model of the Octávio Frias de Oliveira cable-stayed bridge in the wind tunnel Prof. Joaquim Blessmann.

in the return section of the wind tunnel Prof. Joaquim Blessmann. The simulation devices and roughness elements are similar to those described in Section 2.2.

4.4 Study of atmospheric dispersion by means of scale model

The concentration fields in the proximities of a local gas emission source were experimentally analyzed in several combinations of wind incidences and source emissions. Concentration measurements were performed by an aspirating probe in a boundary layer wind tunnel. The analysis included the mean concentration values and the intensity of concentration fluctuations in a neutral atmospheric boundary layer flow [18–20].

To perform atmospheric diffusion studies, it is usual to consider full-scale wind speeds in the range of 5–20 m/s [21]. Thus, in order to fulfill the Froude number similarity, the wind tunnel modeling must be performed at low free-stream mean velocities. Atmospheric boundary layer developed with low mean velocities similar to the full-depth simulations described in Section 3.1 was used in the UFRGS wind tunnel.

The hot-wire anemometer, by incorporating the aspirating probe, becomes a density measurement system, and when binary gas mixtures are used, the system measures instantaneous concentrations. A gas mixer was used to provide known air-helium mixtures to calibrate the probe [22]. This type of probe produces a wide useful bandwidth of frequency response, and it allows the evaluation of the plume fluctuating concentration near the source in a turbulent wind. At each measurement point, a sample of 1 min was taken at a sampling frequency of 1024 Hz.

Different configurations were tested in the wind tunnel of Prof. Joaquim Blessmann of the UFRGS, but in this work only the case of an isolated stack in a homogeneous terrain is shown in partial form (**Figure 17**). The results obtained are presented as profiles of concentration coefficient K and intensity of the concentration fluctuations I_c, being $K = CU_H H^2/Q_0$ and $I_c = \sigma_c/C$, where C and σ_c are the mean concentration and the standard deviation (rms) of the concentration fluctuations, respectively, Q_0 is the total exhaust volume flow rate (m^3/s), U_H is the wind velocity at the emission source height (stack height), and z is the vertical coordinate measured from the wind tunnel floor.

Figure 18 presents vertical profiles of concentration coefficient K and I_c for a specific condition of emission where plume velocity ratio is 0.66, plume momentum is 0.060, and the buoyancy parameter is −0.031. The experimental mean concentration values are contrasted with Gaussian profiles. It was possible to highlight the observation of the plume vertical asymmetry in the case of an isolated emission source and different probabilistic behavior of the concentration fluctuation data in a cross-sectional measurement plane inside the plume.

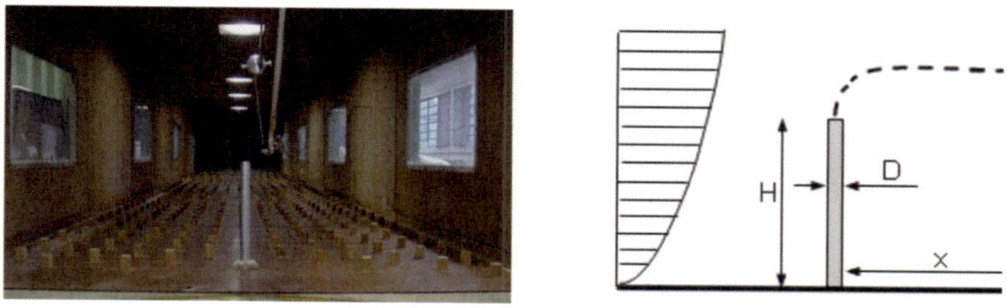

Figure 17.
Isolated emission source model in the test section of the UFRGS wind tunnel.

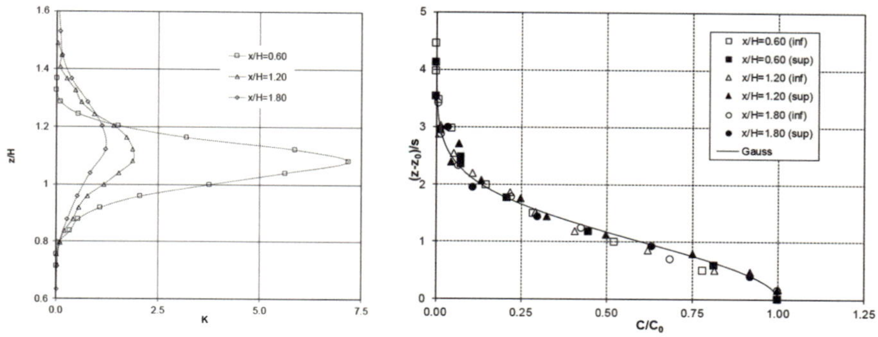

Figure 18.
Concentration profiles K, *at* x/H = *0.60, 1.20, and 1.80 and comparison with the Gaussian profile.*

One practical application of this type of development is presented next. Wind tunnel tests were realized to evaluate some characteristics of the Alcântara Launch Center (ALC), which is the Brazilian gate to the space located at the north coast of Maranhão State, close to the Equator. Topographical local characteristics modify the parameters of incident atmospheric winds, and it can cause great influence on the gas dispersion process.

The topographical scale models were built to measure mean and fluctuating flow characteristics in order to understand the real behavior of ALC winds, and then, physical simulations of the effluent dispersion process were made using these scale models. The wind velocity was measured by a hot-wire anemometer, and the concentration fields in the proximities of a gas emission source were analyzed by an aspirating probe connected to the same anemometer system [23].

The dispersion process of the gases emitted from the launch center is illustrated in **Figure 19**. Different effluent conditions were tested to reproduce the emission caused by a rocket. Helium gas was used at the emission source to simulate the turbulent diffusion process. The results obtained were compared with previous full-scale measurements and computational evaluations considering the emission at ground level. A coherent behavior with the physics of the phenomena was observed [24].

4.5 Wind tunnel tests of the flow in the wake of wind turbines

The interaction between the incident wind and wind turbines in a wind farm causes mean velocity deficit and increased levels of turbulence in the wake. The turbulent flow is characterized by the superposition of wind turbine wakes. A research work that included a series of wind tunnel tests to evaluate experimentally the spectral characteristics of turbulence in the wake of a wind turbine. Longitudinal

Figure 19.
Launch of a space vehicle at the ACL, Maranhão, Brazil, and simulation of the dispersion process in the wind tunnel of the LAC/UFRGS.

velocity fluctuations were measured in the incident flow and in the wake of a wind turbine-reduced model in the test section of the UNNE wind tunnel. In these experiments, the adequacy of spectral technique and changes in the turbulence spectral composition of the incident wind and the wake were analyzed [25].

All longitudinal velocity fluctuation measurements were realized employing a neutral ABL flow obtained by the Counihan method similar to the full-depth simulation described in Section 2.2. The simulated incident wind corresponds to a power law profile with an exponent α = 0.27 and a gradient height zg = 1.20 m. Wind tunnel measurements were made using a hot-wire anemometer system. The wind turbine model corresponds to a three-bladed UNIPOWER wind turbine, with a tower height of 100 m and a rotor diameter of 100 m. The scale of the model is approximately 1/450, and the model height is 0.33 m. **Figure 20** shows the wind incident, making the turbine model rotate.

The rotational velocity of the wind turbine was estimated and remained nearly constant during the measurements, but the values of the dimensionless speed ratio λ ensure the similarity of phenomenon in the range of the proper operation of the generator. Vertical profiles of dimensionless mean longitudinal velocity U/U_{ref} measured in the incident wind and in the wake generated by the wind turbine are

Figure 20.
Wind turbine model spinning during the wind tunnel test.

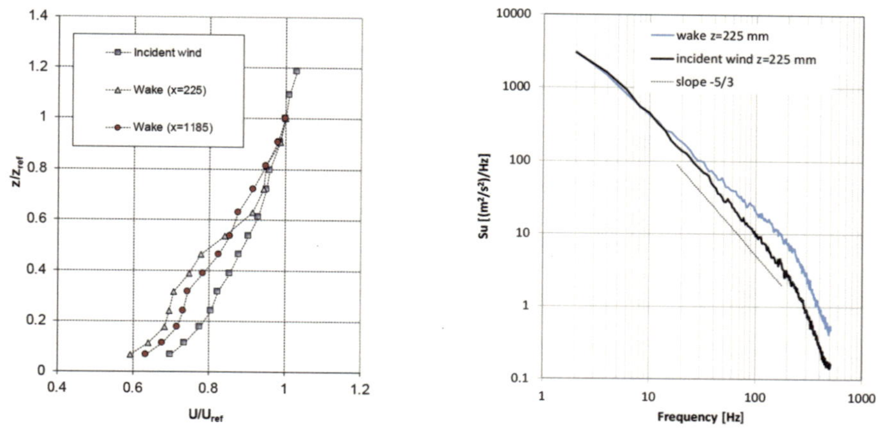

Figure 21.
Vertical profiles of dimensionless mean longitudinal velocity and spectral comparison of the incident wind and the wake flow at z = 225 mm.

indicated in **Figure 21**. Two profiles measured at locations x = 225 and 1185 mm downwind of the plane determined by the rotor blades are included. The comparison of the characteristic spectra of the incident wind and those obtained in the wake are also shown in **Figure 21** and allow observing the changes in the energy fluctuation distribution. These changes are a product of the turbulence introduced by the wind generator. Measurements allowed to analyze the configuration of the spectra in different frequency ranges, the effect of analog signal filtering, and differences in the spectral behavior of the incident wind relative to wind in the wake of the turbine.

5. Concluding remarks

In this work, different boundary layer flows were experimentally analyzed. The BLF developed at the UNNE wind tunnel include a naturally developed boundary layer with the empty wind tunnel, a full-depth ABL generated by the Counihan method, and a part-depth ABL simulated by the Standen method.

A simplified analysis considering the depth of the neutral ABL of about 500 m compared with the gradient height (0.30 m) obtained for the empty tunnel implicates a scale factor of 1/1650. In addition, turbulence intensity values inside the boundary layer are always minor than 10%, concluding that this boundary layer flow is not appropriate to wind engineering experiments.

Full-depth and part-depth simulations developed in the UNNE wind tunnel seem to show adequate performance. It is observed that values of measured turbulence intensity are lower than the values in the neutral atmosphere, mainly in the positions above, but other authors obtained similar results. Dimensionless spectral comparison indicates a deviation of the experimental results with respect to design spectra, but it is possible that parameters used to normalize the spectrum are not fully adequate. Some studies are being developed to verify this behavior.

Comparison of ABL flows obtained with low velocities and the ABL flow obtained with high velocity at the UFRGS wind tunnel indicates an acceptable behavior of the mean velocity and turbulence intensity distribution. Dimensionless spectra were not obtained for measurements with low velocities. However, a poor spectral definition was observed for measurements realized at the lowest velocity.

Five recent applications of ABL simulations in both wind tunnels (UNNE and UFRGS) are presented. A study of local wind loads on a high-rise building considering the urban environment and the surrounding topography realized in the UNNE wind tunnel, where full-depth simulation flows were used. An experimental study of a low structure, specifically an airport where a part-depth boundary layer simulation developed at the UNNE wind tunnel, was utilized. Some wind tunnel applications to the aerodynamic analysis of cable-stayed bridges are shown where different incident flows were used.

Then, a pollutant atmospheric dispersion study realized in the UFRGS wind tunnel was shown. ABL flows obtained with low velocities were used to simulate the gas plume emission. A case study applied to the Brazilian Launch Center of Alcântara to evaluate the emitted gas dispersion process is also shown.

Finally, a recent wind tunnel study of the flow in the wake of wind turbines is presented. Measurements of the flow characteristics upwind and downwind of the turbine rotor were analyzed. Comparison of the turbulence spectra were also developed to evaluate the rotor effects on the turbine wake flow.

Also numerical methods are used mainly for forecasting and studying the dynamics of the airflow over large surfaces, usually with domains of several square kilometers. The Weather Research and Forecasting (WRF) model, which is a numerical weather prediction and atmospheric simulation system, is an example of this type of computational modeling. The size of the domain of the simulation of these models is much larger than the simulated spaces of boundary layer flows in a wind tunnel. However, some efforts are being made to link results from computational model and experimental data. In South America, for example, a WRF model was used by Puliafito et al. [26] to simulate mesoscale events of Zonda winds, and the obtained results were compared with meteorological data. The next objective of this research is to try the physical simulation of these events in a wind tunnel [27].

Acknowledgements

This research was partially funded by Conselho Nacional de Desenvolvimento Científico e Tecnológico (CNPq, Brazil) and Secretaría General de Ciencia y Técnica, Universidad Nacional del Nordeste (SGCYT-UNNE, Argentina).

Author details

Adrián R. Wittwer[1]*, Acir M. Loredo-Souza[2], Mario E. De Bortoli[1] and Jorge O. Marighetti[1]

1 Facultad de Ingeniería, Universidad Nacional del Nordeste (UNNE), Resistencia, Argentina

2 Laboratório de Aerodinâmica das Construções, Universidade Federal de Rio Grande do Sul (UFRGS), Porto Alegre, Brazil

*Address all correspondence to: a_wittwer@yahoo.es

References

[1] Arya S. Atmospheric boundary layers over homogeneous terrain. In: Plate EJ, editor. Engineering Meteorology. Amsterdam: Elsevier Scientific Publishing Company; 1982. pp. 233-266

[2] Blessmann J. O Vento na Engenharia Estrutural. Porto Alegre, Brazil: Editora da Universidade. UFGRS; 1995

[3] Cermak J, Takeda K. Physical modeling of urban air-pollutant transport. Journal of Wind Engineering and Industrial Aerodynamics. 1985;**21**:51-67

[4] Surry D. Consequences of distortions in the flow including mismatching scales and intensities of turbulence. In: Proceedings of the International Workshop on Wind Tunnel Modeling Criteria and Techniques in Civil Engineering Applications; Gaithersburg, Maryland, USA. 1982. pp. 137-185

[5] Cook N. Determination of the model scale factor in wind-tunnel simulations of the adiabatic atmospheric boundary layer. Journal of Industrial Aerodynamics. 1978;**2**:311-321

[6] Vosáhlo L. Computer Programs for Evaluation of Turbulence Characteristics from Hot-wire Measurements. KfK 3743. Karlsruhe: Kernforschungszentrum Karlsruhe; 1984

[7] Möller S. Experimentelle Untersuchung der Vorgänge in engen Spalten zwischen den Unterkanälen von Stabbündeln bei turbulenter Strömung [dissertation]. KfK 4501. Karlsruhe, RFA: Universität Karlsruhe (TH); 1988

[8] Wittwer A, Möller S. Characteristics of the low speed wind tunnel of the UNNE. Journal of Wind Engineering and Industrial Aerodynamics. 2000;**84**:307-320

[9] Counihan J. An improved method of simulating an atmospheric boundary layer in a wind tunnel. Atmospheric Environment. 1969;**3**:197-214

[10] Standen N. A spire array for generating thick turbulent shear layers for natural wind simulation in wind tunnels. National Research Council of Canada. NAE. Report LTR-LA-94. 1972

[11] Blessmann J. The boundary layer wind tunnel of UFRGS. Journal of Wind Engineering and Industrial Aerodynamics. 1982;**10**:231-248

[12] Yassin M, Katob S, Ookab R, Takahashib T, Kounob R. Field and wind-tunnel study of pollutant dispersion in a built-up area under various meteorological conditions. Journal of Wind Engineering and Industrial Aerodynamics. 2005;**93**:361-382

[13] Wittwer A, Loredo-Souza A, Oliveira M, De Bortoli M, Marighetti J. Análisis experimental de cargas de viento localizadas en un edificio de gran altura. In: Memorias de XXXVI Jornadas Sudamericanas de Ingeniería Estructural; Montevideo, Uruguay. 2014

[14] Wittwer A, De Bortoli M, Natalini B, Marighetti J, Natalini M. Estudio de cargas de viento de la nueva terminal de pasajeros del Aeropuerto de Ezeiza mediante ensayos en túnel de viento. In: Memorias de XXI Jornadas Argentinas de Ingeniería Estructural; Buenos Aires, Argentina. 2010

[15] Cook N. The designer's guide to wind loading of building structures. In: Part 1: Background, Damage Survey, Wind Data and Structural Classification. London, UK: Building Research Establishment; 1990

[16] Loredo-Souza AM, Rocha MM. Determinação Experimental,

em Túnel de Vento, do Comportamento Aerodinâmico do Complexo Viário Jornalista Roberto Marinho/Octávio Frias de Oliveira, Fase III; Relatório LAC-UFRGS; Porto Alegre. Brazil; 2005

[17] Loredo-Souza AM, Rocha MM. Determinação experimental, em túnel de vento, do comportamento aerodinâmico do Complexo Viário Jornalista Roberto Marinho/Octávio Frias de Oliveira. Fase VI–Modelo aeroelástico completo; Relatório LAC-UFRGS; Porto Alegre, Brazil; 2007

[18] Wittwer A, Loredo-Souza A, Camaño Schettini E. Analysis of concentration fluctuations in a plume emission model. In: Proceedings of the 11th Americas Conference on Wind Engineering; June 22-26, Puerto Rico. 2009

[19] Wittwer A, Loredo-Souza A, Camaño Schettini E. Laboratory evaluation of the urban effects on the dispersion process using a small-scale model. In: Proceedings of the 13th International Conference on Wind Engineering. ICWE13; Amsterdam, The Netherlands. 2011

[20] Wittwer A, Loredo-Souza A, Camaño Schettini E, Castro H. Wind tunnel study of plume dispersion with varying source emission configurations. Wind and Structures. 2018;**27**(6):417-430

[21] Isyumov N, Tanaka H. Wind tunnel modelling of stack gas dispersion—difficulties and approximations. In: Proceedings of the fifth International Conference on Wind Engineering; Fort Collins, Colorado, USA; 1980

[22] Camaño Schettini E. Etude Expérimentale des Jets Coaxiaux avec Différences de Densité [Thèse de Docteur]. France: Institut National Polytechnique de Grenoble; 1996

[23] Wittwer A, Alvarez y Alvarez G, Demarco G, Martins L, Puhales F, Acevedo O, et al. Employing wind tunnel data to evaluate a turbulent spectral model. American Journal of Environmental Engineering. 2016;**6**(4A):156-159

[24] Wittwer A, Loredo-Souza A, Oliveira M, Fisch G, De Souza B, Goulart E. Study of gas turbulent dispersion process in the Alcântara Launch Center. Journal of Aerospace Technology and Management. 2018;**10**:1-17

[25] Wittwer A, Dorado R, Alvarez y Alvarez G, Degrazia G, Loredo-Souza A, Bodmann B. Flow in the wake of wind turbines: Turbulence spectral analysis by wind tunnel tests. American Journal of Environmental Engineering. 2016;**6**(4A):109-115

[26] Puliafito E, Allende D, Mulena C, Cremades P, Lakkis S. Evaluation of the WRF model configuration for Zonda wind events in a complex terrain. Atmospheric Research. 2015;**166**:24-32

[27] Loredo-Souza A, Wittwer A, Castro H, Vallis M. Characteristics of Zonda wind in south American Andes. Wind and Structures. 2017;**24**(16):657-677

Chapter 7

Dimple Generators of Longitudinal Vortex Structures

Volodymyr Voskoboinick, Andriy Voskoboinick, Oleksandr Voskoboinyk and Volodymyr Turick

Abstract

Visual research of characteristic features and measurement of velocity and pressure fields of a vortex flow inside and nearby of a pair of the oval dimples on hydraulically smooth flat plate are conducted. It is established that depending on the flow regime inside the oval dimples, potential and vortex flows with ejection of vortex structures outside of dimples in the boundary layer are formed. In the conditions of a laminar flow, a vortex motion inside dimples is not observed. With an increase of flow velocity in dimples, boundary layer separation, shear layer, and potential and circulating flows are formed inside the oval dimples. In the conditions of the turbulent flow, the potential motion disappears, and intensive vortex motion is formed. The profiles of longitudinal velocity and the dynamic and wall-pressure fluctuations are studied inside and on the streamlined surface of the pair of oval dimples. The maximum wall-pressure fluctuation levels are pointed out on the aft walls of the dimples. The tonal components corresponding to oscillation frequencies of vortical flow inside the dimples and ejection frequencies of the large-scale vortical structures outside the dimples are observed in velocity and pressure fluctuation spectra.

Keywords: dimple generator, oval dimple, visualization, vortex structure, velocity profile, wall-pressure fluctuations

1. Introduction

Various inhomogeneities of the streamlined surface in the form of cavities or dimples are present in many hydraulic structures and constructions. Under appropriate conditions of the flow, large-scale coherent vortex systems and small-scale vortices are formed inside dimples that generate intense fluctuations of velocity, pressure, temperature, vorticity, and other turbulence parameters [1–3]. Boundary layer control uses these artificial vortex structures for drag reduction, increase of mixing, and noise minimization. Vortex structures of various scales, directions, rotational frequencies, and oscillations are generated in space and in time depending on the flow regime, the geometric parameters, and the shape of the cavities. Experimental and numerical results of aerodynamic and thermophysical studies showed a rather high efficiency of dimple reliefs, which allowed to increase heat and mass transfer for a slight increase in the level of hydrodynamic losses [4–6].

The boundary layer separation from the frontal edge of the cavity and the instability of the shear layer flow generate vortex structures inside the cavity. With the increase of flow velocity, one of the edges of vortex structures, circulating in

 IntechOpen

the cavity, is separated from the streamlined surface of the cavity and is extracted following the flow. These inclined structures have a longitudinal dimension that substantially exceeds their lateral scale. They intensively initiate the interaction of medium of the cavity and the surrounding area [2, 3, 7, 8].

The experience achieved by scientists and engineers when using dimple surfaces indicates that the creation of time and space stable vortex systems generated inside the cavities has a perspective value for boundary layer control. The creation of large-scale coherent vortex structures, with predefined qualities, allows you to change the structure of the boundary layer or the separation flow. It improves the heat and mass transfer, reduces the drag of streamlined structures, or changes the spectral composition of aerohydrodynamic noise, in order to reduce it [3, 9, 10].

In Refs. [11, 12], it was noted that spherical cavities for heat and hydraulic efficiency are not the best for turbulent regime of heat carrier flow and for laminar regime; their use is practically not justified. The presence of a switching mechanism of generation and ejection of vortex structures inside spherical cavities on a streamlined surface [13–15] does not allow to form longitudinal vortex structures that are stable in space and time, which are necessary for boundary layer control. This defect is absent in oval dimples, which are at an angle to the current direction. Asymmetry of the dimple shape due to its lateral deformation allows transforming the vortex structure and intensifying the transverse flow of liquid within its boundaries. Adding a shallow dimple of an asymmetric shape leads to a reorganization of its flow. A two-dimensional vortical structure in the dimple, generated in a symmetrical dimple during its laminar flow, is changed to an inclined monovortex. The high stability of the inclined structure should be noted, which ensures the stability of vortex intensification of heat transfer [16–18].

In this connection, the purpose of this experimental work is to study the characteristic features of the flow of a system of oval dimples on a flat plate and to study the fields of dynamic and wall-pressure fluctuations inside and on the streamlined surface of the inclined oval dimples and in their vicinity.

2. Experimental setup

Experimental research was carried out in a hydrodynamic flume with an open surface of water 16 m long, 1 m wide, and 0.4 m deep. The scheme of the experimental stand and the location of the measuring plate with dimples are given in works [19, 20]. At a distance of about 8 m from the input part of the flume, there were a measuring section equipped with control equipment and means of visual recording of the flow characteristics, coordinate devices, lighting equipment, and other auxiliary tools necessary for conducting experimental research. The design and equipment of the hydrodynamic flume allowed the flow velocity and water depth control in wide limits.

Transparent walls of a hydrodynamic flume, which were made of thick shockproof glass, ensured high-quality visual research.

Hydraulically flat plate made of polished organic glass of 0.01 m thick, 0.5 m in width, and 2 m in length was sharpened from one (front) and from the other (aft) side. End washers are fixed to the lateral sides of the plate. At a distance of X = 0.8 m from the front edge of the plate, there was a hole, where the system of two oval dimples was installed, which was located at an angle of 30 degrees to the direction of flow (**Figure 1**). The diameter of a spherical part of the dimple (d) was 0.025 m. The width and length of the cylindrical part of the dimple were also 0.025 m. Thus, the oval dimples located at a distance of 0.005 m from each other had a width of 0.025 m, a length of 0.05 m, and a depth to width ratio of h/d = 0.22.

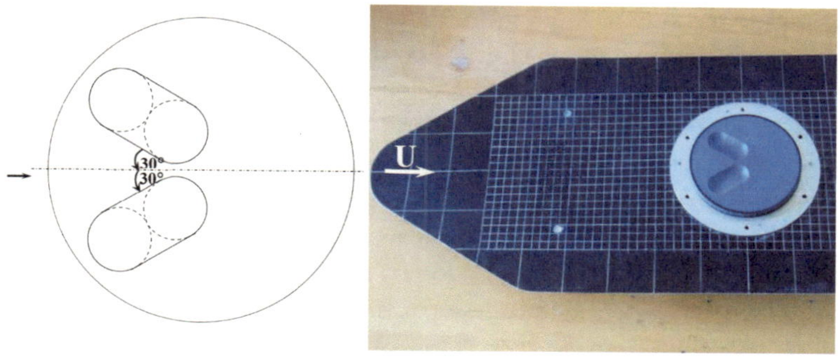

Figure 1.
Scheme and photography of the experimental plate with pair of the oval dimples.

According to the developed program and experimental research methodology, visual studies were initially carried out. Then, in the characteristic points of the vortex generation and the places of interaction of vortices with a streamlined surface, measurements of the fields of velocity and pressure were carried out. Visualization was carried out by drawing of contrasting coatings on the streamlined surface and coloring agents that were introduced into the stream. Paints and labeled particles through a small diameter tube were introduced into the boundary layer before the dimple and/or inside the dimple.

The study of the pressure fluctuation fields on the streamlined surface of the oval dimples and the plate, as well as the velocity fields of the vortex flow over the investigated surfaces, was carried out using miniature piezoceramic and piezoresistive sensors of pressure fluctuations and differential electronic manometers (**Figure 2a**). Specially designed and manufactured pressure sensors were installed in a level with a streamlined surface and measured the absolute pressure and the wall-pressure fluctuations [9, 21, 22]. Inside of the system of oval dimples and in their near wake, 12 sensors of pressure fluctuations were used (**Figure 2b**). The field of velocity fluctuations inside a pair of oval dimples and over a streamlined plate surface was measured by sensors of the dynamic pressure fluctuations or dynamic velocity pressure based on piezoceramic sensing elements.

The degree of the flow turbulence in the hydrodynamic flume did not exceed 10% for the velocity range from 0.03 to 0.5 m/s. The levels of acoustic radiation in the area of the dimples were no more than 90 dB relative to 2×10^{-5} Pa in the frequency range from 20 Hz to 20 kHz, and the vibration levels of the test plate with

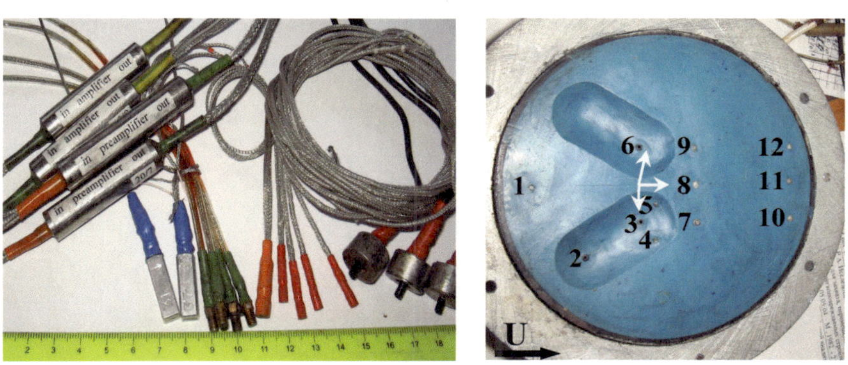

a) b)

Figure 2.
Absolute pressure and pressure fluctuation sensors (a) and their disposition (b).

a pair of dimples and sensor holder did not exceed −55 dB relative to g (gravitation constant) in the frequency range from 2 Hz to 12.5 kHz. The measurement error of the averaged parameters of the fields of velocity and pressure did not exceed 10% (reliability 0.95). The measurement error of the spectral components of the velocity fluctuations did not exceed 1 dB, and the pressure and acceleration fluctuations—no more than 2 dB—in the frequency range from 2 Hz to 12.5 kHz.

3. Research results

The vortical motion in the middle of the dimples is not what was observed (**Figure 3a**) for a laminar flow regime over a pair of oval dimples (U = (0.03...0.06) m/s, $Re_X = UX/\nu$ = (24,000...48,000), and $Re_d = Ud/\nu$ = (750...1500), where ν is the kinematic coefficient of water viscosity). The contrast dye was transferred inside the dimple along its front spherical and cylindrical parts and gradually filled the entire volume of the oval dimples. The separation flow was not observed inside the dimples, and colored dye, which was moved from the front of the dimple to its aft part, made non-intensive oscillatory motion.

When the flow velocity was increased to (0.08...0.12) m/s, then a separation zone of the boundary layer appeared inside the front parts of the oval dimples. A shear layer began to form over the dimple opening, generating a circulating flow and a slow vortex motion inside the dimples (**Figure 3b**). This fluid motion had a kind of longitudinal spirals and was slow and almost symmetrical in each of the dimples. The liquid of the dimples fluctuated in three mutually perpendicular planes. The oscillation frequencies in each of the dimples were practically equal, but the destruction of the vortex sheet did not occur simultaneously. Contrast material went inside the dimples along their front semispherical and cylindrical parts. The separation and circulation areas behind the front edge of the dimple occupied almost half the volume of the dimple. There was a very slow rotation of the fluid inside the dimples, and its direction was coincided with the direction of the flow as well as its fluctuations along the longitudinal and transverse axes of the dimples. The disturbance package was transferred in the direction of the flow at a transfer velocity of approximately (0.4...0.5) U. In this case, the contrast material was ejected into the plate boundary layer over the region of the combination of the aft cylindrical and spherical parts of the dimple (**Figure 3b**). The ejection of a large-scale vortex or spiral-like vortices from the oval dimple was observed at a frequency close to f = (0.16...0.2) Hz, which the Strouhal number corresponded to $St = fd/U$ = (0.04...0.05). A wake of the contrast material into the boundary layer outside the dimples was traced at a distance of about 8–10 in diameter of the dimple.

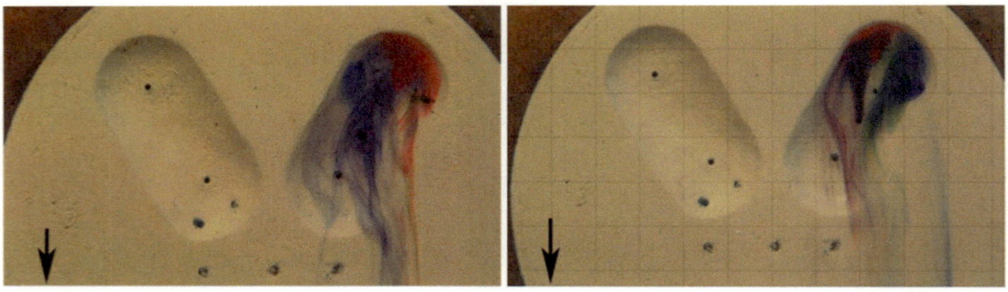

a) b)

Figure 3.
Flow visualization inside the oval dimple for Re_x = 48,000 (a) *and* Re_x = 96,000 (b).

The vortex motion became more intense when the flow velocity over the dimple system was increased up to (0.2...0.3) m/s (Re_X = (160,000...240,000) and Re_d = (5000...7500)). The zone of potential flow near the separation wall of the oval dimple had almost vanished (**Figure 4a**). The entire fluid filling the front spherical part of the dimple is turned in a circulating flow and formed a coherent large-scale spindle-shaped vortex. This vortex had a source near the center of the spherical part of the dimple and made intensive oscillations. During the ejection, the spindle-shaped vortex structures began to lift above the front hemispheres of oval dimples and to stretch along the axis of the dimples. Then they were ejected outside the dimples over their aft parts. These large-scale vortex structures were rotated in the XOZ plane in each of the dimples in opposite directions. For example, in the left dimple in **Figure 4a**, the vortex rotated against the clockwise arrow, and in the right—clockwise. Ejection of vortex structures from the dimples was sometimes observed at the same time, but in most cases, ejections occurred at different time intervals. At the same time, there was no interaction of these vortex structures in the near wake of the dimples. The frequency of ejections of large-scale vortex structures from each of the dimples was estimated as (0.4...0.6)Hz or St = (0.04...0.06). In addition, ejections of the small-scale eddy structures were also observed. These vortices were broken off from the upper part of a large-scale spindle-shaped vortex during its formation, when its transverse scale exceeded the depth of the dimple. Vortex structures retained their identity at a distance (7...9) of the diameter of the dimple.

The contrast dye inside the dimple was concentrated inside the front spherical parts of the dimple (**Figure 4b**) for developed turbulent flow and flow velocity (0.4...0.5) m/s (Re_X = (320,000...400,000) and Re_d = (10,000...12,500)). Here, spindle-shaped vortex structures were generated, and they were ejected from the dimples at a frequency close to 1 Hz (St ~0.05). Color dyes, swirling in a spindle-shaped vortex, were intensively oscillated in three mutually perpendicular planes. When the transverse scale of the spindle-shaped vortex exceeded the depth of the dimple, an intensive ejection of small-scale structures was observed from its upper part. These vortices were flushed over the region of the conjugation of the front spherical part of the dimple and its aft cylindrical part. The ejection frequency of small-scale vortices was estimated as (4...5) Hz or St = (0.2...0.25). As shown by the dye visualization (see **Figure 4b**), in the gap between the dimples, the flow did not undergo significant perturbations, as can be seen on the dye on the axis of the plate, which was not washed.

The intensity of the fluctuations of the longitudinal velocity over the streamlined surface of the oval dimple (u'/U), which was calculated from the fluctuations of the dynamic pressure $\left(u'_{rms} = \sqrt{2\left(p'_{rms}\right)_{dyn}/\rho}\right)$ measured by the piezoceramic

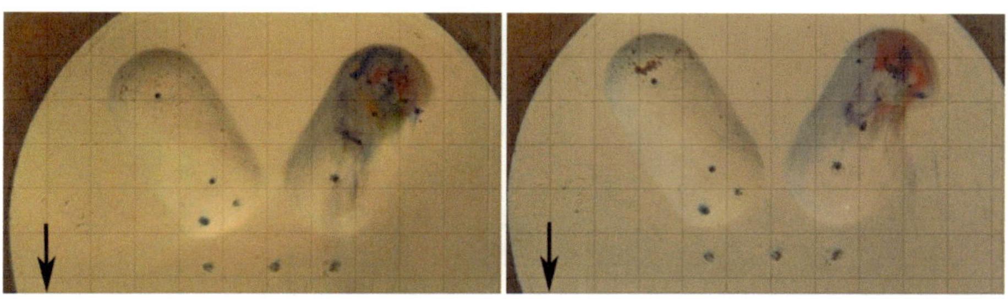

a) b)

Figure 4.
Flow visualization inside the dimple for Re_x = 240,000 (a) and Re_x = 400,000 (b).

sensors, depending on the distance from the streamlined surface (y/δ, where δ is the boundary layer thickness in front of the oval dimple), is presented in **Figure 5a**. These results were obtained for two flow velocities, namely, U = 0.25 m/s (Re_X = 200,000) and U = 0.45 m/s (Re_X = 360,000). The results were measured over the streamlined surface of one of the oval dimples above the wall-pressure fluctuation sensor of No. 2 (see **Figure 1b**). The longitudinal velocity fluctuations are increased with approach to the streamlined plate surface. They have a maximum and then are decreased at the level of the plate surface above the opening of the oval dimple. When the dynamic pressure fluctuation sensors are deepened in the opening of the oval dimple, the velocity fluctuations again increased (the boundary of the shear layer) and then decreased (the core of the circulating flow inside the dimple).

The change of the rms values of the wall-pressure fluctuations measured on the streamlined surface of the oval dimple and in its vicinity is presented in **Figure 5b** depending on the Reynolds number. The normalization of the root mean square values of the wall-pressure fluctuations was carried out by the dynamic pressure ($q = \rho U^2/2$). In this figure, the curve numbers correspond to the numbers of the wall-pressure fluctuation sensors, which are set to the level with the streamlined surface in accordance with **Figure 1b**.

Thus, the wall-pressure fluctuations on a flat surface before the dimples are subjected to a quadratic dependence on the flow velocity. It should be noted that the wall-pressure fluctuations normalized by the dynamic pressure in the undisturbed boundary layer before the oval dimples are approximately 0.01, practically, in the entire range of studied Reynolds numbers.

Consequently, the smallest levels of the wall-pressure fluctuations are observed at the bottom of the oval dimples, in their front parts, especially for low flow velocities and Reynolds numbers (curve 2, **Figure 5b**). Inside the oval dimples, the levels of the wall-pressure fluctuations are greatest in the aft spherical parts of the dimples and in the near wake immediately after the dimples (see curves 4, 7, and 9 in **Figure 5b**).

A spectral analysis of the wall-pressure fluctuations on the streamlined surfaces of the oval dimples and plate was performed. To do this, we used the fast Fourier transform algorithm and the Hanning weighting function, as recommended in [23–25]. Power spectral densities of the wall-pressure fluctuations on a streamlined surface of oval dimples and on a flat plate near the system of these dimples have clearly visible discrete peaks which correspond to the nature of the vortex and jet motion over the investigated surfaces.

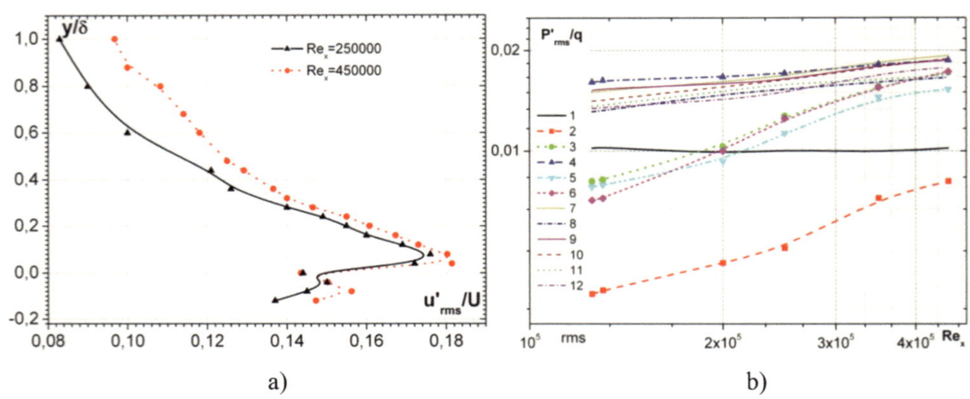

Figure 5.
Velocity profile (a) and root mean square value of wall-pressure fluctuations (b).

Figure 6a shows the power spectral densities of the wall-pressure fluctuations, which were measured inside one of the oval dimples for a flow velocity of 0.25 m/s (Re_d = 6250 and Re_x = 200,000). The spectra were normalized by the dynamic pressure and external variables, namely, the diameter of the oval dimple and the flow velocity ($p_q^*(St) = (p')^2(St)U/q^2d$). Frequency ($f$) was normalized and presented as the number of Strouhal St. The numbers of curves correspond to the numbers of pressure fluctuation sensors, which are shown in **Figure 1b**. The maximum levels of the wall-pressure fluctuations occur at the aft wall of the dimple, where there are the intense interactions of the vortex flow ejected from the dimple and the shear layer formed above the streamlined surface of the plate. The smallest spectral levels of wall-pressure fluctuations occur at the bottom of the front spherical part of the oval dimple (curve 2). The ejections of the large-scale vortex structures observed during visual investigations occur at a frequency of (0.37...0.45) Hz or St = (0.04...0.05).

Oscillations of the vortex flow inside the oval dimple are observed at frequencies (0.035...0.037) Hz or St = (0.003...0.004) and (0.13...0.15) Hz or St = (0.013...0.015) in the longitudinal and transverse directions relative to the axes of the oval dimple, respectively. In this case, the oscillations of the vortex motion and, respectively, the field of the wall-pressure fluctuations inside the dimple correspond to the subharmonics and harmonics of higher orders of these frequencies, as it is clearly illustrated in **Figure 6a**.

The results of the measurements of the power spectral densities of the wall-pressure fluctuations along the middle section of the oval dimple system, as well as inside the dimples, are shown in **Figure 6b**. It should be noted that under the boundary layer on a flat surface of a hydraulically smooth plate, the spectral levels of the wall-pressure fluctuations (curve 1) are minimal and do not have the tonal or discrete peaks observed inside and near the dimples. Behind the oval dimples, these discrete peaks are clearly visible on the spectra, but the tone frequencies near the system of oval dimples and at a distance of 2d differ from them (see curves 8 and

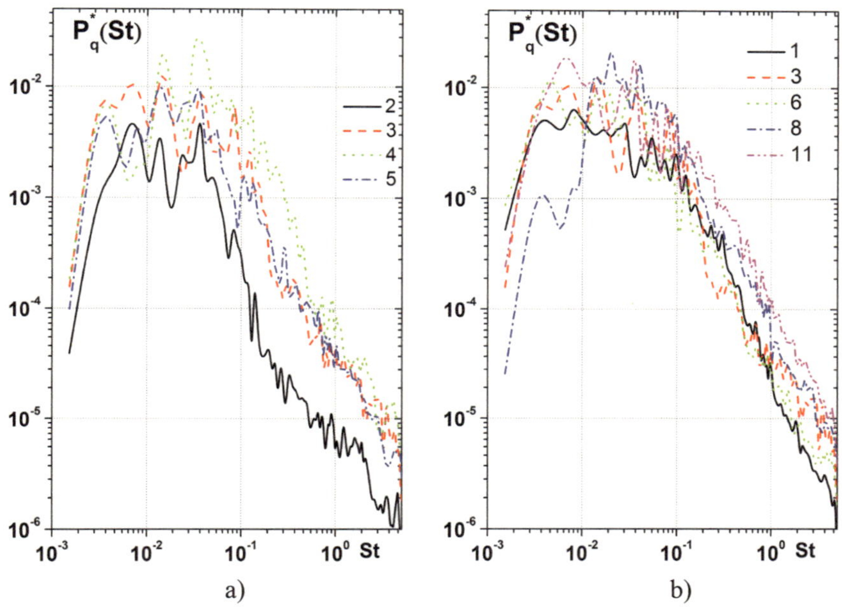

Figure 6.
Power spectral densities of the wall-pressure fluctuations inside (a) and near (b) the oval dimple for the Reynolds number Re_x = *200,000.*

11 in **Figure 6b**). In the middle section of the oval dimple system, where the sensor number 8 is located, the character of the pressure fluctuation spectrum differs from that which occurs on the aft wall of the dimple. Here, the maximum of spectral levels is observed at a frequency of 0.2 Hz (St = 0.02), and in the frequency range of the order of 0.03 Hz (St = 0.003), the intensity of the wall-pressure fluctuations is negligible. At a distance of 2*d* from the dimples, the tonal peaks appear in the spectra corresponding to the ejection frequencies of large-scale vortex structures outside of the dimples. Thus, the traces of the vortex flows ejected from the oval dimples are intersected at the location of the sensor No. 11.

Experiments have shown that all sensors located at a distance of 2*d* from the system of oval dimples record the field of the wall-pressure fluctuations with tonal peaks in the spectra corresponding to the ejection frequencies of large-scale vortex structures from the dimples, the frequencies of oscillatory motion inside the dimples, and their subharmonics and harmonics of higher orders. In this case, the spectral levels at such a distance from the dimples are of lesser value than in the near wake of the dimples. Thus, with the distance from the system of oval dimples, the boundary layer is gradually restored, which was observed during the visualization of the flow.

In the conditions of developed turbulent flow (Re_d > 11,000 and Re_X > 350,000), the spectral characteristics of the wall-pressure fluctuation field are similar to those observed for Reynolds numbers Re_X = 200,000 and Re_d = 6250. But the spectral levels become higher (**Figure 7a**). The highest spectral levels of wall-pressure fluctuations, as well as tonal peaks, are observed on the aft spherical part of the oval dimple as for the lower flow velocity. The smallest spectral levels are generated in the forward spherical part of the dimple (curve 2). Discrete peaks are observed at frequencies (0.05...0.06) Hz or St = (0.003...0.004), (0.11 ...0.13) Hz or St = (0.006...0.007), and (0.8...0.9) Hz or St = (0.04...0.05). The first two low-frequency ranges correspond to the oscillation frequency in the oval dimple, and the frequency range St = (0.04...0.05) is due to the ejection of large-scale vortex structures from the dimple. Also in the spectral characteristics of the field of the

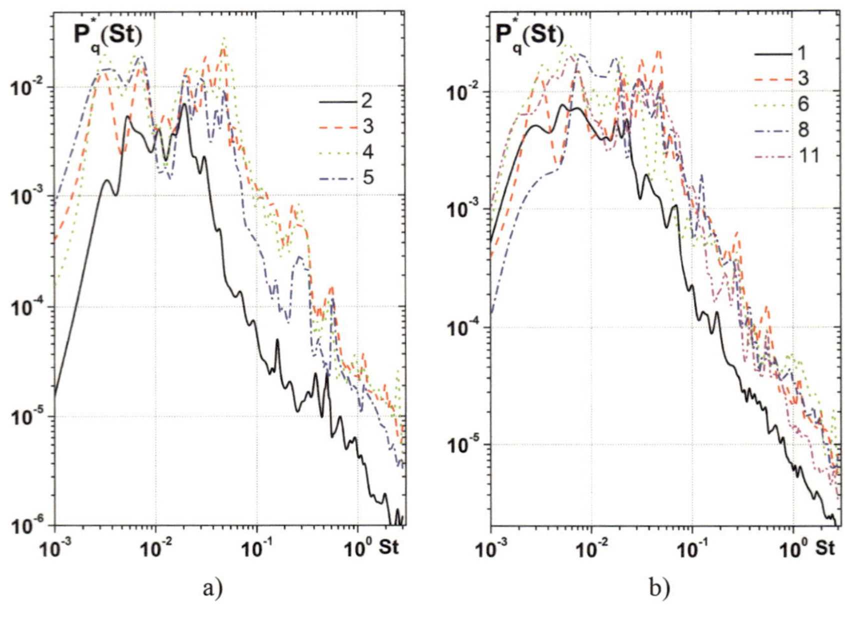

Figure 7.
Power spectral densities of the wall-pressure fluctuations inside (a) and near (b) the oval dimple for the Reynolds number Re_x *= 400,000.*

wall-pressure fluctuations inside the oval dimple, there are discrete peaks that correspond to subharmonics and harmonics of higher orders of dominant frequencies of the vortex motion.

The features of the vortex motion, as well as wall-pressure fluctuation field, which it generates, in the near wake of the oval dimple system, in its middle section and at a distance of $2d$ from the system of oval dimples, are shown in **Figure 7b**. The spectral levels of the wall-pressure fluctuations in the wake behind the aft spherical part of the dimple are similar to those obtained inside the oval dimple as for a lower flow velocity. At the same time, the spectra in the middle section of the system of oval dimples (in their near wake) have a specific character with a maximum at 0.13 Hz (curve 8 in **Figure 7b**). Behind the aft spherical part of the oval dimple in the spectral dependences of the field of the wall-pressure fluctuations, there are tone peaks which are characteristic of the vortical motion inside the oval dimples. In the middle section of the oval dimple system at a distance $2d$ from the dimples, discrete peaks appear in the spectral levels of wall-pressure fluctuations. They are characteristic for the low-frequency oscillations of the vortex motion inside the dimples, as well as for the ejection of large-scale vortex structures from the dimples. The intensity, for example, of the wall-pressure fluctuations at the ejection frequency, is much lower for this flow regime than that observed for the velocity flow 0.25 m/s (**Figures 6b** and **7b**). This is due to the fact that for a large flow velocity, the interaction between the vortices of each of the dimples takes place more distant from the dimples, and the distance $2d$ from the pair of oval dimples for this regime is in the initial stage of this interaction.

4. Conclusions

1. The visual images of the vortex flow formed inside the oval dimple system are obtained, and the characteristic features of vortex formation for different flow regimes are determined. It has been experimentally established that the separation flow was not observed inside the dimples for laminar regime. For transient flow regime and small flow velocities within the oval dimples, the formation of very intense longitudinal spirals is observed, which are rotated and slowly fluctuated along the longitudinal and transverse axes of the dimples. For a turbulent flow regime inside the oval dimples, the spindle-shaped vortices are formed, which, with increasing velocity, are pressed against the front spherical parts of the dimples. These spindle-shaped vortices, reaching the scales of the dimples, are ejected from the oval dimples, disturbing the structure of the boundary layer. Inside the oval dimples, there is a low-frequency oscillatory motion in mutually perpendicular planes relative to the axes of the dimples, whose frequency is increased with increasing flow velocity.

2. It is shown that the intensity of the field of the velocity fluctuations has maximum values near the streamlined surface and also on the boundary of the shear layer in the opening of the oval dimple. The intensity of the wall-pressure fluctuation field is greatest in the interaction region between vortex structures of the shear layer and large-scale vortex systems ejected from the dimples with the aft wall of the oval dimple. The smallest intensity of the wall-pressure fluctuations occurred at the bottom of the oval dimple in its forward spherical part.

3. It has been established that depending on the flow regimes in the spectral characteristics of the field of the wall-pressure fluctuations measured on the

streamlined surface, characteristic features appear in the form of discrete peaks corresponding to the frequencies of low-frequency oscillations of the vortex flow inside the oval dimples and the ejection frequencies of large-scale vortex systems from the dimples. In the middle section of the system of oval dimples (in their near wake), there is no interaction of vortex structures that are ejected from the dimples. With a distance of more than two diameters of the dimple, intensive tone peaks are observed in the spectral dependences. They correspond to the ejection frequencies of large-scale vortices and the frequency of oscillations of the vortex motion inside the dimples, both in the middle section of the system of the dimples and behind their aft spherical parts. With the distance from the system of oval dimples, the intensity of the tonal oscillations, which are characteristic for the vortical motion inside the dimples, is decreased, and the boundary layer is restored.

Author details

Volodymyr Voskoboinick[1*], Andriy Voskoboinick[1], Oleksandr Voskoboinyk[1] and Volodymyr Turick[2]

1 Institute of Hydromechanics of NAS of Ukraine, Kyiv, Ukraine

2 NTUU Igor Sikorsky Kyiv Polytechnic Institute, Kyiv, Ukraine

*Address all correspondence to: vlad.vsk@gmail.com

References

[1] Kiknadze GI, Gachechiladze IA, Oleinikov VG, Alekseev VV. Mechanisms of the self-organization of tornado-like jets flowing past three-dimensional concave reliefs. Heat Transfer Research. 2006;**37**(6):467-494

[2] Ligrani PM, Harrison JL, Mahmmod GI, Hill ML. Flow structure due to dimple depressions on a channel surface. Physics of Fluids. 2001;**13**(11):3442-3451

[3] Khalatov AA. Heat transfer and fluid mechanics over surface indentations (dimples). National Academy of Sciences of Ukraine, Institute of Engineering Thermophysics, Kyiv; 2005

[4] Leontiev AI, Kiselev NA, Vinogradov YA, Strongin MM, Zditovets AG, Burtsev SA. Experimental investigation of heat transfer and drag on surfaces coated with dimples of different shape. International Journal of Thermal Sciences. 2017;**118**:152-167

[5] Li P, Luo Y, Zhang D, Xie Y. Flow and heat transfer characteristics and optimization study on the water-cooled microchannel heat sinks with dimple and pin-fin. International Journal of Heat and Mass Transfer. 2018;**119**:152-162

[6] Voropayev GA, Voskoboinick VA, Rozumnyuk NV, Voskoboinick AV. Vortical flow features in a hemispherical cavity on a flat plate. Papers of the Sixth International Symposium on Turbulence and Shear Flow Phenomena, TSFP-6, June 22-24, 2009, Seoul, Korea; 2009. pp. 563-568

[7] Douay CL, Pastur LR, Lusseyran F. Centrifugal instabilities in an experimental open cavity flow. Journal of Fluid Mechanics. 2016;**788**:670-694

[8] Borchetta CG, Martin A, Bailey SCC. Examination of the effect of blowing on the near surface flow structure over a dimpled surface. Experiments in Fluids. 2018;**59**:36-1-13

[9] Voskoboinick V, Kornev N, Turnow J. Study of near wall coherent flow structures on dimpled surfaces using unsteady pressure measurements. Flow, Turbulence and Combustion. 2013;**90**(4):709-722

[10] Zhang F, Wang X, Li J. Flow and heat transfer characteristics in rectangular channels using combination of convex-dimples with grooves. Applied Thermal Engineering. 2017;**113**:926-936

[11] Isaev SA, Leont'ev AI, Baranov PA, Metov KT, Usachov AE. Numerical analysis of the effect of viscosity on the vortex dynamics in laminar separated flow past a dimple on a plane with allowance for its asymmetry. Journal of Engineering Physics and Thermophysics. 2001;**74**(2):339-346

[12] Isaev SA, Leontiev AI, Mityakov AV, Pyshnyi IA, Usachov AE. Intensification of tornado turbulent heat exchange in asymmetric holes on a plane wall. Journal of Engineering Physics and Thermophysics. 2003;**76**(2):266-270

[13] Picella F, Loiseau J-C, Lusseyran F, Robinet J-C, Cherubini S, Pastur L. Successive bifurcations in a fully three-dimensional open cavity flow. Journal of Fluid Mechanics. 2018;**844**:855-877

[14] Turnow J, Kornev N, Isaev S, Hassel E. Vortex mechanism of heat transfer enhancement in a channel with spherical and oval dimples. Heat and Mass Transfer. 2011;**47**(3):301-313

[15] Sun Y, Taira K, Cattafesta LN III, Ukeiley LS. Biglobal instabilities of compressible open-cavity flows. Journal of Fluid Mechanics. 2017;**826**:270-301

[16] Isaev SA, Baranov PA, Leontiev AI, Popov IA. Intensification of a laminar flow in a narrow microchannel with single-row inclined oval-trench dimples. Technical Physics Letters. 2018;**44**(5):398-400

[17] Isaev S, Voropaiev G, Grinchenko V, Sudakov A, Voskoboinick V, Rozumnyuk N. Drag reduction of lifting surfaces at the use of oval dimples as vortex generators. Abstract of the European Drag Reduction and Flow Control Meeting "EDRFCM 2010" 2-4 September, 2010, Kyiv, Ukraine;2010. pp. 32-33

[18] Isaev SA, Schelchkov AV, Leontiev AI, Gortyshov Yu F, Baranov PA, Popov IA. Vortex heat transfer enhancement in the narrow plane-parallel channel with the oval-trench dimple of fixed depth and spot area. International Journal of Heat and Mass Transfer. 2017;**109**:40-62

[19] Voskoboinick VA, Voskoboinick AV, Voskoboinick AA. Pressure fluctuations inside and nearby the pair of oval dimples on a flat plate. Acoustic Bulletin. 2015;**17**(1):23-33

[20] Voskoboinick VA, Voskoboinick AV, Voskoboinick AA. Vortex generation by a couple of oval cavities on flowing plane surface. Applied Hydromechanics. 2015;**17**(3):10-17

[21] Voskoboinick VA, Turick VN, Voskoboinyk OA, Voskoboinick AV, Tereshchenko IA. Influence of the deep spherical dimple on the pressure field under the turbulent boundary layer. Proc. Intern. Conf. on Computer Science, Engineering and Education Applications (ICCSEEA2018); 18-20 January 2018, Kiev, Ukraine 1-10;2018

[22] Vinogradnyi GP, Voskoboinick VA, Grinchenko VT, Makarenkov AP. Spectral and correlation characteristics of the turbulent boundary layer on an extended flexible cylinder. Journal of Fluid Dynamics. 1989;**24**(5):695-700

[23] Bendat JS, Piersol AG. Random data: Analysis and measurement procedures. NY: John Willey & Sons, Inc; 1986

[24] Voskoboinick VA, Makarenkov AP. Spectral characteristics of the hydrodynamical noise in a longitudinal flow around a flexible cylinder. International Journal of Fluid Mechanics Research. 2004;**31**(1):87-100

[25] Blake WK. Mechanics of Flow-induced Sound and Vibration. Vol. 2. New York: Academic Press; 1986

Section 4

Laminar Velocity and Thermal Boundary Layers

Chapter 8

Thermal-Hydrodynamic Characteristics of Turbulent Flow in Corrugated Channels

Nabeel S. Dhaidana and Abdalrazzaq K. Abbas

Abstract

The heat transfer-flow characteristics of turbulent flow inside corrugated channels heated by constant heat flux are numerically investigated. The rate of heat transfer, pressure drop, and performance evaluation criterion is determined for smooth channel and various designs of corrugated channels at the Reynolds number ranged from 5000 to 60,000. The effect of rib arrangement distributions of inward, outward, and inward-outward ribs are examined. The various rib configurations of corrugated channels are also tested. In addition, the influences of rib roughness parameters (height, pitch, and width) and rib shapes (semicircular, trapezoidal, and rectangular) are researched. The Reynolds-averaged Navier-Stokes equations (RANS) are used to model the governing flow equations. The computational model is validated through a reasonable agreement between the present numerical results and the outcomes of related works. For different geometrical and operating conditions, the results revealed that the rate of heat exchange in corrugated channels exceeds higher than that of smooth ones but with additional pressure loss. Moreover, the rib arrangements, rib configuration, and rib roughness parameters exhibit a relatively significant effect on the performance of the corrugated channels. On the other hand, the influence of the rib shapes seems to be small.

Keywords: thermal-flow performance, corrugated channel, rib distribution, rib configuration, rib shapes

1. Introduction

The reliable efficient heat exchangers transfer the maximum rate of heat with minimum friction losses. The rate of heat transfer of most fluids is restricted by their low thermal conductivity. Thus, the thermal systems adopt techniques of heat transfer enhancement to reduce the effect of this issue. There are three techniques of enhancing heat transfer, namely, active methods (require external power) [1], passive methods (fins, corrugation, ribs, etc.) [2], and compound techniques (simultaneous use of active and passive techniques) [3]. Corrugation of tubes and channels is considered an efficient passive method to augment the rate of heat exchange. The thermal-flow features of turbulent flow in corrugated tubes are reported extensively in many articles (for example [4–8]).

Corrugated channels are widely utilized in industrial applications as they are the major components in plate heat exchangers. Naphon [9] conducted experiments to show the performance of a turbulent flow inside a two-sided corrugated channel with an in-line and staggered arrangements. He showed the important effect of corrugation on the augmentation of heat transfer and pressure loss. Eiamsa-ard and Promvonge [10] experimentally examined the thermal-hydrodynamic performance of the three types of ribbed-grooved ducts. They reported that the maximum rate of heat exchange and pressure drop exist in the ducts with a rectangular rib and a triangular groove. Elshafei et al. [11] conducted experiments to examine the thermal-hydraulic performance of corrugated channels under the influence of variations of phase shift and channel spacing. The corrugated channels exhibit a compound increase in heat transfer and pressure loss. Mohammed et al. [12] performed a computational model to investigate the effects of wavy tilt angle, channel height, and channel height on the flow-thermal fields in a corrugated channel. A three-dimensional numerical model to investigate the employing baffles on the heat transfer-flow in the corrugated channels was presented by Li and Gao [13]. Increasing the baffle height enhances heat transfer effectively but leads to dramatic penalty in pressure drop. Pehlivan et al. [14] experimentally investigated the rate of heat exchange for sharp corrugation peak fins of corrugated channel for three different types and sinusoidal converging–diverging channels. It is reported that the rate of heat transfer increases with the corrugated angle. The numerical results showed that the wavy channel is an efficient method to increase the heat transfer. Ravi et al. [15] numerically studied the impact of different rib configurations on the heat transfer-flow characteristics of the turbulent flow inside corrugated channels. Shubham et al. [16] numerically investigated the thermal-hydrodynamic transport characteristics of non-Newtonian fluids in corrugated channels. It was found that using of shear thinning fluids is more convenient for maximum augmentation of thermal performance with a minimum penalty in pressure drop.

The present study offers a numerical model to investigate the thermal flow attributes of turbulent flow in corrugated channels. The performance of corrugated channels are examined under the effects of corrugation arrangement (inward, outward, and inward-outward rib distribution), corrugation configuration, corrugation roughness parameters (rib pitch, rib width, and rib height), and rib shapes (rectangular, trapezoidal, and semicircular). The comparisons between the predicted thermal flow performance of corrugated channels and that of smooth ones are fulfilled under a large range of Reynolds number (5000–60,000).

2. Numerical model

The two-dimensional corrugated channel with a width (b) of 10 mm is described schematically in **Figure 1**. The water as heat transfer fluid enters the computational domain at a temperature of 27°C and intensity of turbulent of 5%. Also, 5% of turbulent intensity is considered at the exit. The end effects and viscous dissipation terms are ignored. The constant heat flux of 600 W/cm^2 is applied on the channel wall. The consideration of an axisymmetric situation reduces the size of the numerical domain for saving computational time.

The flow-thermal behavior is modeled by the governing conservation equations (continuity, momentum, and energy) in a RANS technique as

$$\frac{\partial u_i}{\partial x_i} = 0 \tag{1}$$

$$\frac{\partial}{\partial x_j}(\rho\, u_i u_j) = -\frac{\partial P}{\partial x_i} + \frac{\partial}{\partial x_j}\left[\mu\left(\frac{\partial u_i}{\partial x_j} + \frac{\partial u_j}{\partial x_i} - \frac{2}{3}\delta_{ij}\frac{\partial u_j}{\partial x_j}\right)\right] + \frac{\partial}{\partial x_j}\left(-\rho\, \overline{u_i' u_j'}\right) \tag{2}$$

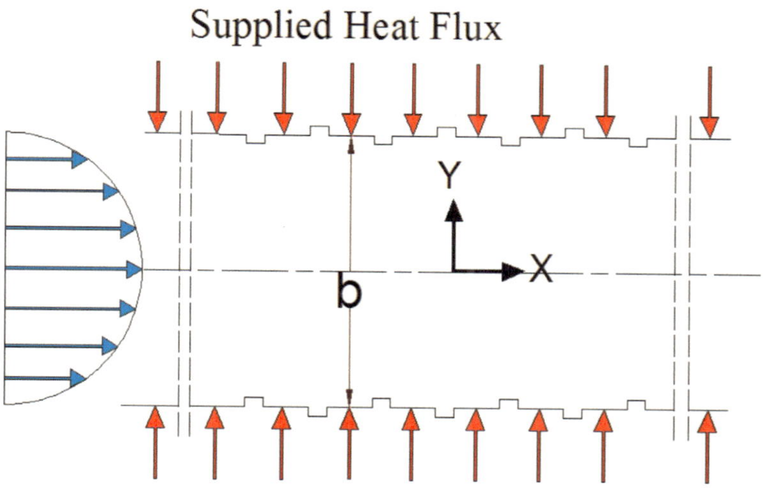

Figure 1.
Schematic representation of the computational domain.

in which ρ, μ, u', and $\rho\overline{u_i'u_j'}$ are density, viscosity, fluctuated velocity, and turbulent shear stress, respectively.

$$\frac{\partial}{\partial x_i}[u_i(\rho E + P)] = \frac{\partial}{\partial x_j}\left[\frac{\partial T}{\partial x_j}\left(kt + \frac{C_p \mu_t}{\mathrm{Pr}_t}\right) + u_i\left(\tau_{ij}\right)_{eff}\right] \tag{3}$$

where Pr_t is the turbulent Prandtl number and $(\tau_{ij})_{eff}$ is the deviatoric stress tensor which is evaluated as

$$\left(\tau_{ij}\right)_{eff} = \mu_{eff}\left(\frac{\partial u_i}{\partial x_j} + \frac{\partial u_j}{\partial x_i}\right) - \frac{2}{3}\mu_{eff}\frac{\partial u_i}{\partial x_j}\delta_{ij} \tag{4}$$

The transport equations in k-e model are presented as [17]

$$\frac{\partial}{\partial x_i}(\rho k u_i) = \frac{\partial}{\partial x_j}\left[\left(\mu + \frac{\mu_t}{\sigma_k}\right)\frac{\partial k}{\partial x_j}\right] + G_k - \rho\varepsilon \tag{5}$$

$$\frac{\partial}{\partial x_i}(\rho \varepsilon u_i) = \frac{\partial}{\partial x_j}\left[\left(\mu + \frac{\mu_t}{\sigma_\varepsilon}\right)\frac{\partial \varepsilon}{\partial x_j}\right] + C_{1\varepsilon}(\varepsilon/k)G_k - C_{2\varepsilon}\rho\left(\varepsilon^2/k\right) \tag{6}$$

and μ_t is the eddy viscosity which is modeled as

$$\mu_t = \frac{\rho\, C_\mu\, k^2}{\varepsilon} \tag{7}$$

The model constants C_μ, $C_{1\varepsilon}$, $C_{2\varepsilon}$, σ_k, and σ_ε are 0.09, 1.44, 1.92, 1.0, and 1.3, respectively.

No-slip condition and constant wall heat flux are assumed as boundary conditions.

The thermal-hydrodynamic performance of the corrugated channels is assessed by dimensionless parameters which are the Nusselt number, friction factor, and performance evaluation criterion (*PEC*).

The average Nusselt number is presented as

$$Nu = \frac{q''d}{kt}\int_0^x \frac{1}{T_w(x) - T_b(x)}dx \tag{8}$$

where q'' and $T_w(x)$ and $T_b(x)$ act as the supplied heat flux and wall and local bulk temperatures, respectively.

The friction factor is defined as

$$f = \frac{2\,\Delta P\, d}{L\,\rho\, {u_m}^2} \tag{9}$$

The comparison between the enhancement in thermal performance and a penalty in the pressure drop is assessed by introducing the performance evaluation criteria (*PEC*) of corrugated channels with different roughness dimensions. The *PEC* can be calculated as

$$PEC = \frac{Nu/Nu_s}{(f/f_s)^{1/3}} \tag{10}$$

where f_s and Nu_s are the friction factor and the Nusselt number of smooth channel, respectively.

The performance of corrugated channels is estimated according to different values of the Reynolds number which is introduced as

$$\mathrm{Re} = \frac{\rho\, u_m\, d_h}{\mu} \tag{11}$$

where μ, ρ, d_h, and u_m are dynamic viscosity, density, hydrodynamic diameter, and mean fluid velocity.

The ANSYS Fluent CFD package-based control volume method is adopted to discretize the governing equations and simulate thermal flow behavior of corrugated channels. The SIMPLE algorithm is utilized for solving the flow field. The diffusion terms and other resulting terms are discretized by employing the first-order upwind scheme. The residuals lower than 10^{-6} is chosen to achieve the convergence criterion for all variables. A fine grid discretization close to the wall is adopted. Also, the meshing system of 23,964 grids is sufficient for solution accuracy. On the other hand, the numerical code that is validated through a reasonable agreement is shown (**Figure 2a**) between the Nusselt number of the present work and the same number which is obtained from the well-known Gnielinski correlation [18]. Furthermore, good agreement is indicated for the friction factor (**Figure 2b**) between the present work and the work of San and Huang [5].

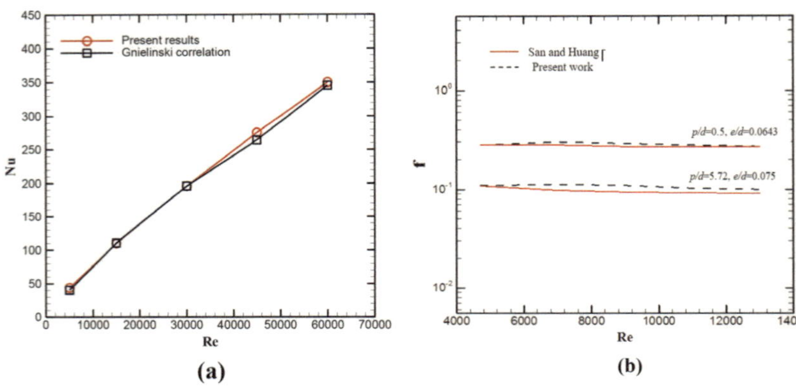

Figure 2.
(a) Numerical Nu of the present work and that obtained from Gnielinski's correlation [17] and (b) Numerical f and that of San and Huang [5].

3. Results and discussion

The flow-thermal features of turbulent flow in corrugated channels are evaluated numerically. The enhanced heat transfer and an accompanied pressure loss are assessed for corrugated channels under the influences of rib arrangement, rib configuration, rib roughness parameters, and rib shapes. The dimensionless parameters Nu, f, and *PEC* through a wide range of *Re* are presented to assess the performance of corrugated channels.

3.1 The effect of rib arrangements

Corrugated channels exist in three layouts depending on rib arrangements, IOCC, ICC, and OCC, as described in **Figure 1a**. The variations of *Nu* and f with the *Re* number of all rectangular rib arrangements of corrugated channels and smooth one are presented in **Figure 3a** and **b**, respectively. The rate of heat that is transferred in corrugated channels is higher than that of the smooth channel. The heat transfer varies insignificantly with the rib distribution at the low *Re*. The rib distribution experiences a pronounced influence on the Nusselt number when *Re* increases. The ICC shows a maximum ability to exchange the heat, while the OCC has a lower thermal performance. On the other hand, there is an additional pressure loss associated with corrugated channels compared with smooth ones as exhibited in **Figure 3b**. The friction factor decreases slightly with the *Re*. Also, the OCC has a

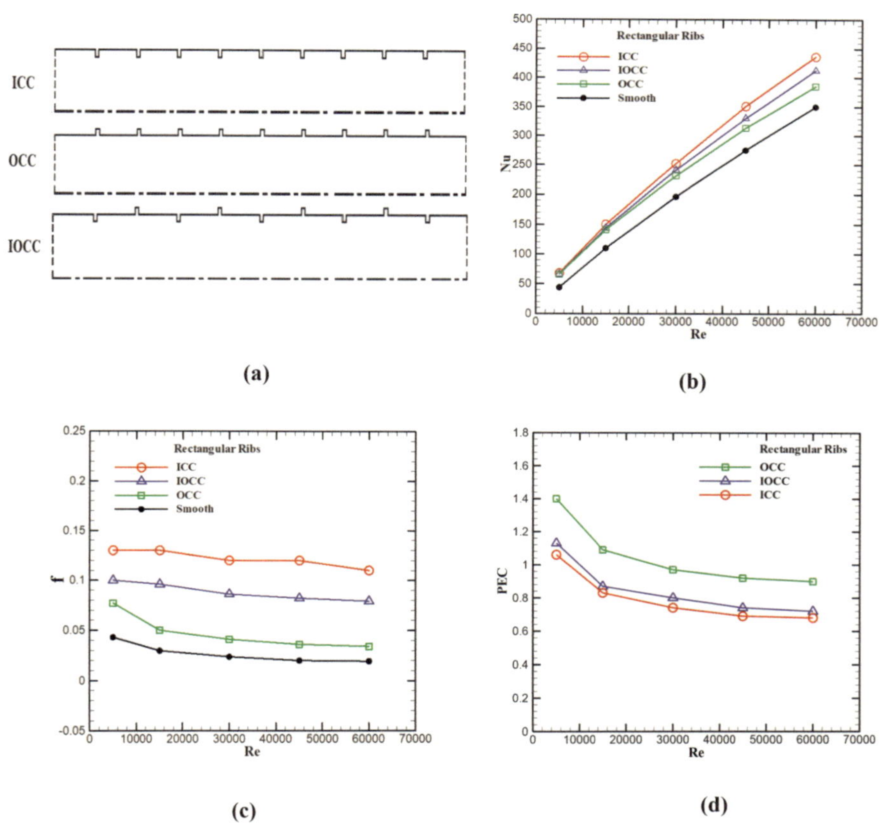

Figure 3.
(a) Different rib arrangements of corrugated channels and the influence of rib configuration on Nu, f, and PEC as described in (b), (c), and (d), respectively, for the different values of Re.

minimum friction factor, while the ICC owns a maximum pressure loss. Moreover, the performance evaluation criterion (PEC) varies inversely with the Re as exhibited in **Figure 3c**. The increase in pressure loss exceeds the enhancement in the heat transfer for all corrugated channel layouts. Also, OCC has higher PEC than both IOCC and ICC channels. This is due to the increase in *f* of OCC is lower than that of ICC and IOCC. Even though, both ICC and IOCC have higher *Nu* than IOCC.

3.2 The influence of rib configurations

Seven configurations of rib trapezoidal corrugated channels are denoted (B1, B2, C1, C2, C3, D1, and D2) which are presented in **Figure 4a**. Also, the smooth channel is indicated by A. The variation of the Nusselt number for all channels is depicted in **Figure 4b**. The increase in *Re* and flow velocity causes enhancement in mixing the

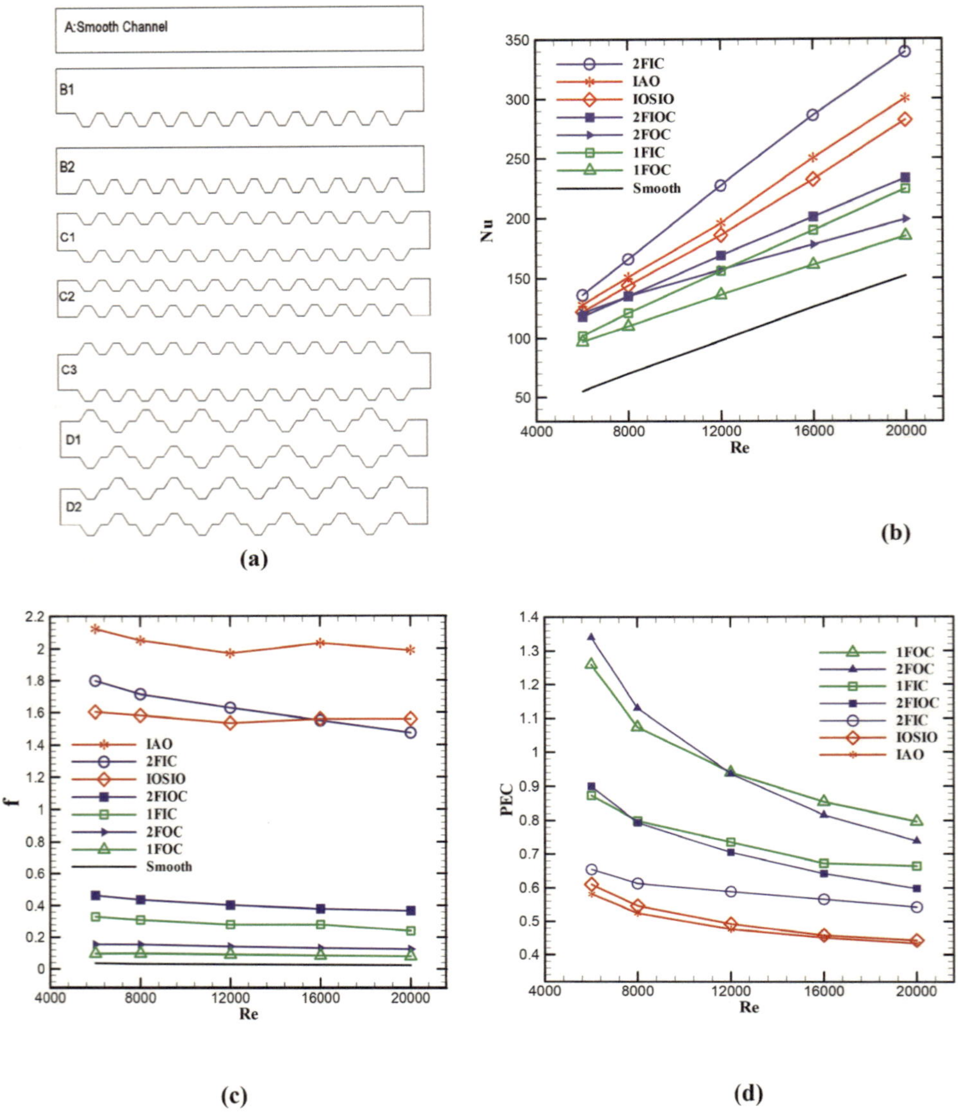

Figure 4.
(a) Different configurations of corrugated channels and the influence of rib configuration on Nu, f, and PEC as depicted in (a), (b), and (c), respectively, for the different values of Re.

rate between the core flow and recirculating flow. Thus, the heat exchange between the heating wall and the flow is enhanced. On the other hand, f is higher for corrugated channels than the smooth one as revealed in **Figure 4c**. In one side, the results revealed that the heat is transferred more effectively in the corrugated channel than the smooth one due to the additional surface area, suppressing the boundary layer thickness associated with corrugated channels. On the other side, the corrugation results in a substantial flow recirculation and separation and an extra surface area, and thus it creates higher pressure drop. The corrugated channel C1 registers the highest *Nu*, while the minimum *Nu* is achieved for corrugated channel B1. Conversely, the results exhibit that the minimum pressure drop is registered for B1 configuration channel among other corrugated channels. Moreover, the influence of rib configuration of corrugated channels on the *PEC* is presented in **Figure 4d**. The results reveal that there is a monotonic decrease of *PEC* with the *Re*. The optimum performance is accomplished at the lower *Re*. As *Re* increases the conflict between the augmentation in thermal performance and degradation in pressure drop is initiated. The higher values of *PEC* are obtained for C3 and B1 corrugated channels, whereas D1 and D2 configurations have the minimum values of *PEC*.

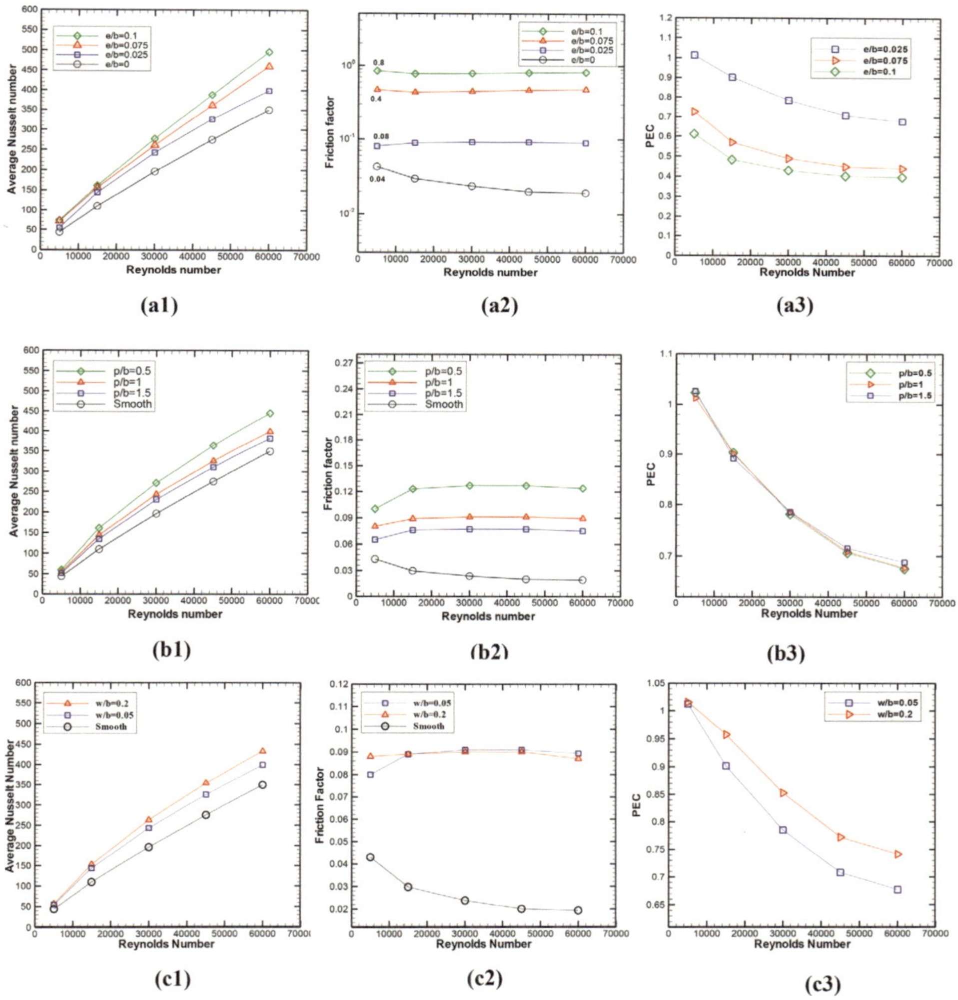

Figure 5.
Nu, f, and PEC for different (a) rib heights, (b) rib pitches, and (c) rib widths.

3.3 The impact of rib roughness parameters

The roughness parameters of corrugated channels involve relative rib height (*e/b*), relative rib pitch (*p/b*), and relative rib width (*w/b*) as illustrated later in **Figure 6a**. The impact of roughness parameters on the thermal-flow behavior of corrugated channels is presented in **Figure** 5. The computed *Nu*, f, and PEC are tested for different relative roughness heights which are presented in **Figure 5a1, 5a2**, and **5a3**, respectively, with constant values of *p/b* and *w/b*. Generally, corrugated channels have higher *Nu* than a smooth channel. It is observed that the Nusselt number increases monotonically with both rib height and *Re*. But there is a relatively small effect of rib height on the *Nu* at lower values of *Re*. At the same time, the friction factor varies positively with the relative rib height. While, there is an insignificant effect of *Re* on *f*, the variation of *PEC* (**Figure 5a3**) confirms that the diverse effect of friction factor exceeds the enhancement in transferred heat especially with an increase of *Re*. The influence of rib pitch of corrugation on *Nu*, *f*, and *PEC* of corrugated channels is illustrated in **Figure 5b1, 5b2**, and **5b3**, respectively, for constant corrugation height and width. Decreasing the pitch results in an increase in the number of ribs for unit length and excites the secondary flow. Therefore, the thickness of boundary layer is decreased, and the rate of heat transfer is augmented. However, the flow impedance is increased due to the increase in the number of roughness elements which add extra friction to the flow stream. It appears that the influence of corrugation pitch is insignificant on the *PEC* as

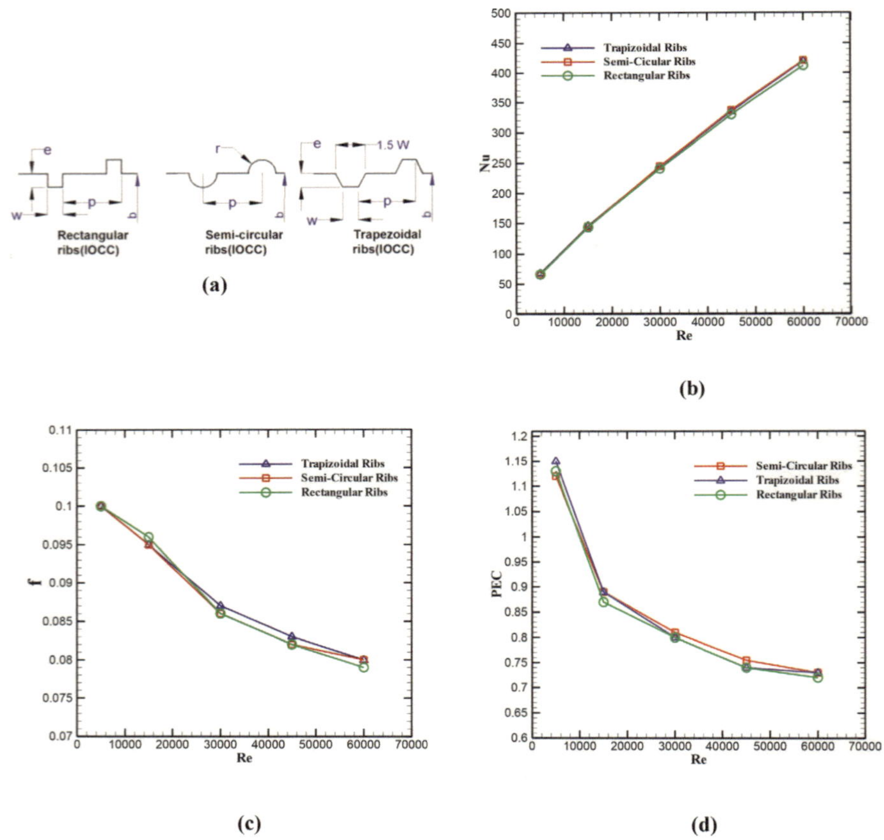

Figure 6.
(a) Different rib shapes of IOCC channels and the influence of rib shapes on the Nu, f, and PEC as presented in (b), (c), and (d), respectively, for the different values of the Re.

presented in **Figure 5b3**. In a similar way, the influences of two values of rib width on the performance of corrugated channel are shown in **Figure 5c**. As the rib width increases, the secondary flow becomes more intense. Therefore, there is a mutual increase in *Nu* and *f* as depicted in **Figure 5c1** and **5c2**, respectively. Furthermore, the *PEC* shows a monotonic decrease with the rib width and *Re* as described by **Figure 5c3**.

3.4 The influence of rib shape

The heat transfer-flow behavior of IOCC channel, for example, is examined for rectangular, semicircular, and trapezoidal rib shapes. The different shapes of the rib are illustrated in **Figure 6a**, while the *Nu*, f, and *PEC* for various rib shapes are presented in **Figure 6b, c** and **d**, respectively, for (p/b = 1, e/b = 0.025, and w/b = 0.05). It is found that the influence of the roughness shape is small on the performance of corrugated channels.

4. Conclusion

The computational investigation of thermal-flow performance of turbulent flow in corrugated channels is carried out for the Reynolds number from 5000 to 60,000. The effects of rib arrangements, rib configurations, rib roughness parameters, and rib shapes are investigated. All layouts of corrugated channels showed a superior ability of exchange heat than that experienced by smooth channel. However, the pressure loss associated with corrugated channels is higher than that of the smooth ones. Furthermore, it is inferred that the arrangement of rib distribution, rib configuration, and rib roughness parameters has a pronounced effect on the thermal-flow performance of corrugated channels, while the influence of rib shapes seems to be small.

Author details

Nabeel S. Dhaidana* and Abdalrazzaq K. Abbas
Department of Mechanical Engineering, College of Engineering, Kerbala University, Kerbala, Iraq

*Address all correspondence to: engnab74@yahoo.com

References

[1] Léal L, Miscevic M, Lavieille P, Amokrane M, Pigache F, Topin F, et al. An overview of heat transfer enhancement methods and new perspectives: Focus on active methods using electroactive materials. International Journal of Heat and Mass Transfer. 2013;**61**:505-524. DOI: 10.1016/j.ijheatmasstransfer.2013.01.083

[2] Liu S, Sakr M. A comprehensive review on passive heat transfer enhancements in pipe exchangers. Renewable and Sustainable Energy Reviews. 2013;**19**:64-81. DOI: 10.1016/j.rser.2012.11.021

[3] Alamgholilou A, Esmaeilzadeh E. Experimental investigation on hydrodynamics and heat transfer of fluid flow into channel for cooling of rectangular ribs by passive and EHD active enhancement method. Experimental Thermal and Fluid Science. 2012;**38**:61-73. DOI: 10.1016/j.expthermflusci.2011.11.008

[4] Vicente PG, Garcia A, Viedma A. Experimental investigation on heat transfer and frictional characteristics of spirally corrugated tubes in turbulent flow at different Prandtl numbers. International Journal of Heat and Mass Transfer. 2004;**47**:671-681. DOI: 10.1016/j.ijheatmasstransfer.2003.08.005

[5] San JY, Huang WC. Heat transfer enhancement of transverse ribs in circular tubes with consideration of entrance effect. International Journal of Heat and Mass Transfer. 2006;**49**: 2965-2971. DOI: 10.1016/j.ijheatmasstransfer.2006.01.046

[6] Li XW, Meng JA, Li ZX. Roughness enhanced mechanism for turbulent convective heat transfer. International Journal of Heat and Mass Transfer. 2011; **54**:1775-1781. DOI: 10.1016/j.ijheatmasstransfer.2010.12.039

[7] Dizaji HS, Jafarmadar S, Mobadersani F. Experimental studies on heat transfer and pressure drop characteristics for new arrangements of corrugated tubes in a double pipe heat exchanger. International Journal of Thermal Sciences. 2015;**6**:211-220. DOI: 10.1016/j.ijthermalsci.2015.05.009

[8] Dhaidan NS, Abbas AR. Turbulent forced convection flow inside inward-outward rib corrugated tubes with different rib-shapes. Heat Transfer - Asian Research. 2018:1-13. DOI: 10.1002/htj.21365

[9] Naphon P. Heat transfer characteristics and pressure drop in the channel with V corrugated upper and lower plates. Energy Conversion and Management. 2007;**48**:1516-1524. DOI: 10.1016/j.enconman.2006.11.020

[10] Eiamsa-ard S, Promvonge P. Thermal characteristics of turbulent rib–grooved channel flows. International Communications in Heat and Mass Transfer. 2009;**36**:705-711. DOI: 10.1016/j.icheatmasstransfer.2009.03.025

[11] Elshafei EAM, Awad MM, El-Negiry E, Ali AG. Heat transfer and pressure drop in corrugated channels. Energy. 2010;**35**:101-110. DOI: 10.1016/j.energy.2009.08.031

[12] Mohammed HA, Abed AM, Wahid MA. The effects of geometrical parameters of a corrugated channel within out-of-phase arrangement. International Communications in Heat and Mass Transfer. 2013;**40**:47-57. DOI: 10.1016/j.icheatmasstransfer.2012.10.022

[13] Li Z, Gao Y. Numerical study of turbulent flow and heat transfer in cross-corrugated triangular ducts with delta-shaped baffles. International Journal of Heat and Mass Transfer. 2017;

108:658-670. DOI: 10.1016/j.ijheatmasstransfer.2016.12.054

[14] Pehlivan H, Taymaz I, İslamoğlu Y. Experimental study of forced convective heat transfer in a different arranged corrugated channel. International Communications in Heat and Mass Transfer. 2013;**46**:106-111. DOI: 10.1016/j.icheatmasstransfer.2013.05.016

[15] Ravi BV, Singh P, Ekkad SV. Numerical investigation of turbulent flow and heat transfer in two-pass ribbed channels. International Journal of Thermal Sciences. 2017;**112**:31-43. DOI: 10.1016/j.ijthermalsci.2016.09.034

[16] Shubham SA, Dalala A, Pati S. Thermo-hydraulic transport characteristics of non-Newtonian fluid flows through corrugated channels. International Journal of Thermal Sciences. 2018;**129**:201-208. DOI: 10.1016/j.ijthermalsci.2018.02.005

[17] Launder BE, Spalding DB. Mathematical Models of Turbulence. New York: Academic Press; 1972

[18] Gnielinski V. New equations for heat and mass transfer in turbulent pipe and channel flow. International Chemical Engineering. 1976;**16**:359-368

Section 5

Numerical Methods in Boundary Layer Theory

Chapter 9

Singularly Perturbed Parabolic Problems

Asan Omuraliev and Ella Abylaeva

Abstract

The aim of this work is to construct regularized asymptotic of the solution of a singularly perturbed parabolic problems. Namely, in the first paragraph, we consider the case when the scalar equation contains a free term consisting of a finite sum of the rapidly oscillating functions. In the first paragraph, it is shown that the asymptotic solution of the problem contains parabolic, power, rapidly oscillating, and angular boundary layer functions. Angular boundary layer functions have two components: the first one is described by the product of a parabolic boundary layer function and a boundary layer function, which has a rapidly oscillating change. The second section is devoted to a two-dimensional equation of parabolic type. Asymptotic of the scalar equation contains a rapidly oscillating power, parabolic boundary layer functions, and their product; then, the multidimensional equation additionally contains a multidimensional composite layer function.

Keywords: singularly perturbed parabolic problem, asymptotic, stationary phase, power boundary layer, parabolic boundary layer, angular boundary layer

1. Asymptotics of the solution of the parabolic problem with a stationary phase and an additive-free member

1.1 Introduction

Singularly perturbed problems with rapidly oscillating free terms were studied in [1–3]. Ordinary differential equations with a rapidly oscillating free term whose phase does not have stationary points are studied in [1]. Using the regularization method for singularly perturbed problems [4], differential equations of parabolic type with a small parameter were studied in [2, 3] when fast-oscillating functions as a free member. The asymptotic solutions constructed in [1–3] contain a boundary layer function having a rapidly oscillating character of change. In addition to such a boundary layer function, ordinary differential equations contain an exponential [1], and parabolic equations - parabolic [2, 3] and angular boundary layer [2, 5] functions. If the phase of the free term has stationary points, then boundary layers arise additionally, having a power character of change. In this case, the asymptotic solution consists of regular and boundary layer terms. The boundary layer members are parabolic, power, rapidly oscillating boundary layer functions, and their products, which are called angular boundary layer functions [4]. In this chapter we used the methods of [4, 5].

1.2 Statement of the problem

In this chapter we study the following problem:

$$L_\varepsilon u\ (x,t,\varepsilon) \equiv \partial_t u - \varepsilon^2 a(x)\partial_x^2 u - b(x,t)u = \sum_{k=1}^{N} f_\kappa(x,t)\exp\left(\frac{i\theta_\kappa(t)}{\varepsilon}\right),\ (x,t)\in\Omega, \tag{1}$$

$$u(x,t,\varepsilon)|_{t=0} = u(x,t,\varepsilon)|_{x=0} = u(x,t,\varepsilon)|_{x=1} = 0$$

where $\varepsilon > 0$ is a small parameter and $\Omega = \{(x,t): x\in(0,1), t\in(0,T]\}$.
The problem is solved under the following assumptions:

1. $a(x) > 0,\ a(x)\in C^\infty[0,1],\ b(x,t),\ f\ (x,t)\in C^\infty(\overline{\Omega})$.

2. $\forall x\in[0,1]$ function $a(x) > 0$.

3. $\theta_k'(t)\big|_{t=0} = 0$ is the phase function.

1.3 Regularization of the problem

For the regularization of problem (Eq. (1)), we introduce regularizing independent variables using methods [5, 6]:

$$\eta = \frac{t}{\varepsilon^2},\ r_k = \frac{i\,[\theta_k(t) - \theta_k(0)]}{\varepsilon},\ \xi_\nu = \frac{\varphi_\nu(x)}{\varepsilon},\ i = \sqrt{-1},$$

$$\zeta_\nu = \frac{\varphi_\nu(x)}{\varepsilon^2},\ \varphi_\nu(x) = (-1)^{\nu-1}\int_{\nu-1}^{x}\frac{ds}{\sqrt{a(s)}},\quad \nu = 1,2, \tag{2}$$

$$\sigma_k = \int_0^t \exp\left(\frac{i\,[\theta_\kappa(s) - \theta_\kappa(0)]}{\varepsilon}\right)ds \equiv p_\kappa(t,\varepsilon),\ l = \overline{0,r},\ j = \overline{0,k_l - 1}$$

Instead of the desired function $u(x,t,\varepsilon)$, we will study the extended function

$$\check{u}(M,\varepsilon),\ M = (x,t,r,\eta,\sigma,\xi,\zeta),\ \sigma = (\sigma_1,\sigma_2 ...\sigma_N),\ r = (r_1,r_2 ...r_N),\ \xi = (\xi_1,\xi_2),\ \zeta = (\zeta_1,\zeta_2)$$

such that its restriction by regularizing variables coincides with the desired solution:

$$\check{u}(M,\varepsilon)|_{\gamma=p(x,t,\varepsilon)} \equiv u(x,t,\varepsilon)$$
$$\gamma = (r,\sigma,\eta,\xi,\zeta) \tag{3}$$

Taking into account (Eqs. (21)) and ((3)), we find the derivatives
On the basis of (Eqs. (1)–(4)) for the extended function $\check{u}(M,\varepsilon)$, we set the problem:

$$\partial_t u(x,t,\varepsilon) \equiv (\partial_t\check{u}\ (M,\varepsilon) + \tfrac{1}{\varepsilon^2}\partial_\eta\check{u}(M,\varepsilon) + \sum_{k=1}^{N}[\frac{i\theta_k{}'(t)}{\varepsilon}\partial_{r_k}\check{u}(M,\varepsilon) + \exp(r_k)\partial_{\sigma_k}\check{u}\ (M,\varepsilon))\Big|_{\gamma=p(x,t,\varepsilon)},$$

$$\partial_x u(x,t,\varepsilon) \equiv \left((\partial_x\check{u}\ (M,\varepsilon) + \sum_{\nu=1}^{2}\left\{\frac{\varphi_\nu'(x)}{\varepsilon}\partial_{\xi_\nu}\check{u}\ (M,\varepsilon) + \frac{\varphi_\nu'(x)}{\varepsilon^2}\partial_{\zeta_\nu}\check{u}\ (M,\varepsilon)\right\}\right)\Big|_{\gamma=p(x,t,\varepsilon)}, \tag{4}$$

$$\partial_x^2 u(x,t,\varepsilon) \equiv \left((\partial_x^2 \breve{u}\,(M,\varepsilon) + \sum_{\nu=1}^{2}\left\{ \left(\frac{\varphi_\nu'(x)}{\varepsilon}\right)^2 \partial_{\xi_\nu}^2 \breve{u}\,(M,\varepsilon) + \left(\frac{\varphi_\nu'(x)}{\varepsilon^2}\right)^2 \partial_{\zeta_\nu}^2 \breve{u}\,(M,\varepsilon) + \frac{1}{\varepsilon} D_{\xi,v}\breve{u}\,(M,\varepsilon) + \frac{1}{\varepsilon^2} D_{\zeta,v}\breve{u}(M,\varepsilon) \right\} \right)\Bigg|_{\gamma=p(x,t,\varepsilon)},$$

$$D_{\xi,v} \equiv 2\varphi_\nu'(x)\partial^2_{x,\xi_\nu} + \varphi_\nu''(x)\partial_{\xi_\nu},$$

$$D_{\zeta,v} \equiv 2\varphi_\nu'(x)\partial^2_{x,\zeta_\nu} + \varphi_\nu''(x)\partial_{\zeta_\nu}.$$

$$\widetilde{L}_\varepsilon \breve{u}(M,\varepsilon) \equiv \frac{1}{\varepsilon^2} T_0\,\breve{u}(M,\varepsilon) + \sum_{k=1}^{N} \frac{i\theta_k'(t)}{\varepsilon}\partial_{r_k}\breve{u}(M,\varepsilon) + T_1\,\breve{u}(M,\varepsilon)$$

$$= \sum_{k=1}^{N} f_\kappa(x,t)\exp\left(r_k + \frac{i\theta_\kappa(0)}{\varepsilon}\right) + L_\zeta\,\breve{u}(M,\varepsilon) + \varepsilon L_\xi \breve{u}(M,\varepsilon) + \varepsilon^2 L_x \breve{u}(M,\varepsilon)$$

$$\breve{u}(M,\varepsilon)\big|_{t=r_k=\eta=0} = \breve{u}(M,\varepsilon)\big|_{x=0,\,\xi_1=\zeta_1=0} = \breve{u}(M,\varepsilon)\big|_{x=1,\,\xi_2=\zeta_2=0} = 0,$$

$$T_1 \equiv \partial_\eta - \sum_{\nu=1}^{2}\partial^2_{\zeta_\nu},$$

$$T_2 \equiv \partial_t - \sum_{\nu=1}^{2}\partial^2_{\xi_\nu} - b(x,t) + \sum_{k=1}^{N}\exp(r_k)\partial_{\sigma_k}, \tag{5}$$

$$L_\xi \equiv a(x)\sum_{v=1}^{2} D_{\xi,v},$$

$$L_\zeta \equiv a(x)\sum_{v=1}^{2} D_{\zeta,v},$$

$$L_x \equiv a(x)\partial_x^2.$$

The problem (Eq. (5)) is regular in ε as $\varepsilon \to 0$:

$$\left(\widetilde{L}_\varepsilon \breve{u}(M,\varepsilon)\right)\Big|_{q=q(x,t,\varepsilon)} \equiv L_\varepsilon \breve{u}(x,t,\varepsilon). \tag{6}$$

1.4 Solution of iterative problems

The solution of problem (Eq. (5)) will be determined in the form of a series:

$$\breve{u}(\,M,\varepsilon) = \sum_{v=0}^{\infty} \varepsilon^v u_v(M). \tag{7}$$

For the coefficients of this series, we obtain the following iterative problems:

$$T_1 u_0(M) = 0,\ T_1 u_1(M) = -i\sum_{k=1}^{N}\theta_k'(t)\partial_{r_k}u_0(M),$$

$$T_1 u_2(M) == -i\sum_{k=1}^{N}\theta_k'(t)\partial_{r_k}u_1(M) - T_2 u_0(M) + \sum_{k=1}^{N} f_\kappa(x,t)\exp\left(r_k + \frac{i\theta_\kappa(0)}{\varepsilon}\right) + L_\zeta\,u_0(M),$$

$$T_1 u_v(M) = -i\sum_{k=1}^{N}\theta_k'(t)\partial_{r_k}u_{v-1}(M) - T_2 u_{v-2}(M) + L_\zeta u_{v-2} + L_\xi u_{v-3}(M) + L_x u_{v-4}(M). \tag{8}$$

The solution of this problem contains parabolic boundary layer functions; internal power boundary layer functions which are connected with a rapidly oscillating free term in a phase which are vanished at $t = t_l$, $l = 0, 1, \ldots\ldots, n$ in addition; and the asymptotic also contain angular boundary layer functions. We introduce a class of functions in which the iterative problems will be solved:

$$G_0 \cong C^\infty(\overline{\Omega}),\ G_1 = \left\{u(M) : u(M) = \oplus_{l=1}^{2} G_0 \otimes \operatorname{erfc}\left(\frac{\xi_l}{2\sqrt{t}}\right)\right\},$$

$$G_2 = \left\{u(M) : u(M) = \oplus_{k=1}^{N} G_0 \otimes \exp(r_k)\right\},$$

$$G_3 = \left\{u(M) : u(M) = \oplus_{k=1}^{N} \oplus_{l=1}^{2} Y_k^l(N_l) \otimes \exp(r_k), \left\|Y_k^l(N_l)\right\| < c\ exp\left(-\frac{\varsigma_l^2}{8\eta}\right)\right\},$$

$$G_4 = \left\{u(M) : u(M) = \oplus_{k=1}^{N} G_0\left(\oplus_{l=1}^{2} G_0 \otimes \operatorname{erfc}\left(\frac{\xi_l}{2\sqrt{t}}\right)\right)\sigma_k\right\}, N_l = (x, t, \eta, \varsigma_1, \varsigma_2).$$

From these spaces we construct a new space:

$$G = \oplus_{l=0}^{4} G_l.$$

The element $u(M) \epsilon G$ has the form:

$$\begin{aligned} u(M) &= v(x,t) + \sum_{l=1}^{2} w^l(x,t) \operatorname{erfc}\left(\frac{\xi_l}{2\sqrt{t}}\right) \\ &+ \sum_{k=1}^{N}\left[c_k(x,t) + \sum_{l=1}^{2} Y_k^l(N_l)\right]\exp(r_k) \\ &+ \sum_{k=1}^{N}\left[z_k(x,t) + \sum_{l=1}^{2} q_k^l(x,t) \operatorname{erfc}\left(\frac{\varsigma_l^2}{2\sqrt{t}}\right)\right]\sigma_k. \end{aligned} \tag{9}$$

1.5 Solvability of intermediate tasks

The iterative problems (Eq. (9)) in general form will be written:

$$T_1 u(M) = H(M). \tag{10}$$

Theorem 1. Suppose that the conditions (1)–(3) and $H(M) \epsilon G_3$ are satisfied. Then, equation (Eq. (10)) is solvable in G.

Proof. Let the free term $H(M) \epsilon G_3$ be representable in the form:

$$H(M) = \sum_{k=1}^{N}\sum_{l=1}^{2} H_k^l(N_l),\ \left\|H_k^l(N_l)\right\| < c\ \exp\left(\frac{\varsigma_l^2}{8\eta}\right).$$

Then, by directly substituting function $u(M) \epsilon\ G$ from (Eq. (9)) in (Eq. (10)), we see that this function is a solution if and only if the function $Y_k^l(N_l)$ will be a solution of equation:

$$\partial_\eta Y_k^l(N_l) = \partial_{\varsigma_l}^2 Y_k^l(N_l) + H_k^l(N_l), l = 1, 2, k = 1, 2, \ldots, N. \tag{11}$$

With the corresponding boundary conditions, this equation has a solution which have the estimate:

$$\left\|Y_k^l(N_l)\right\| < c\ exp\left(\frac{\varsigma_l^2}{8\eta}\right).$$

The theorem is proven.

Theorem 2. Suppose that the conditions of *Theorem 1* are satisfied. Then, under additional conditions:

1.
$$u(M)\big|_{t=\eta=0}=0,\, u(M)\big|_{x=l-1,\,\xi_l=0,\,\varsigma_l=0}=0,\, l=1,2.$$

2.
$$L_{\varsigma}u(M)=0,\, L_{\xi}u(M)=0.$$

3.
$$i\sum_{k=1}^{N}\theta_k'(t)\partial_{r_k}u_v(M)+T_2u_{v-1}(M)+h(M)\in G_3.$$

Eq. (10) is uniquely solvable.

Proof. By *Theorem 1* equation (Eq. (10)) has a solution that is representable in the form (Eq. (9)). With satisfying condition (1), we obtain

$$v(x,t)\Big|_{t=0}=-\sum_{k=1}^{N}c_k(x,0),\, w^l(x,t)\Big|_{t=0}=\overline{w}^l(x), \tag{12}$$

$$Y_k^l(N_l)\big|_{t=\eta=0}=0,\, q_k^l(x,t)\big|_{t=0}=\overline{q}_k^l(x),\;\, d_k^l(x,t)\Big|_{t=0}=\overline{d}_k^l(x),$$

$$w^l(x,t)\big|_{x=l-1}=-c_k(l-1,t),\, q_k^l(x,t)\big|_{x=l-1}=-z_k(l-1,t),\, l=1,2.$$

Due to the fact that the function $\operatorname{erfc}\left(\frac{\theta}{2\sqrt{t}}\right)$ is zero at $\theta=0$, the values for $w^l(x,t)\big|_{t=0}, q_k^l(x,t)\big|_{t=0}$ are chosen arbitrarily.

We calculate

$$\begin{aligned}
&i\sum_{k=1}^{N}\theta_k'(t)\partial_{r_k}u_v(M)+T_2u_{v-1}(M)+h(M)\\
&=i\sum_{k=1}^{N}\theta_k'(t)\left[c_{k,v}(x,t)+\sum_{l=1}^{2}Y_{k,v}^l(N_l)\right]\exp(r_k)+\left[\partial_t v_{v-1}(x,t)-b(x,t)v_{v-1}(x,t)\right]\\
&+\sum_{l=1}^{2}\left[\partial_t w_{v-1}^l(x,t)-b(x,t)w_{v-1}^l(x,t)\right]\operatorname{erfc}\left(\frac{\xi_l}{2\sqrt{t}}\right)\\
&+\sum_{k=1}^{N}\left[\partial_t c_{k,v-1}(x,t)-b(x,t)c_{k,v-1}(x,t)+\sum_{l=1}^{2}\left(\partial_t Y_{k,v-1}^l(N_l)-b(x,t)Y_{k,v-1}^l(N_l)\right)\right]\exp(\tau_k)\\
&+\sum_{k=1}^{N}\left\{\partial_t z_{k,v-1}(x,t)-(x,t)z_{k,v-1}(x,t)+\sum_{l=1}^{2}\left[\partial_t q_{k,v-1}^l(x,t)-b(x,t)q_{k,v-1}^l(x,t)\right]\operatorname{erfc}\left(\frac{\xi_l}{2\sqrt{t}}\right)\right\}\sigma_k\\
&+\sum_{k=1}^{N}\left[z_{k,v-1}(x,t)+\sum_{l=1}^{2}q_{k,v-1}^l(x,t)\operatorname{erfc}\left(\frac{\xi_l}{2\sqrt{t}}\right)\right]\exp(\tau_k)+h_0(x,t)+\sum_{l=1}^{2}h_1^l(x,t)\operatorname{erfc}\left(\frac{\xi_l}{2\sqrt{t}}\right)\\
&+\sum_{k=1}^{N}\left[h_2^k(x,t)+\sum_{l=1}^{2}h_2^{l,k}(x,t)\right]\exp(\tau_k)+\sum_{k=1}^{N}\left[h_3^k(x,t)+\sum_{l=1}^{2}h_3^{l,k}(x,t)\operatorname{erfc}\left(\frac{\xi_l}{2\sqrt{t}}\right)\right]\sigma_k.
\end{aligned} \tag{13}$$

Condition (3) of the theorem will be ensured, if we choose arbitrarily (Eq. (9)) as the solutions of the following equations:

$$\begin{gathered}\partial_t v_{v-1}(x,t) - b(x,t)v_{v-1}(x,t) = -h_0(x,t),\\ \partial_t w^l_{v-1}(x,t) - b(x,t)w^l_{v-1}(x,t) = -h^l_1(x,t),\\ \partial_t Y^l_{k,v-1}(N_l) - b(x,t)Y^l_{k,v-1}(N_l) = -\left(h^{l,k}_2(x,t) + q^l_{k,v-1}(x,t)\operatorname{erfc}\left(\frac{\varsigma_l}{2\sqrt{\eta}}\right)\right),\\ \partial_t z_{k,v-1}(x,t) - b(x,t)z_{k,v-1}(x,t) = -h^k_3(x,t),\\ \partial_t q^l_{k,v-1}(x,t) - b(x,t)q^l_{k,v-1}(x,t) = -h^{l,k}_3(x,t),\\ i\theta'_k(t)c_{k,v}(x,t) = -z_{k,v-1}(x,t) - [\partial_t c_{k,v-1}(x,t) - b(x,t)c_{k,v-1}(x,t)] - h^k_2(x,t).\end{gathered} \tag{14}$$

After this choice of arbitrariness, expression (Eq. (13)) is rewritten:

$$i\sum_{k=1}^{N}\theta'_k(t)\partial_{r_k}u_v(M) + T_2u_{v-1}(M) + h(M) = \sum_{k=1}^{N}\sum_{l=1}^{2}\left[i\theta'_k(t)Y^l_{k,v}(N_l)\right]\exp(\tau_k)\in G_3$$

In (Eq. (14)), transition was made from $\xi_l/2\sqrt{t}$ to variable $\varsigma_l/2\sqrt{\eta}$. The function $Y^l_k(N_l)$ is defined as the solution of equation (Eq. (30)) under the boundary conditions from (Eq. (12)) in the form:

$$Y^l_k(N_l) = d^l_k(x,t)\operatorname{erfc}\left(\frac{\varsigma_l}{2\sqrt{\eta}}\right) + \frac{1}{2\sqrt{\pi}}\int_0^{\eta}\int_0^{\infty}\frac{H^l_k(\cdot)}{\sqrt{\eta-\tau}}\left[\exp\left(-\frac{(\varsigma_l-y)^2}{4(\eta-\tau)}\right) - \exp\left(-\frac{(\varsigma_l+y)^2}{4(\eta-\tau)}\right)\right]dyd\tau. \tag{15}$$

We substitute this function in the corresponding equation from (Eq. (14)); then with respect to $d^l_k(x,t)$, we obtain a differential equation, which is solving under the initial condition $d^l_k(x,t)\big|_{t=0} = \overline{d}^l_k(x)$, and we find

$$d^l_k(x,t) = \overline{d}^l_k(x,t)B(x,t) + P^l_k(x,t),\ B(x,t) = \exp\left(\int_0^t b(x,s)ds\right),$$

where $P^l_k(x,t)$ is known as the function.

By substituting the obtained function into condition for $d^l_k(x,t)\big|_{x=l-1}$ from (Eq. (12)), we define the value of $\overline{d}^l_k(x)\big|_{x=l-1}$. The obtained value is used as an initial condition for a differential equation with respect to $\overline{d}^l_k(x)$, which is obtained after substitution $d^l_k(x,t)$ into the first condition of (2). With that we ensure fulfillment of this condition and uniqueness of the function $Y^l_k(N_l)$. The last equation from (Eq. (14)) due to the fact that $\theta'_k(t_k) = 0$ is solvable if

$$z^l_{k,v-1}(x,0) = -h^k_2(x,0) - [\partial_t c_{k,v-1}(x,t) - b(x,t)c_{k,v-1}(x,t)]\Big|_{t=0}.$$

The obtained ratio is used as the initial condition for the differential equation with respect to $z^l_{k,v-1}(x,t)$ from (Eq. (14)).

The equation with respect to $v_{v-1}(x,t)$ under the initial condition from (12) determines this function uniquely. Equations with respect to $w^l_{k,v-1}(x,t)$, $q^l_{k,v-1}(x,t)$ under the corresponding condition from (Eq. (12)) have solutions representable in the form:

$$w^l_{k,v-1}(x,t) = \overline{w}^l_{k,v-1}(x)B(x,t) + H^l_{1,v-1}(x,t),$$
$$q^l_{k,v-1}(x,t) = \overline{q}^l_{k,v-1}(x)B(x,t) + H^l_{2,v-1}(x,t) \tag{16}$$

where $H^l_{1,v-1}(x,t)$, $H^l_{2,v-1}(x,t)$- are known functions.

With substituting (Eq. (16)) into the conditions under $x = l-1$ from (Eq. (12)), we define values of $\overline{w}^l_{k,v-1}(x)\big|_{x=l-1}, \overline{q}^l_{k,v-1}(x)\big|_{x=l-1}$. These conditions are used in solving differential equations which are obtained from the second condition of (Eq. (21)):

$$L_\xi\left(w^l_{k,v-1}(x,t)\text{erfc}\left(\frac{\xi_l}{2\sqrt{t}}\right)\right) = 0,\ L_\xi\left(q^l_{k,v-1}(x,t)\text{erfc}\left(\frac{\xi_l}{2\sqrt{t}}\right)\right) = 0.$$

Thus, function $u(M)$ is determined uniquely. The theorem is proven.

1.6 Solution of iterative problems

Eq. (8) is homogeneous for k = 0; therefore, by *Theorem 1*, it has a solution in G, representable in the form:

$$u_0(M) = v_0(x,t) + \sum_{l=1}^{2} w^l(x,t)\text{erfc}\left(\frac{\xi_l}{2\sqrt{t}}\right)$$
$$+ \sum_{k=1}^{N}\left\{\left(c_{k,0}(x,t) + \sum_{l=1}^{2} Y^l_{k,0}(N_l)\right)e^{r_k} + \left[z_{k,0}(x,t) + \sum_{l=1}^{2} q^l_{k,0}(x,t)\text{erfc}\left(\frac{\xi_l}{2\sqrt{t}}\right)\right]\sigma_k\right\} \tag{17}$$

If the function $Y^l_{k,0}(N_l)$ is the solution of the equation $\partial_\eta Y^l_{k,0}(N_l) = \partial^2_{\varsigma_l} Y^l_{k,0}(N_l)$ which is satisfying that

$$Y^l_{k,0}(N_l)\big|_{t=\eta=0} = 0,\ Y^l_{k,0}(N_l)\big|_{x=l-1,\,\varsigma_l=0} = -c_{k,0}(l-1,t).$$

from the last problem, we define

$$Y^l_{k,0}(N_l) = d^l_{k,0}(x,t)\ \text{erfc}\left(\frac{\varsigma_l}{2\sqrt{\eta}}\right),\ d^l_{k,0}(x,t)\big|_{x=l-1} = -c_{k,0}(l-1,t),\ \text{where}\ d^l_{k,0}(x,t)\big|_{t=0} = \overline{d}^l_{k,0}(x).$$

$\overline{d}^l_{k,0}(x)$ is the arbitrary function. In the next step, equation (Eq. (8)) for k = 1 takes the form:

$$T_1u_1(M) = -i\sum_{k=1}^{N}\theta'_k(t)\left[c_{k,0}(x,t) + \sum_{l=1}^{2} Y^l_{k,0}(N_l)\right]e^{r_k}.$$

According to *Theorem 1*, this equation is solvable in U, if $c_{k,0}(x,t)=0$; the function $Y^l_{k,0}(N_l)$ is the solution of the differential equation $\partial_\eta Y^l_{k,0}(N_l) = \partial^2_{\varsigma_l} Y^l_{k,0}(N_l) + H^l_{k,0}(N_l)$, and its solution is representable in the form (Eq. (14)), where $H^l_k(0) = i\theta'_k(t)Y^l_{k,0}(N_l)$. Satisfying condition (1)–(3) of *Theorem 1*, we obtain (see (Eq. (14)))

$$\begin{gathered}\partial_t v_0 - b(x,t)v_0(x,t) = 0, \partial_t w_0^l(x,t) - b(x,t)w_0^l(x,t) = 0,\\ \partial_t d_{k,0}^l(x,t) - b(x,t)d_{k,0}^l(x,t) = -q_{k,0}^l(x,t),\\ \partial_t z_{k,0}(x,t) - b(x,t)z_{k,0}(x,t) = 0,\\ \partial_t q_{k,0}^l(x,t) - b(x,t)q_{k,0}^l(x,t) = 0,\\ i\theta_k'(t)c_{k,1}(x,t) = -z_{k,0}(x,t) + f_k(x,t)\exp\left(\frac{i\theta_\kappa(0)}{\varepsilon}\right),\\ L_\varsigma\left(d_{k,0}^l(x,t)\mathrm{erfc}\left(\frac{\varsigma_l}{2\sqrt{\eta}}\right)\right) = 0.\end{gathered} \tag{18}$$

When the equation is obtained with respect to $d_{k,0}^l(x,t)$ in the $q_{k,0}^l(x,t)\mathrm{erfc}\left(\frac{\xi_l}{2\sqrt{t}}\right)$, a transition $\frac{\xi_l}{2\sqrt{t}} = \frac{\varsigma_l}{2\sqrt{\eta}}$ occurs.

The initial conditions for equation (Eq. (18)) are determined from (Eq. (12)). Functions $w_0^l(x,t)$, $d_{k,0}^l(x,t)$, $q_{k,0}^l(x,t)$ are expressed through arbitrary functions $\overline{w}_0^l(x)$, $\overline{d}_{k,0}^l(x)$, $\overline{q}_{k,0}^l(x)$. These arbitrary functions provide the condition:

$$L_\xi u_k(m) = 0, L_\varsigma u_k(m) = 0,$$

ensuring the solvability of the equation with respect to $c_{k,1}^l(x,t)$. Suppose that

$$Z_{k,0}(x,t)|_{t=0} = f_k(x,t)\exp\left(\frac{i\theta_k(0)}{\varepsilon}\right).$$

This relation is used by the initial condition for determining $Z_{k,0}(x,t)$ from the equation entering into (Eq. (18)).

Further repeating this process, we can determine all the coefficients of $u_k(m)$ of the partial sum:

$$u_{\varepsilon n}(m) = \sum_{i=0}^{n}\varepsilon^i u_i(m).$$

In each iteration with respect to $v_i(x,t)$, $w_i^l(x,t)$, $d_{k,i}^l(x,t)$, $z_{k,i}(x,t)$, $q_{k,i}^l(x,t)$, we obtain inhomogeneous equations.

1.7 Assessment of the remainder term

For the remainder term

$$R_{\varepsilon n}(x,t,\varepsilon) \equiv R_{\varepsilon n}(m,\varepsilon)\Big|_{\gamma=\rho(x,t,\varepsilon)} = u(x,t,\varepsilon) - \sum_{i=0}^{n}\varepsilon^i u_i(m)\Big|_{\gamma=\rho(x,t,\varepsilon)},$$

taking into account (Eqs. (3) and (6)), we obtain the equation

$$L_\varepsilon R_{\varepsilon n}(x,t,\varepsilon) = \varepsilon^{n+1}g_n(x,t,\varepsilon)$$

with homogeneous boundary conditions. Using the maximum principle, like work of [7], we get the estimate:

$$|R_{\varepsilon n}(x,t,\varepsilon)| < c\varepsilon^{n+1}. \tag{19}$$

Theorem 3. Suppose that conditions (1)–(3) are satisfied. Then, the constructed solution is an asymptotic solution of problem (Eq. (1)), i.e., $\forall n = 0, 1, 2, \ldots$; the estimate is fair (Eq. (18)).

2. Two-dimensional parabolic problem with a rapidly oscillating free term

2.1 Introduction

In the case when a small parameter is also included as a multiplier with a temporal derivative, the asymptotic of the solution acquires a complex structure.

Different classes of singularly perturbed parabolic equations are studied in [2]. There, regularized asymptotics of the solution of these equations are constructed, when a small parameter is in front of the time derivative and with one spatial derivative. It is shown that the constructed asymptotic contains exponential, parabolic, and angular products of exponential and parabolic boundary layer functions. The equations are studied when the limiting equation has a regular singularity. Such equations have a power boundary layer. If a small parameter is entering as the multiplier for all spatial derivatives, then the asymptotic solution contains a multidimensional parabolic boundary layer function. When entering into the equation, as free terms of rapidly oscillating functions, then the asymptotic of the solution additionally contains fast-oscillating boundary layer functions. If it is additionally assumed that the phase of this free term has a stationary point, in addition to the rapidly oscillating boundary layer function that arises as a power boundary layer.

This section is devoted to a two-dimensional equation of parabolic type.

2.2 Statement of the problem

Consider the problem:

$$L_\varepsilon u(x,t,\varepsilon) \equiv \partial_t u - \varepsilon^2 \Delta_a u - b(x,t)u = f(x,t)\exp\left(\frac{i\theta(t)}{\varepsilon}\right), (x,t)\epsilon E,$$

$$u|_{t=0} = 0, u|_{\partial\Omega=0} = 0, \tag{20}$$

where $\varepsilon > 0$ is the small parameter, $x = (x_1, x_2)$, $\Omega = (0 < x_1 < 1)x$ $(0 < x_2 < 1)$, $E = (0 < t \leq T)x\Omega$, $\Delta_a \equiv \sum_{l=1}^{2} a_l(x_l)\partial_{x_l}^2$.

The problem is solved under the following assumptions:

1. $\forall x_l \in [0, 1]$ the function $a_l(x_l) \in C^\infty[0, 1]$, $l = 1, 2$.

2. $b(x,t), f(x,t) \in C^\infty[E]$.

3. $\theta'(0) = 0$.

2.3 Regularization of the problem

Following the method of regularization of singularly perturbed problems [1, 2], along with the independent variables (x, t), we introduce regularizing variables:

$$\mu = \frac{t}{\varepsilon}, \xi_l = \frac{(-1)^{l-1}}{\sqrt{\varepsilon^3}} \int_{l-1}^{x_1} \frac{ds}{\sqrt{a_1(s)}}, \eta_l = \frac{\varphi_l(x_1)}{\varepsilon^2}$$
$$\xi_{l+2} = \frac{(-1)^{l-1}}{\sqrt{\varepsilon^3}} \int_{l-1}^{x_2} \frac{ds}{\sqrt{a_2(s)}}, \eta_{l+2} = \frac{\varphi_{l+2}(x_2)}{\varepsilon^3} \tag{21}$$
$$\sigma = \int_0^t e^{\frac{i[\theta(s)-\theta(0)]}{\varepsilon}} ds, \tau_2 = \frac{i[\theta(t)-\theta(0)]}{\varepsilon}, \tau_1 = \frac{t}{\varepsilon^2},$$
$$\varphi_l(x_r) = (-1)^{l-1} \int_{l-1}^{x_r} \frac{ds}{\sqrt{a_r(s)}},$$

For extended function $\widetilde{u}(M,\varepsilon), M = (x,t,\tau,\xi,\eta)$ such that

$$\widetilde{u}(M,\varepsilon)|_{\mu=\psi(x,t,\varepsilon)} \equiv u(x,t,\varepsilon),$$
$$\chi = (\tau,\xi,\eta), \tau = (\tau_1,\tau_2), \xi = (\xi_1,\xi_2,\xi_3,\xi_4),$$
$$\eta = (\eta_1,\eta_2,\eta_3,\eta_4),$$
$$\psi(x,t,\varepsilon) = \left(\frac{t}{\varepsilon^2}, \frac{t}{\varepsilon}, \frac{i[\theta(t)-\theta(0)]}{\varepsilon}, \frac{\varphi(x)}{\varepsilon}, \frac{\varphi(x)}{\varepsilon^2}\right),$$
$$\varphi(x) = (\varphi_1(x_1), \varphi_2(x_1), \varphi_3(x_2), \varphi_4(x_2))$$
$$\widetilde{u}(M,\varepsilon)|_{\mu=\psi(x,t,\varepsilon)} \equiv u(x,t,\varepsilon), \chi = (\tau,\xi,\eta), \tag{22}$$
$$\tau = (\tau_1,\tau_2), \xi = (\xi_1,\xi_2,\xi_3,\xi_4),$$
$$\eta = (\eta_1,\eta_2,\eta_3,\eta_4),$$
$$\psi(x,t,\varepsilon) = \left(\frac{t}{\varepsilon^2}, \frac{t}{\varepsilon}, \frac{i[\theta(t)-\theta(0)]}{\varepsilon}, \frac{\varphi(x)}{\varepsilon}, \frac{\varphi(x)}{\varepsilon^2}\right),$$
$$\varphi(x) = (\varphi_1(x_1), \varphi_2(x_1), \varphi_3(x_2), \varphi_4(x_2)).$$

Find from (Eq. (22)) the derivatives based on

$$\partial_t u \equiv \left(\partial_t \widetilde{u} + \tfrac{1}{\varepsilon}\partial_\mu \widetilde{u} + \tfrac{1}{\varepsilon^2}\partial_{\tau_1}\widetilde{u} + \tfrac{i\theta'(t)}{\varepsilon}\partial_{\tau_2}\widetilde{u} + \exp(\tau_2)\partial_\sigma \widetilde{u}\right)\Big|_{\chi=\psi(x,t,\varepsilon)},$$
$$\partial_{x_r} u \equiv \left(\partial_{x_r}\widetilde{u} + \sum_{l=2r-1}^{2r}\left[\frac{\varphi_l'(x_r)}{\sqrt{\varepsilon^3}}\partial_{\xi_l}\widetilde{u} + \frac{\varphi_l'(x_r)}{\varepsilon^2}\partial_{\zeta_l}\widetilde{u}\right]\right)\Big|_{\chi=\psi(x,t,\varepsilon)},$$
$$\partial_{x_r}^2 u \equiv \left(\partial_{x_r}^2\widetilde{u} + \sum_{l=2r-1}^{2r}\left[\frac{\varphi_l'2(x_r)}{\varepsilon^3}\partial_{\xi_l}^2\widetilde{u} + \frac{\varphi_l'2(x_r)}{\varepsilon^4}\partial_{\zeta_l}^2\widetilde{u}\right]\right.$$
$$\left.+\sum_{l=2r-1}^{2r}\left[\frac{2\varphi_l'(x_r)}{\sqrt{\varepsilon^3}}\partial_{x_r\xi_l}^2\widetilde{u} + \frac{\varphi_l''(x_r)}{\sqrt{\varepsilon^3}}\partial_{\xi_l}\widetilde{u} + \frac{1}{\varepsilon^2}\left(\varphi_l'(x_r)\partial_{x_r\eta_l}^2\widetilde{u} + \varphi_l''(x_r)\partial_{\eta_l}\widetilde{u}\right)\right]\right)\Big|_{\chi=\psi(x,t,\varepsilon)} \tag{23}$$

Below, it is shown that the solution of the iterative problems does not contain terms depending on $(\xi_1,\xi_2), (\xi_3,\xi_4), (\zeta_1,\zeta_2), (\zeta_3,\zeta_4), (\xi_l,\zeta_k), l,k = 1,2$. Therefore, to simplify recording, the mixed derivatives of these variables are omitted. Based on (Eq. (20)), (Eq. (22)), and (Eq. (23)) for extended function $\widetilde{u}(M,\varepsilon)$, set the problem:

$$\widetilde{L}_\varepsilon\widetilde{u} \equiv \frac{1}{\varepsilon^2}T_0\widetilde{u} + \frac{1}{\varepsilon}i\theta'(t)\partial_{\tau_2}\widetilde{u} + \frac{1}{\varepsilon}T_1\widetilde{u} + D_\sigma\widetilde{u} - L_\eta\widetilde{u} - \sqrt{\varepsilon}L_\xi\widetilde{u} - \varepsilon^2\Delta_a\widetilde{u} = f(x,t)\exp\left(\tau_2 + \frac{i\theta(0)}{\varepsilon}\right),$$
$$\widetilde{u}|_{t=\tau_1=\tau_2=0} = 0, \widetilde{u}|_{x_l=r-1,\ \xi_k=\eta_k=0} = 0, r = 1, 2,\ \ l = 1, 2, k = \overline{1, 4}$$
$$T_0 \equiv \partial_{\tau_1} - \Delta_\eta, T_1 \equiv \partial_\mu + \Delta_\xi, D_\sigma \equiv D_t + \exp(\tau_2)\partial_\sigma, D_t \equiv \partial_t + b(x,t),$$
$$L_\eta \equiv \sum_{r=1}^{2}\sum_{l=2r-1}^{2r} a_r(x_r)D_{x,\eta}^{r,l},$$
$$D_{x,\xi}^{r,l} \equiv \left[2\varphi_l'(x_r)\partial^2_{x_r\xi_l} + \varphi_l''(x_r)\partial_{\eta_l}\right],$$
$$\Delta_\eta \equiv \sum_{k=1}^{4}\partial^2_{\eta_k},\ \ E_1 = Ex(0,\infty)^{10} \tag{24}$$

In this case, the identity is satisfied:

$$\left(\widetilde{L}_\varepsilon\widetilde{u}\right)\Big|_{\chi=\psi(x,t,\varepsilon)} \equiv L_\varepsilon u(x,t,\varepsilon). \tag{25}$$

2.4 Solution of iterative problems

For the solution of the extended function (Eq. (24)), we search in the form of series

$$\widetilde{u}(M,\varepsilon) = \sum_{i=0}^{\infty}\varepsilon^{\frac{i}{2}}u_i(M). \tag{26}$$

Then, for the coefficients of this series, we get the following problems:

$$\begin{aligned}
&T_0u_v(M) = 0, v = 0, 1,\\
&T_0u_q = -i\theta'(t)\partial_{\tau_2}u_{q-2} - T_1u_{q-2}, q = 2, 3.\\
&T_0u_4 = f(x,t)\exp\left(\tau_2 + \frac{i\theta(0)}{\varepsilon}\right) - T_1u_2 - D_\sigma u_0 + L_\eta u_0,\\
&T_0u_i = -i\theta'(t)\partial_{\tau_2}u_{i-2} - T_1u_{i-2} - D_\sigma u_{i-4} + L_\eta u_{i-4} + L_\xi u_{i-5} + \Delta_a u_{i-8},\\
&u_i|_{t=\tau=0} = 0, u_i|_{x_l=r-1,\ \xi_k=\eta_k=0} = 0, l, r = 1, 2.k = \overline{1, 4}
\end{aligned} \tag{27}$$

We introduce a class of functions:

$$U_0 = \{V_0(N) = [c(x,t) + F_1(N) + F_2(N)]\exp(\tau_2), F_1(N) \in U_4, F_2(N) \in U_5, c(x,t) \in C^\infty(\overline{E})\},$$
$$U_1 = \{V_1(M) : V_1(M) = v(x,t) + F_1(M) + F_2(M), F_1(M) \in U_4, F_2(M) \in U_5, v(x,t) \in C^\infty(\overline{E})\},$$
$$U_2 = \{V_2(M) : V_2(M) = [z(x,t) + F_1(M) + F_2(M)]\sigma, F_1(M) \in U_4, F_2(M) \in U_5, z(x,t) \in C^\infty(\overline{E})\},$$
$$U_4 = \left\{V_4(M) : V_1(M) = \sum_{l=1}^{4}Y^l(N_l), |Y^l(N_l)| < cexp\left(-\frac{\eta_l^2}{8\tau_1}\right)\right\},$$
$$U_5 = \left\{V_5(M) : V_2(M) = \sum_{r,\,l=1}^{4}Y^{r+2,l}(N_{r+2,l}), |Y^{r+2,l}(N_{r+2,l})| < cexp\left(-\frac{|\eta^{r,l}|^2}{8\tau_1}\right), |\eta^{r,l}| = \sqrt{\eta_r^2 + \eta_l^2}\right\}.$$

From these classes we will construct a new one, as a direct sum:

$$U = U_0 \bigoplus U_1 \bigoplus U_2.$$

Any item u(M) ∈ U is representable in the form:

$$u(M) = v(x,t) + c(x,t)\exp(\tau_2) + z(x,t)\sigma + \left[\sum_{l=1}^{4} Y^l(N_l) + \sum_{r,\,l=1}^{2} Y^{r+2,\,l}(N_{r+2,\,l})\right]\exp(\tau_2)$$

$$+\sum_{l=1}^{4} w^l(x,t)\operatorname{erfc}\left(\frac{\xi_l}{2\sqrt{\mu}}\right) + \sum_{l,\,r=1}^{2} w^{r+2,\,l}(M_{r+2,\,l}) + \left[\sum_{l=1}^{4} q^l(x,t)\operatorname{erfc}\left(\frac{\xi_l}{2\sqrt{\mu}}\right) + \sum_{l,\,r=1}^{2} z^{r+2,\,l}(M_{r+2,\,l})\right]\sigma,$$

$$N_l = (x,t,\tau_1,\eta_l),\, N_{r+2,\,l} = (x,t,\tau_1,\eta_l,\eta_{r+2}),$$

$$M_l = (x,t,\mu,\xi_l),\, M_{r+2,\,l} = (x,t,\mu,\xi_l,\xi_{r+2}). \tag{28}$$

Let's satisfy this function to the boundary conditions:

$$v(x,0) = -c(x,0),\, Y^l(N_l)\big|_{t=\tau_1=0} = 0 \tag{29}$$

$$Y^{r+2,\,l}(N_{r+2,\,l})\big|_{t=\tau_1=0} = 0,\, w^l\big|_{t=0} = \overline{w}^l(x),$$

$$q^l\big|_{t=0} = \overline{q}^l(x),$$

$$w^{r+2,\,l}(M_{r+2,\,l})\big|_{t=\mu=0} = 0,$$

$$z^{r+2,\,l}(M_{r+2,\,l})\big|_{t=\mu=0} = 0,$$

$$w^l(x,t)\big|_{x_1=l-1} = -v(l-1,x_2,t),$$

$$q^l(x,t)\big|_{x_1=l-1} = -z(l-1,x_2,t),$$

$$Y^l\big|_{x_1=l-1,\,\eta_l=0} = -c(l-1,x_2,t),\;\; Y^{r+2,\,l}\big|_{x_1=l-1,\,\eta_l=0} = -Y^{r+2,\,l}(N_{r+2,\,l})\big|_{x_1=l-1},$$

$$w^{r+2,\,l}\big|_{x_1=l-1,\,\xi_l=0} = -w^{r+2}(l-1,x_2,t)\operatorname{erfc}\left(\frac{\xi_{r+2}}{2\sqrt{t}}\right),$$

$$z^{r+2,\,l}\big|_{x_1=l-1,\,\xi_l=0} = -q^{r+2}(l-1,x_2,t)\operatorname{erfc}\left(\frac{\xi_{r+2}}{2\sqrt{t}}\right),$$

$$w^l(x,t)|_{x_r=l-1} = -v(x,t)\big|_{x_r=l-1},$$

$$q^l(x,t)|_{x_r=l-1} = -z(x,t)\big|_{x_r=l-1},$$

$$Y^{r+2}\big|_{x_2=l-1,\,\eta_{r+2}=0} = -c(x_1,l-1,t),$$

$$Y^{r+2,\,l}\big|_{x_2=l-1,\,\eta_{r+2}=0} = -Y^l\big|_{x_2=l-1},$$

$$w^{r+2,\,l}\big|_{x_2=l-1,\,\xi_{r+2}=0} = -w^l\big|_{x_2=l-1}\operatorname{erfc}\left(\frac{\xi_l}{2\sqrt{t}}\right),$$

$$z^{r+2,\,l}\big|_{x_2=l-1,\,\xi_{r+2}=0} = -q^l\big|_{x_2=l-1}\operatorname{erfc}\left(\frac{\xi_l}{2\sqrt{t}}\right),\, l,r = 1,2.$$

We compute the action of the operators T_0, T_1, L_η, L_ξ on function $u(M) \in U$, and we have

$$T_1 u(M) = \sum_{r,\,l=1}^{2}\left\{\partial_\mu w^{r+2,\,l} - \Delta_\xi w^{r+2,\,l} + \sigma\left[\partial_\mu z^{r+2,\,l} - \Delta_\xi z^{r+2,\,l}\right]\right\},$$

$$L_\eta u = \sum_{r=1}^{2}\sum_{l=2r-1}^{2r} D_{x,\eta}^{r,l} Y^l(N_l) + \sum_{v=1r,}^{2}\sum_{l=1}^{2} D_{x,\eta}^{v,l} Y^{r+2,l}(N_{r+2,l}),$$

$$L_\xi u = \sum_{r=1}^{2}\sum_{l=2r-1}^{2r} D_{x,\xi}^{r,l} w^l(x,t)\operatorname{erfc}\left(\frac{\xi_l}{2\sqrt{\mu}}\right) + \sum_{v=1r,}^{2}\sum_{l=1}^{2} D_{x,\xi}^{v,l} w^{r+2,l}(M_{r+2,l})$$

$$+\sigma\left[\sum_{r=1}^{2}\sum_{l=2r-1}^{2r} D_{x,\xi}^{r,l} q^l(x,t)\operatorname{erfc}\left(\frac{\xi_l}{2\sqrt{\mu}}\right) + \sum_{v=1r,}^{2}\sum_{l=1}^{2} D_{x,\xi}^{v,l} z^{r+2,l}(M_{r+2,l})\right],$$

$$D_\sigma u(M) = D_t v(x,t) + \sum_{l=1}^{4} D_t w^l(x,t)\operatorname{erfc}\left(\frac{\xi_l}{2\sqrt{\mu}}\right) + \sum_{r,\,l=1}^{2} D_t w^{r+2,l}(M_{r+2,l})$$

$$+\left[D_t c(x,t) + \sum_{l=1}^{4} D_t Y^l(N_l) + \sum_{r,\,l=1}^{2} D_t Y^{r+2,l}(N_{r+2,l})\right]\exp(\tau_2)$$

$$+\sigma\left[D_t z(x,t) + \sum_{l=1}^{4} D_t q^l(x,t)\operatorname{erfc}\left(\frac{\xi_l}{2\sqrt{\mu}}\right) + \sum_{r,\,l=1}^{2} D_t z^{r+2,l}(M_{r+2,l})\right]$$

$$+\left[z(x,t) + \sum_{l=1}^{4} q^l(x,t)\operatorname{erfc}\left(\frac{\xi_l}{2\sqrt{\mu}}\right) + \sum_{r,\,l=1}^{2} z^{r+2,l}(M_{r+2,l})\right]\exp(\tau_2)$$

We write iterative equation (8) in the form:

$$T_0 u(M) = H(M). \tag{31}$$

Theorem 1. Let be $H(M)\in U_4 \bigoplus U_5$ and condition (1) is satisfied. Then, Eq. (31) is solvable in U, if the equations are solvable:

$$T_0 Y^l(N_l) = H_1(N_l),\, l = \overline{1,4},\, T_0 Y^{r+2,l}(N_{r+2,l}) = H_2(N_{r+2,l}),\, r, l = 1, 2.$$

Theorem 2. Let be $H_1(N_l)\in U_4$. Then, the problem $\partial_{\tau_1} Y^l(N_l) = \Delta_\eta Y^l(N_l) + H_1(N_l)$, $Y^l(N_l)\big|_{\tau_1=0} = 0$, $Y^l(N_l)\big|_{\eta_l=0} = d^l(x,t)$, $l = \overline{1,4}$ (Eq. (32)) has a solution $Y^l(N_l)\in U_4$.

Theorem 3. Let be $H_2(N_{r+2,l})\in U_5$, $Y^l(N_l)\in U_4$, and then the problem $\partial_{\tau_1} Y^{r+2,l}(N_{r+2,l}) = \Delta_\eta Y^{r+2,l}(N_{r+2,l}) + H_2(N_{r+2,l})$, $Y^{r+2,l}(N_{r+2,l})\big|_{\eta_l=0} = -Y^{r+2}(N_{r+2})$, $Y^{r+2,l}(N_{r+2,l})\big|_{\eta_{r+2}=0} = -Y^l(N_l)$, $r, l = 1, 2$ has a solution $Y^{r+2,l}(N_{r+2,l})\in U_5$.

The proof of these theorems is given in [2].

2.5 The decision of the iterative problems

Eq. (27) under $v = 0, 1$ is homogeneous. By *Theorem 1*, it has a solution representable in the form $u_0(M)\in U$ if functions $Y^l(N_l)$ and $Y^{r+2,l}(N_{r+2,l})$ – are solutions of the following equations:

$$T_0 Y_v^l(N_l) = 0,\, T_0 Y_v^{r+2,l}(N_{r+2,l}) = 0.$$

Based on the boundary conditions from (Eq. (29)), the solution is written:

$$Y_v^l(N_l) = d_v^l(x,t)\operatorname{erfc}\left(\frac{\eta_l}{2\sqrt{\tau_1}}\right), l = 1, 2, 3, 4. \tag{32}$$

$$Y_v^{r+2,l}(N_{r+2,l}) = -\int_0^{\tau_1}\int_0^{\infty} Y_v^l(*)\left[\frac{\partial}{\partial_\xi}\ G(N_l,\xi,\eta,\tau_1-\tau\ )\right]|_{\xi=0}d\eta d\tau$$

$$-\int_0^{t}\int_0^{\infty} Y_v^{r+2}(*)\left[\frac{\partial}{\partial_\eta}\ G(N_{r+2,l},\xi,\eta,\tau_1-\tau\ )\right]|_{\eta=0}d\xi d\tau,$$

where $d^l(x,t)$– is arbitrary function such as

$$d_v^p(x,t)\Big|_{t=0} = -\overline{d}_v^p(x), d_v^l(x,t)\Big|_{x_1=l-1} = -c_v(l-1,x_2,t),$$

$$G(\eta_l,\eta_{r+2,l},\xi,\eta,\tau_1) = \frac{1}{4\pi\tau_1}\left\{\exp\left(-\frac{(\eta_l-\xi)^2}{4\tau_1}\right)-\exp\left(-\frac{(\eta_l+\xi)^2}{4\tau_1}\right)\right\} \quad (33)$$

$$\left\{\exp\left(-\frac{(\eta_{r+2}-\eta)^2}{4\tau_1}\right)-\exp\left(-\frac{(\eta_{r+2}+\eta)^2}{4\tau_1}\right)\right\}.$$

Due to the fact that the function $d_v^l(x,t)$ при $t=\tau_1=0$ multiplied by the function becomes as $d_0^l(x,t)\Big|_{t=0} = -\overline{d}_0^l(x)$, an arbitrary function is accepted, and its values under $x_1 = l-1$ are determined from the second relation. According to *Theorems 2* and *3*, the functions found by the formula (Eq. (33)) satisfy the estimates:

$$|Y_v^l(N_l)| < c\exp\left(-\frac{\eta_l^2}{8\tau_1}\right), |Y_v^{r+2,l}(N_{r+2,l})| < c\exp\left(-\frac{\eta_{r+2}^2+\eta_l^2}{8\tau_1}\right), r,l=1,2. \quad (34)$$

Free member of equation (Eq. (27)) under $v = 2, 3$ has a form

$$F_{v-2}(M) \equiv T_1u_{v-2}(M) + i\theta'(t)\partial_\sigma u_{v-2}(M) = i\theta'(t)\left[c_{v-2}(x,t) + \sum_{l=1}^{4} Y_{v-2}^l(N_l) + \sum_{r,l=1}^{2} Y_{v-2}^{r+2,l}(N_{r+2,l})\right]$$

$$\exp(\tau_2) + \sum_{l,r=1}^{2}\left\{\partial_\mu w_{v-2}^{r+2,l} - \Delta_\xi w_{v-2}^{r+2,l} + \sigma\left[\partial_\mu z_{v-2}^{r+2,l} - \Delta_\xi z_{v-2}^{r+2,l}\right]\right\},$$

so that equation (Eq. (27)), under $v = 2, 3$, has a solution in U; we set

$$c_{v-2}(x,t) = 0, T_1w_{v-2}^{r+2,l} = 0, T_1z_{v-2}^{r+2,l} = 0.$$

Solutions of the last equations under the boundary conditions from (Eq. (29)) have a form (Eq. (33)) for which estimates of the form (Eq. (35) are fair. Eq. (27), i=4, has a free term:

$$F_4(M) = -i\theta'(t)\partial_{\tau_2} - T_1u_2 + f(x,t)\exp\left(\frac{i\theta(0)}{\varepsilon}\right) - D_\sigma u_0 + L_\eta u_0$$

$$= -i\theta'(t)\left[c_2(x,t) + \sum_{l=1}^{4} Y_2^l(N_l) + \sum_{r,l=1}^{2} Y_2^{r+2,l}(N_{r+2,l})\right]\exp(\tau_2)$$

$$-\sum_{l,r=1}^{2}\left[T_0w_2^{r+2,l}(M_{r+2,l}) + \sigma T_0z_2^{r+2,l}\right] - D_tv_0(x,t) - \sum_{l=1}^{4} D_tw_0^l(x,t)\mathrm{erfc}\left(\frac{\xi_l}{2\sqrt{\mu}}\right)$$

$$-\sum_{l,r=1}^{2} D_tw_0^{r+2,l}(x,t) - \exp(\tau_2)\left[\partial_tc_0(x,t) + \sum_{l=1}^{4}\partial_tY_0^l + \sum_{l,r=1}^{2} D_tY_0^{r+2,l}\right]$$

$$-\sigma\left[D_t z_0(x,t)+\sum_{l=1}^{4}D_t q_0^l(x,t)\operatorname{erfc}\left(\frac{\xi_l}{2\sqrt{\mu}}\right)+\sum_{r,\,l=1}^{2}D_t z_0^{r+2,\,l}(M_{r+2,\,l})\right]$$

$$-\left[z_0(x,t)+\sum_{l=1}^{4}q_0^l(x,t)\operatorname{erfc}\left(\frac{\xi_l}{2\sqrt{\mu}}\right)+\sum_{r,\,l=1}^{2}z_0^{r+2,\,l}(M_{r+2,\,l})\right]\exp(\tau_2)$$

$$+\sum_{r=1}^{2}\sum_{l=2r-1}^{2r}D_{x,\,\xi}^{r,\,l}w_0^p(x,t)\operatorname{erfc}\left(\frac{\xi_l}{2\sqrt{\mu}}\right)+\sum_{v=1}^{2}\sum_{r,\,l=1}^{2}D_{x,\,\eta}^{v,\,l}Y_0^{r+2,\,l}(N_{r+2,\,l}).$$

By providing $F_4(M)\in U_4\bigoplus U_5$ with regard to $c_v(x,t)=0,\,v=0,1$, we set

$$\begin{gathered}-i\theta'(t)c_2(x,t)+f(x,t)\exp\left(\frac{i\theta(0)}{\varepsilon}\right)-z_0(x,t)=0,\\ D_t v_0(x,t)=0,\,D_t z_0(x,t)=0,\\ D_t Y_0^l(N_l),\,T_0 w_2^{r+2,\,l}=0,\,T_0 z_2^{r+2,\,l}=0,\\ D_t w_0^l=0,\,D_t w_0^{r+2,\,l}=0,\,D_t Y_0^{r+2,\,l}=0,\\ D_t q_0^l\,(x,t)=0,\,D_t z_0^{r+2,\,l}\,(x,t)=0,\\ D_{x,\,\xi}^{r,\,l}w_0^l(x,t)=0,\,D_{x,\,\eta}^{v,\,l}Y_0^{r+2,\,l}=0,\,D_{x,\,\eta}^{r,\,l}Y_0^l=0,\end{gathered}\tag{35}$$

then

$$F_4(M)=-i\theta'(t)\left[\sum_{l=1}^{4}Y_2^l(N_l)+\sum_{r,\,l=1}^{2}Y_2^{r+2,\,l}(N_{r+2,\,l})\right]\exp(\tau_2)$$

$$-\left[\sum_{l=1}^{4}q_0^l(x,t)\operatorname{erfc}\left(\frac{\eta_l}{2\sqrt{\tau_2}}\right)+\sum_{r,\,l=1}^{2}z_0^{r+2,\,l}(N_{r+2,\,l})\right]\exp(\tau_2).$$

In the last bracket, the transition is from the variables $\frac{\xi_l}{2\sqrt{\mu}}$ to the variables $\frac{\eta_l}{2\sqrt{\tau_2}}$.

Substituting the value $Y_0^l(N_l)=d_0^l(x,t)\operatorname{erfc}\left(\frac{\eta_l}{2\sqrt{\tau_1}}\right)$ into equation $D_t Y_0^l(N_l)=0$, with respect to $d_0^l(x,t)$, we get the equation $D_t d_0^l(x,t)=0$, which is solved under an arbitrary initial condition $d_0^l(x,t)\Big|_{t=0}=\overline{d}_0^l(x)$. This arbitrary function provides the condition $L_\eta\,Y_0^l=0$; therefore, $D_{x,\,\eta}\,Y_0^l=0$.The initial condition for this equation is determined from the relation:

$$d_0^l(x,t)\Big|_{x_1=l-1}=-c_0(l-1,x_2,t),d_0^{l+2}(x,t)\Big|_{x_2=l-1}=-c_0(x_1,l-1,t),$$

which comes out from (Eq. (29)) and (Eq. (33)). The function $Y_0^{r+2,\,l}(N_{r+2,\,l})$ expresses through $Y_0^l(N_l)$ therefore provided that

$$D_t Y_0^{r+2,\,l}=0,\,D_{x,\,\eta}^{v,\,l}Y_0^{r+2,\,l}=0.$$

The same is true for functions $w_0^{r+2,\,l}(M_{r+2,\,l}),\,z_0^{r+2,\,l}(M_{r+2,\,l})$; in other words, the following relations hold: $D_t w_0^{r+2,\,l}=0,\,D_t z_0^{r+2,\,l}=0,\,D_{x,\,\xi}^{v,\,l}w_0^{r+2,\,l}=0,\,D_{x,\,\xi}^{v,\,l}z_0^{r+2,\,l}=0.$

Solutions of equations with respect $w_0^{r+2,\,l},z_0^{r+2,\,l}$ under appropriate boundary conditions from (Eq. (29) are representable as (Eq. (33)), and they are expressed through $w_2^l(x,t),\,q_2^l(x,t)$. The first equation (Eq. (36)) is solvable, if $z_0(x,t)|_{t=0} = f(x,0)\exp\left(\frac{i\theta(0)}{\varepsilon}\right)$. This ratio is used by the initial condition for the equation

$D_t z_0(x,t) = 0$. The remaining equations from (Eq. (36)) are solvable under the initial conditions from (Eq. (29)).

Thus, the main term of the asymptotics is uniquely determined. As can be seen from the representation (Eq. (28)) and the estimates (Eq. (35)), we note that the asymptotics of the solution have a complex structure. In addition to regular members, it contains various boundary layer functions. Parabolic boundary layer functions have an estimate:

$$|Y^l(N_l)| < c\exp\left(-\frac{\eta_l^2}{8\tau_1}\right), \left|w^l(x,t)\mathrm{erfc}\left(\frac{\xi_l}{2\sqrt{\mu}}\right)\right| < c\exp\left(-\frac{\xi_l^2}{8\mu}\right).$$

Multidimensional and angular parabolic boundary layer functions have an estimate:

$$|Y^{r+2,l}(N_{r+2,l})| < c\exp\left(-\frac{\eta_{r+2}^2 + \eta_l^2}{8\tau_1}\right),$$

$$|w^{r+2,l}(M_{r+2,l})| < c\exp\left(-\frac{\xi_{r+2}^2 + \xi_l^2}{8\mu}\right).$$

The boundary layer functions with rapidly oscillating exponential and power type of change:

$$c(x,t)\exp(\tau_2), \sigma = \int_0^t e^{\frac{i[\theta(s)-\theta(0)]}{\varepsilon}} ds.$$

In addition, the asymptotic contains the product of the abovementioned boundary layer functions.

Repeating the above process, we construct a partial sum:

$$\widetilde{u}_{\varepsilon n}(M) = \sum_{i=0}^{n} \varepsilon^{\frac{i}{2}} u_i(M). \tag{36}$$

2.6 Assessment of remainder term

Substituting the function $\widetilde{u}(M,\varepsilon) = u_{\varepsilon n}(M) + \varepsilon^{n+\frac{1}{2}} R_{\varepsilon n}(M)$ into problem (Eq. (24)), then taking into account the iterative tasks of (Eq. (27)) and (Eq. (29)), we obtain the following problem for the remainder term $R_{\varepsilon n}(M)$:

$$\widetilde{L}_\varepsilon R_{\varepsilon n}(M) = g_n(M,\varepsilon), R_{\varepsilon n}(M)|_{t=0} = R_{\varepsilon n}(M)|_{x_l=r-1,\,\xi_r=0,\,\eta_k} = 0, r = 1,2; k = \overline{1,4}, \tag{37}$$

where $g_n(M,\varepsilon) = -i\theta'(t)\partial_{\tau_2} u_{n-1} - \varepsilon^{\frac{1}{2}} i\theta'(t)\partial_{\tau_2} u_n(M) - T_1 u_{n-1}(M) - \varepsilon^{\frac{1}{2}} T_1 u_n(M) - (D_\sigma - L_\eta)\sum_{k=0}^{3} \varepsilon^{\frac{k}{2}} u_{n-3+k}(M) + L_\eta \sum_{k=0}^{5} \varepsilon^{\frac{k}{2}} u_{n-5+k}(M) + \Delta_a \sum_{k=0}^{7} \varepsilon^{\frac{k}{2}} u_{n-7+k}(M)$.

We put in both parts (Eq. (38)) $\chi = \psi(x,t,\varepsilon)$ considering (Eq. (25)), with respect to

$L_\varepsilon R_{\varepsilon n}(x,t,\varepsilon) = g_{\varepsilon n}(x,t,\varepsilon), R_{\varepsilon n}|_{t=0} = 0, R_{\varepsilon n}|_{\partial\Omega=0}$.

By virtue of the above constructions, the function is $|g_{\varepsilon n}(x,t,\varepsilon)| < c, \forall (x,t) \in$; therefore, applying the maximum principle, an estimate is established:

$$|R_{\varepsilon n}(x, t, \varepsilon)| < c.$$

Thus, we have proven the following:

Theorem 4. Suppose that the conditions (1)–(3) are satisfied. Then, using the above method for solving $u(x, t, \varepsilon)$ of the problem (Eq. (20)), a regularized series (Eq. (26)) such that $\forall n = 0, 1, 2, \ldots$ can be constructed, and for small enough $\varepsilon > 0$, inequality is fair:

$$|u(x, t, \varepsilon) - u_{\varepsilon n}(x, t, \varepsilon)| = |R_{\varepsilon n}\ (x, t, \varepsilon)| < c\varepsilon^{n+\frac{1}{2}},$$

where $\boldsymbol{c}$ is independent of $\boldsymbol{\varepsilon}$.

Author details

Asan Omuraliev* and Ella Abylaeva
Kyrgyz-Turkish Manas University, Bishkek, Kyrgyzstan

*Address all correspondence to: asan.omuraliev@mail.ru

References

[1] Feschenko S, Shkil N, Nikolaenko L. Asymptotic Methods in the Theory of Linear Differential Equations. Kiev: Naukova Dumka; 1966

[2] Omuraliev AS, Sadykova DA. Regularization of a singularly perturbed parabolic problem with a fast-oscillating right-hand side. Khabarshy –Vestnik of the Kazak National Pedagogical University. 2007;**20**:202-207

[3] Omuraliev AS, Sheishenova ShK. Asymptotics of the solution of a parabolic problem in the absence of the spectrum of the limit operator and with a rapidly oscillating right-hand side, investigated on the integral-differential equations. 2010:**42**:122-128

[4] Butuzov VF. Asymptotics of the solution of a difference equation with small steps in a rectangular area. Computational Mathematics and Mathematical Physics. 1972;**3**:582-597

[5] Omuraliev A. Regularization of a two-dimensional singularly perturbed parabolic problem. Journal of Computational Mathematics and Mathematical Physics. 2006;**46**(/8): 1423-1432

[6] Lomov S. Introduction to the General Theory of Singular Perturbations. Moscow: Nauka; 1981

[7] Ladyzhenskaya OA, Solonnikov VA, Uraltseva NN. Linear and Quasilinear Equations of Parabolic Type. Moscow: Nauka; 1967

Chapter 10

Solving Partial Differential Equation Using FPGA Technology

Vu Duc Thai and Bui Van Tung

Abstract

This chapter introduces the method of using CNN technology on FPGA chips to solve differential equation with large space, with lager computing space, while limitation of resource chip on FPGA is needed, we have to find solution to separate differential space into several subspaces. Our solution will do: firstly, division of the computing space into smaller areas and combination of sequential and parallel computing; secondly, division and combination of boundary areas that are required to be continuous to avoid losing temporary data while processing (using buffer memory to store); and thirdly, real-time data exchange. The control unit controls the activities of the whole system set by the algorithm. We have configured the CNN chip for solving Navier-Stokes equation for the hydraulic fluid flow successfully on the Virtex 6 chip XCVL240T-1FFG1156 by Xilinx and giving acceptance results as well.

Keywords: Navier-Stokes equation, cellular neural network, field programmable gate array, boundary processing, separating computing space

1. Introduction

Solving the partial differential equation (PDE) has been investigated by many researchers, implementing digital decoding on PCs successfully. However, with the problem of large computing space, the resolution on the PC is difficult to meet the requirements of speed and accuracy calculations; in some cases, the problem cannot be solved because of the calculation. Cellular Neural Network technology (CNN) researchers have applied cellular neural network (CNN) technology successfully to perform analysis of the problem, design CNN chip, and solve some PDEs.

Using CNN technology for solving PDE, we have to analyze and difference the original particular equations of problem, find templates, design CNN architecture, and then configure FPGA to make a CNN chip. It means that there is no CNN chip for every equation, but for each problem (consist of some equations), there is need to design appropriate CNN chip. When solving large problems, computing resources are needed to configure blocks of CNN chips. In order to save resources, we have proposed a solution for dividing computing space into smaller subspaces and composite parallel and sequential calculations, which ensures high computing rates but saves resources of FPGA chips used.

Because the architecture of CNN chips varies depending on each problem, making the CNN chip is very difficult and costly with traditional methods. Using the FPGA technology, users can use hardware programming languages, such as Verilog and VHDL, to configure the logic elements in the FPGA to produce the electronic circuit of a CNN chip. The recent FPGA architectures (Virtex 7; Stratix 10) have many tools support to test, optimize, and coordinate data exchange. The CNN designer should use FPGA for making a CNN chip.

2. CNN and FPGA technology

2.1 Cellular neural network technology

Cellular neural network (CNN) was introduced by Chua and Yang at Berkeley University, California (USA), in 1988, which combined both analog spatial temporal dynamic and logic [1–3]. The CNN paradigm is a natural framework to describe the behavior of locally interconnected dynamic systems, which has an arrayed structure, so it is very useful in solving the partial differential equations [3–7]. Today, visual microprocessors based on this processing type can perform at TeraOPs computing power and approximately 50,000 fps. The possibility of developing algorithms and programs based on CNN was quickly exploited worldwide. Up to now, there are several CNN models for processing images, solving PDE, recognizing pattern, gene analysis, etc. Depending on problems, the designer can make a CNN chip having size of millions cells. The common CNN architectures are 1D, 2D, and 3D.

The standard CNN 2D is the dynamic system of autonomous cells that are connected locally with its neighbor forming a two-dimensional array [2, 18]. Each cell in the array C(i,j) contains one independent voltage source, one independent current source, a linear capacitor, resistors, and linear voltage-controlled current sources which are coupled to its neighbor cells via the controlling input voltage, and the feedback from the output voltage of each neighbor cell C(k,l). The templates A(i,j;kl) and B(i,j;k,l) are the parameters linking cell C(i,j) to neighbor C(k,l). The effective range of Sr(I,j) on radius r of cell C(I,j) is identified by the set of neighbor cells which satisfies (**Figure 1**).

$$\mathrm{Sr}(\mathrm{i},\ \mathrm{j}) = \{\mathrm{C}(\mathrm{k},\mathrm{l}) \mid \max\{|\mathrm{k}-\mathrm{i}|, |\mathrm{l}-\mathrm{j}|\} \le \mathrm{r}\}$$
$$\text{with } 1 \le \mathrm{k} \le \mathrm{M}, 1 \le \mathrm{l} \le \mathrm{N}.$$

The state equation of cell C(i,j) is given by the following equation:

$$C\frac{\partial x_{ij}}{\partial t} = -\frac{1}{R}x_{ij} + \sum_{C(k,l)\in S_r(i,\ j)} A(i,\ j;\ k,l)\,y_{kl} + \sum_{C(k,l)\in S_r(i,\ j)} B(i,\ j;\ k,l)u_{kl} + z_{ij} \quad (1)$$

With R, C is the linear resistor and capacitor; A(i,j;kl) is the feedback operator parameter; B(i,j;kl) is the control parameter; and zij is the bias value of the cell C(i,j). On the CNN chip, (A, B, z) are the local connective weight values of each cell C(i,j) to its neighbors. The output of the cell C(i,j) is presented by Y_{ij} as:

$$Y_{ij} = f\left(x_{ij}\right) = \frac{1}{2}\ |x_{ij} + 1| + \frac{1}{2}\ |x_{ij} - 1| \quad (2)$$

The characteristic of the CNN output function $Y_{i,j} = f(x_{ij})$ is presented in **Figure 2**.

On the CNN 3D, beside connection with neighbors, the cell has other connection to upper and lower layer in the three-dimensional space [18] as shown in **Figure 3**.

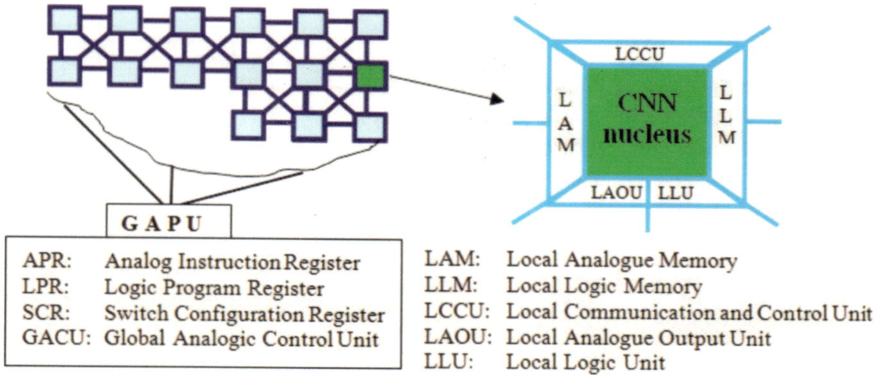

Figure 1.
The architecture of a CNN chip

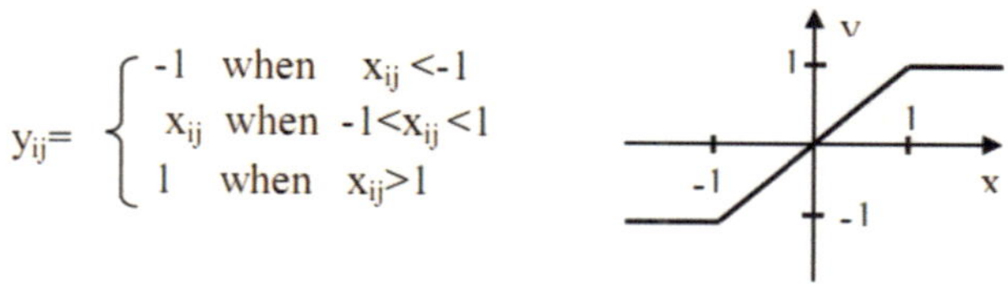

Figure 2.
CNN circuit output function.

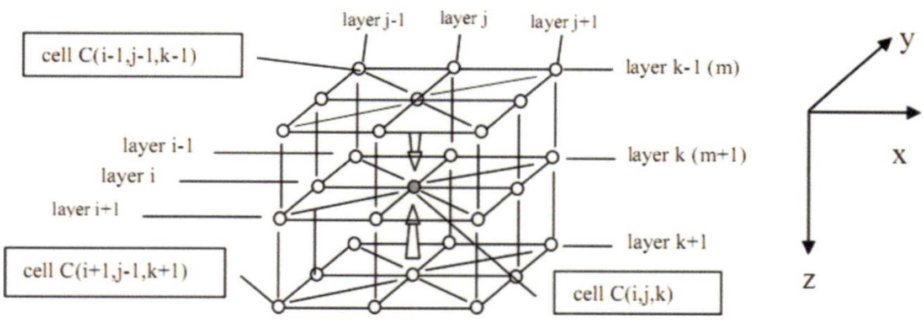

Figure 3.
The 3D CNN, with r = 1, (having 26 neighbors) in three dimensions coordinates x,y,z.

Thus, if radius r = 1, the cell C(i,j,k) has 26 neighbors; hence, the templates A and B have more three coefficients A(i,j,k) and B(i,j,k).

The state equation of CNN 3D takes the form:

$$C\frac{\partial x_{ijk}}{\partial t} = -\frac{1}{R}x_{ijk} + \sum_{C(l,m,n)\in S_r(i,\,j,k)} A(i,j,k;l,m,n)\,y_{lmn} + \sum_{C(l,m,n)\in S_r(i,\,j,k)} B(i,j,k;l,m,n)\,y_{lmn} + z_{ijk} \tag{3}$$

The output function is similar to CNN 2D:

$$y_{ijk} = f(\mathbf{x}_{ijk}) = \frac{1}{2}\left(|x_{ijk}+1| + |x_{ijk}-1|\right)$$

For the problem-solving of three-dimensional PDE, the CNN 3D must be used. The original PDE is differentiated and from that the appropriate templates (A,B,z) of the CNN 3D are generated.

2.2 Field-programmable gate array technology

Field-programmable gate array (FPGA) is the technology in which the blank blocks have available resources of logic gates and RAM blocks are used to implement complex digital computations. FPGAs can be used to implement any logical function. The FPGA block is able to update the functionality after shipping, partial reconfiguration of a portion of the design, and the low nonrecurring engineering costs relative to an ASIC design [13–16].

A recent trend has been to take the coarse-grained architectural approach by combining the logic blocks and interconnects of traditional FPGAs with embedded chips and related peripherals to form a complete "system on a programmable chip" [17–19].

Users like teachers and students could use FGGA for making prototypes for testing application system, with VHDL or Verilog users easily design and test and then reconfigure the system until it has desired results.

2.3 Using FPGA to make CNN chip for solving PDE

Because the CNN architecture is not the same for every application, based on the standard model, the designer develops a particular chip for each problem. FPGA is the most useful for configuring a blank chip to make a CNN chip using programming language like Verilog or VHDL. For solving PDE, firstly, one needs to analyze (differencing) the original model of partial differential equations for finding appropriate template, then base on template found designing architecture CNN chip, finally, using VHDL to configure FPGA following designed hardware making CNN chip.

Some PDEs have been solved using the CNN technology:

Burger equation [3]:

$$\frac{\partial u(x,t)}{\partial t} = \frac{1}{R}\frac{\partial^2 u(x,t)}{\partial x^2} - u(x,t)\frac{\partial u(x,t)}{\partial x} + F(x,t)$$

Klein-Gordon equation [19]:

$$\frac{\partial^2 u(x,t)}{\partial t^2} = \nabla^2 u(x,t) - \sin u(x,t)$$

Heat diffusion equation [3]:

$$\frac{\partial u(x, y,t)}{\partial t} = c\nabla^2 u(x, y,t)$$

Black-Scholes equation [9]:

$$\frac{\partial V(x,t)}{\partial t} = rV(x,t) - \frac{1}{2}\sigma^2 S^2 \frac{\partial^2 V(x,t)}{\partial S^2} - \mathrm{rS}\frac{\partial V(x,t)}{\partial S}$$

Air pollution equation [4]:

$$\frac{\partial \varphi}{\partial t} + div\boldsymbol{v}\varphi + \sigma\varphi - \gamma\frac{\partial^2 \varphi}{\partial z^2} - \mu\nabla^2\varphi = f(x, y,z)$$

Saint venant 2D equation [5]:

$$\frac{\partial H}{\partial t}+\frac{\partial u}{\partial x}+\frac{\partial v}{\partial y}=0$$

$$\frac{\partial u}{\partial t}+\frac{\partial u^2}{\partial x}+g\frac{\partial H}{\partial x}+\frac{\partial uv}{\partial y}=-gu\frac{\left(u^2+v^2\right)^{1/2}}{K_x^2H^2}$$

$$\frac{\partial v}{\partial t}+\frac{\partial v^2}{\partial y}+g\frac{\partial H}{\partial y}+\frac{\partial uv}{\partial x}=-gv\frac{\left(u^2+v^2\right)^{1/2}}{K_y^2H^2}$$

Saint venant 1D equation [6]:

$$b\frac{\partial h(x,t)}{\partial t}+\frac{\partial Q(x,t)}{\partial x}=q \tag{4}$$

$$\frac{\partial Q(x,t)}{\partial t}+\frac{\partial\left[\frac{Q(x,t)^2}{bh(x,t)}\right]}{\partial x}+gbh(x,t)\frac{\partial h(x,t)}{\partial x}-gIbh(x,t)+gJbh(x,t)=k_q q \tag{5}$$

Example of making a CNN chip for solving Saint venant 1D:

- Designing the templates

First, changing the original equation (4)

$$b\frac{\partial h(x,t)}{\partial t}+\frac{\partial Q(x,t)}{\partial x}=q$$

$$\Leftrightarrow\frac{\partial h(x,t)}{\partial t}=\frac{-\partial Q(x,t)}{b\partial x}+\frac{q}{b} \tag{6}$$

and then choosing the difference space of variables x with step Δx for right part of (6). After differencing only the right side of (6) for space variable x by Taylor expansion, one has equation for cell at position (i):

$$\frac{\partial h}{\partial t}=-\frac{1}{2b\Delta x}\left(Q_{i+1}-Q_{i-1}\right)+\frac{q}{b} \tag{7}$$

Note that, following the CNN algorithm, on the left, we do use symbol $(\partial h/\partial t)$. From (7), one has found templates:

$$A^{hQ}=\left[\frac{1}{2b\Delta x}\ \frac{1}{R^h}\ \frac{-1}{2b\Delta x}\right];\mathrm{B^h}=[0\ 1\ 0];\mathrm{z^h}=0;$$

where R^h is the linear resistance on cell circuit of h.
For Eq. (5), changing slightly with assumptions above:

$$\frac{\partial Q(x,t)}{\partial t}+\frac{\partial\left[\frac{Q(x,t)^2}{bh(x,t)}\right]}{\partial x}+gbh(x,t)\frac{\partial h(x,t)}{\partial x}-gIbh(x,t)+gJbh(x,t)=k_q q \qquad (8)$$

Assume that $q > 0$, then $k_q = 0$. After differencing, applying the template design algorithm of CNN, one can has templates for (8):

$$A^Q = \frac{Q_{i+1}}{2b\Delta x h_{i+1}} \quad \frac{1}{R^Q} \quad -\frac{Q_{i-1}}{2b\Delta x h_{i-1}}];$$

$$A^{Qh} = \left[\frac{gbh_i}{2\Delta x} \quad gb(I-J) \quad -\frac{gbh_i}{2\Delta x}\right]; B^Q = 0; z^Q = 0;$$

From template found, we can design the CNN architecture for problem as (1) two layered-1D CNN chip (**Figure 4**) and (2) the h, Q processing block (**Figure 5**).

The cell is mixed both of h, Q in one block to make the physical architecture of a CNN cell.

In general, for each calculation, we need some basic computing block like ADDITION, SUBTRACT, MULTIPLE, DIVIDE. When designing a CNN cell using FPGA, one has to design many separate blocks of them to perform arithmetical

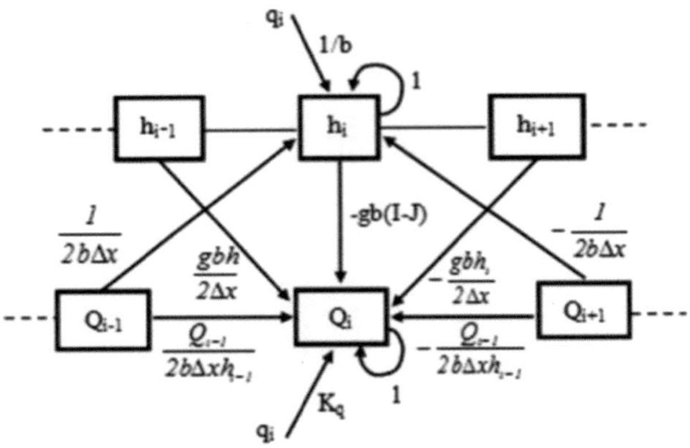

Figure 4.
Logical architecture of a CNN cell.

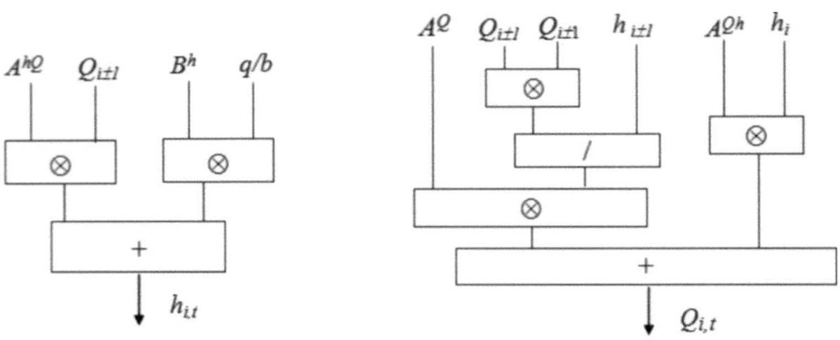

Figure 5.
Logical architecture of a h, Q cell.

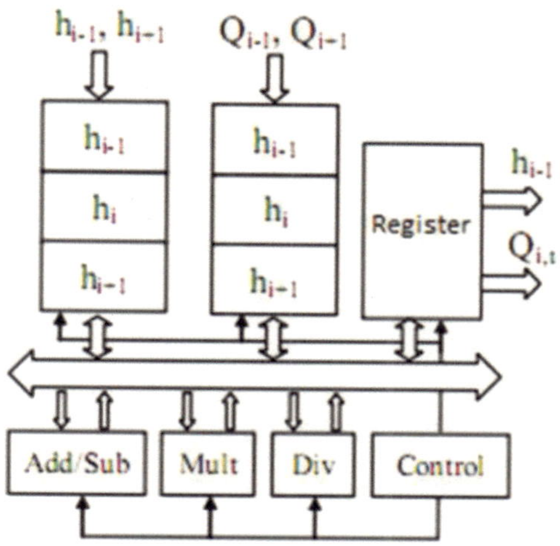

Figure 6.
Physical architecture of CNN cell.

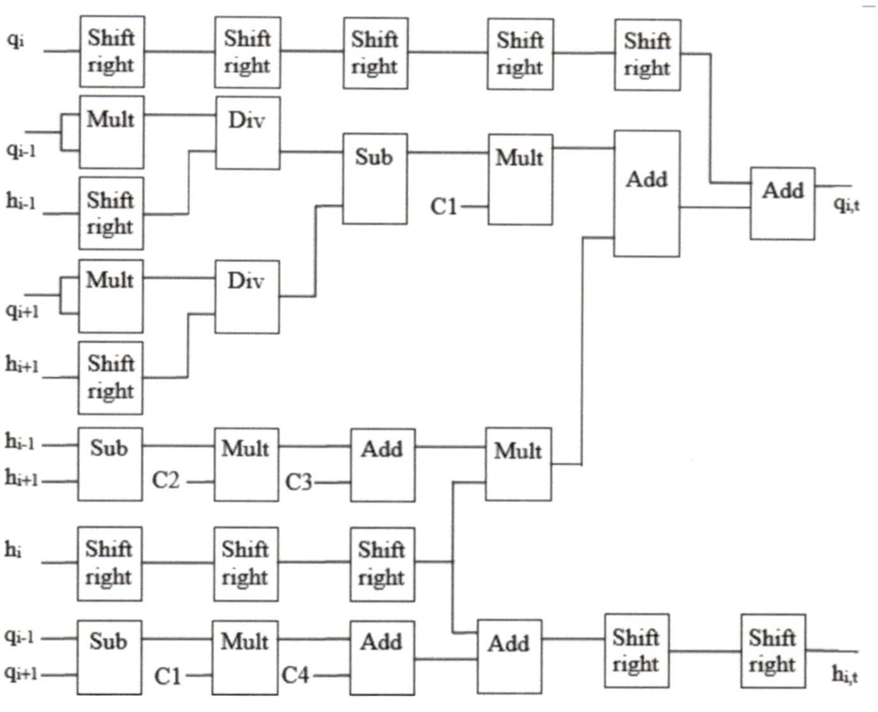

Figure 7.
Solution for physical architecture CNN chip.

processing for each input. In order to save computing resource in FPGA, the method that shares basic block in one cell leading to sequential calculating can be used (**Figure 6**). In this case, the processing time of each cell will be high. To reduce the processing time of each cell, we can use a pipeline mechanism shown in **Figure 7**, but it needs more computing resource for each cell. Finally, for cells in a CNN chip, we process parallel as in **Figure 8**.

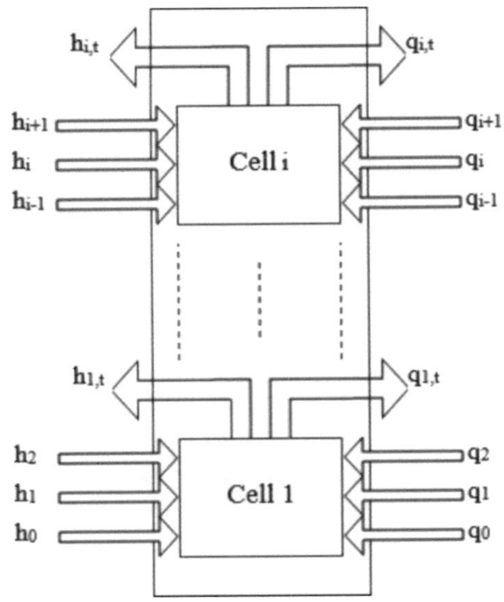

Figure 8.
A core architecture for CNN chip.

C1, ..., C4 are the coefficients as shown in **Figure** 7, (C1= $\frac{1}{2b\Delta x}\Delta t$; C2= $\frac{gb}{2\Delta x}\Delta t$; C3= $gb(I-J)\Delta t$; C4= $\frac{q}{b}\Delta t$).

If each cell is uses a pipeline mechanism shown in **Figure** 7. With the length of a pipeline is 6, the first calculation pays 6 clock pulse (clk), and each calculation after that only needs 1 clk.

3. Solving Navier-Stokes equations

3.1 Physico-mathematical model of Navier-Stokes equations

In hydraulics, many flow models have been researched, such as flows in channels, streams, or rivers, for controlling the flow for preventing disasters, saving water, and exploiting energy of the flow as well. Most of mathematical models of those phenomena are partial differential equations like Saint venant equations and Navier-Stokes equations [8, 9]. Some types of Navier-Stokes equations have various parameters and constraints. Using CNN technology, we could solve some of them which have clear values of boundary conditions; it means we do not research boundary problems deeply. The effectiveness of the CNN technology is making a physical parallel computing chip to increase the computing speed for satisfying a real-time system.

Navier-Stokes equations here consist of three partial differential equations, with functional variables representing water height, and flow velocity in x- and y-directions. The empirical model is a flow through a small port, which diffuses in two directions Ox and Oy.

Solving Navier-Stokes equations by using CNN requires the discretion of continuity model by difference method, the smaller difference intervals the higher accuracy. However, if difference intervals are too small, then it leads to increasing the calculation complexity and time. The CNN chip with parallel physically processing abilities, the above difficulties will be overcome.

3.2 Description equations in Navier-Stokes equations

- Equations describing the water level

$$\frac{\partial\rho z_w}{\partial t} + \frac{\partial\rho q_x}{\partial x} + \frac{\partial\rho q_y}{\partial y} = \rho q_A \tag{9}$$

Assume that the height of water is taken from the bottom of the flow, which is regarded as the origin of the coordinate system, so z_w has no negative values.

- Momentum equations in x-direction:

$$\frac{\partial\rho q_x}{\partial t} + \frac{\partial}{\partial x}\left(\rho\beta\frac{q_x^2}{d}\right) + \frac{\partial}{\partial y}\left(\rho\beta\frac{q_x q_y}{d}\right) + \rho g d\frac{\partial z_w}{\partial x} + \rho g d S_{fx} - \tau_{wx} - \frac{\partial}{\partial x}\left(\rho K_L\frac{\partial q_x}{\partial x}\right) - \frac{\partial}{\partial y}\left(\rho K_T\frac{\partial q_x}{\partial y}\right) = 0 \tag{10}$$

- Momentum equations in y-direction:

$$\frac{\partial\rho q_y}{\partial t} + \frac{\partial}{\partial y}\left(\rho\beta\frac{q_y^2}{d}\right) + \frac{\partial}{\partial x}\left(\rho\beta\frac{q_y q_x}{d}\right) + \rho g d\frac{\partial z_w}{\partial y} + \rho g d S_{fy} - \tau_{wy} - \frac{\partial}{\partial y}\left(\rho K_L\frac{\partial q_y}{\partial y}\right) - \frac{\partial}{\partial x}\left(\rho K_T\frac{\partial q_y}{\partial x}\right) = 0 \tag{11}$$

Explain the meanings of quantities in the equations:

- $\frac{\partial\rho q_x}{\partial t}$ and $\frac{\partial\rho q_y}{\partial t}$: quantities characterizing the momentum variation over time in x-axis and y-axis, respectively.
- $\frac{\partial}{\partial x}\left(\rho\beta\frac{q_x^2}{d}\right)$ and $\frac{\partial}{\partial y}\left(\rho\beta\frac{q_y^2}{d}\right)$: kinetic energy variations of flow in x- and y-directions.
- $\rho g d\frac{\partial z_w}{\partial x}$ and $\rho g d\frac{\partial z_w}{\partial y}$: potential energy variations of flow in x- and y-directions.
- $\rho g d S_{fx}$ and $\rho g d S_{fy}$: influence of friction by bottom and walls of channel on flow in x- and y-directions. Values of S_{fx} and S_{fy} are determined based on physical properties of bottom and walls of hydraulic channels according to the following formulas:

$$S_{fx} = q_x\frac{n^2\left(q_x^2+q_y^2\right)^{1/2}}{d^{1/3}};S_{fy} = q_y\frac{n^2\left(q_y^2+q_x^2\right)^{1/2}}{d^{1/3}} \quad \text{(n is Manning coefficient)}$$

- τ_{wx} and τ_{wy}: wind pressure on free surface of hydraulic flow in x-and y-directions are calculated as follows:

$$\tau_{wx} = c_s\rho_a W^2\cos(\Psi);\tau_{wy} = c_s\rho_a W^2\sin(\Psi),$$

where:

$$c_x = \begin{Bmatrix} 10^{-3};khi\ W \le W_{min} \\ [c_{s1} + c_{s2}(W\text{-}W_{min})].10^{-3};khi\ W > W_{min} \end{Bmatrix};$$

With c_{s1}; c_{s2}; W_{min} are values get from practical, for example: $W_{min} = 4$ m/s; wind speed is 10 m/s, then $c_{s1} = 1.0$; $c_{s2} = 0.067$;

- ρ_a is the air density at free surface (kgm^{-3}); W is wind speed at free surface; and Ψ is the angle between wind direction and x-axis.
- Expressions, $\frac{\partial}{\partial x}\left(\rho K_L \frac{\partial q_x}{\partial x}\right) - \frac{\partial}{\partial y}\left(\rho K_T \frac{\partial q_x}{\partial y}\right)$ and $\frac{\partial}{\partial y}\left(\rho K_L \frac{\partial q_y}{\partial y}\right) - \frac{\partial}{\partial x}\left(\rho K_T \frac{\partial q_y}{\partial x}\right)$, are the impact of turbulence in hydraulic flow caused between x- and y-directions, where: $K_L = \frac{q_x l}{P_e}$ with Pe as the Peclet coefficient with the value of 15–40; l as the length of flow; K_L as coefficient varying according to locations along flow; and $K_T = 0.3–0.7\ K_L$.

3.3 Analyzing and designing CNN to solve the equations

To simplify, change parameters as: the water level $z_w = h$; and the velocity in x-axis $q_x = u$, in y-axis $q_y = v$. Assume that $q_A = 0$; the kinetic influence of turbulent values between velocity in the direction from 0y to 0x (or 0x to 0y) is trivial since horizontal velocity is small enough to be considered as zero; then (9)–(11) are rewritten:

$$\frac{\partial h}{\partial t} + \frac{\partial u}{\partial x} + \frac{\partial v}{\partial y} = 0 \Leftrightarrow \frac{\partial h}{\partial t} = -\frac{\partial u}{\partial x} - \frac{\partial v}{\partial y} \tag{12}$$

$$\frac{\partial v}{\partial t} + \frac{\partial}{\partial y}\left(\beta\frac{v^2}{d}\right) + \frac{\partial}{\partial x}\left(\beta\frac{vu}{d}\right) + gd\frac{\partial h}{\partial y} + gdS_{fy} - \frac{\tau_{wy}}{\rho} - \frac{\partial}{\partial y}\left(K_L\frac{\partial v}{\partial y}\right) = 0$$

$$\Leftrightarrow \frac{\partial v}{\partial t} = \frac{\partial}{\partial y}\left(K_L\frac{\partial v}{\partial y}\right) - \frac{\partial}{\partial y}\left(\beta\frac{v^2}{d}\right) - \frac{\partial}{\partial x}\left(\beta\frac{vu}{d}\right) - gd\frac{\partial h}{\partial y} + \left(\frac{\tau_{wy}}{\rho} - gdS_{fy}\right) \tag{13}$$

$$\frac{\partial u}{\partial t} = \frac{\partial}{\partial x}\left(\beta\frac{u^2}{d}\right) + \frac{\partial}{\partial y}\left(\beta\frac{uv}{d}\right) + gd\frac{\partial h}{\partial x} + gdS_{fx} - \frac{\tau_{wx}}{\rho} - \frac{\partial}{\partial x}\left(K_L\frac{\partial u}{\partial x}\right)$$

$$\Leftrightarrow \frac{\partial u}{\partial t} = \frac{\partial}{\partial x}\left(K_L\frac{\partial u}{\partial x}\right) - \frac{\partial}{\partial x}\left(\beta\frac{u^2}{d}\right) - \frac{\partial}{\partial y}\left(\beta\frac{uv}{d}\right) - gd\frac{\partial h}{\partial x} + \left(\frac{\tau_{wx}}{\rho} - gdS_{fx}\right) \tag{14}$$

Step 1: Differencing equations following Taylor formula

Using finite difference grid with difference interval in x-axis as Δx and in y-axis as Δy and apply Taylor difference formulas for Eqs. (12)–(14); we have difference equations corresponding to the equations:

$$\frac{\partial h_{ij}}{\partial t} = \frac{u_{i+1,j} - u_{i-1,j}}{2\Delta x} - \frac{v_{i,j+1} - v_{i,j-1}}{2\Delta y} \tag{15}$$

$$\frac{\partial u_{i,j}}{\partial t} = -\frac{\beta}{d}\left[\frac{u_{i+1,j}}{2\Delta x}u_{i+1,j} - \frac{u_{i-1,j}}{2\Delta x}u_{i-1,j}\right] - \frac{\beta}{d}\left[\frac{v_{i,j+1}}{2\Delta y}u_{i+1,j} - \frac{v_{i,j-1}}{2\Delta y}u_{i-1,j}\right]$$

$$-gd\frac{h_{i+1,j} - h_{i-1,j}}{2\Delta x}gdS_{fx} + \frac{1}{\rho}\tau_{wx}K_L\frac{u_{i+1,j} - 2u_{i,j} + u_{i-1,j}}{\Delta x^2}] \tag{16}$$

$$\frac{\partial v_{i,j}}{\partial t} = -\frac{\beta}{d}\left[\frac{v_{\mathrm{i,\,j+1}}}{2\Delta y}v_{i,j+1} - \frac{v_{\mathrm{i,\,j\text{-}1}}}{2\Delta y}v_{i,j-1}\right] - \frac{\beta}{d}\left[\frac{u_{\mathrm{i+1,\,j}}}{2\Delta x}v_{i,j+1} - \frac{u_{\mathrm{i\text{-}1,\,j}}}{2\Delta x}v_{i,j-1}\right]$$
$$-gd\frac{h_{i,j+1} - h_{i,j-1}}{2\Delta x} - \mathrm{gdS_{fy}} + \frac{1}{\rho}\tau_{\mathrm{wy}}K_L\frac{v_{i,j+1} - 2v_{i,j} + v_{i,j-1}}{\Delta y^2}] \tag{17}$$

Step 2: Designing a sample of CNN

Based on CNN state equations and difference equations (15)–(17), we can have CNN templates for layers h, u, v:

- Layer h:

$$A^{hu} = \begin{bmatrix} 0 & 0 & 0 \\ \frac{1}{2\Delta x} & 0 & \frac{-1}{2\Delta x} \\ 0 & 0 & 0 \end{bmatrix} \quad A^{hv} = \begin{bmatrix} 0 & \frac{1}{2\Delta y} & 0 \\ 0 & 0 & 0 \\ 0 & \frac{-1}{2\Delta y} & 0 \end{bmatrix} \tag{18}$$

- Layer u:

$$A^{uv} = \begin{bmatrix} 0 & \frac{\beta u_{i,j-1}}{2d\Delta y} & 0 \\ 0 & 0 & 0 \\ 0 & \frac{-\beta u_{i,j+1}}{2d\Delta y} & 0 \end{bmatrix}; A^{uh} = \begin{bmatrix} 0 & 0 & 0 \\ \frac{gd}{2\Delta x} & 0 & \frac{-gd}{2\Delta x} \\ 0 & 0 & 0 \end{bmatrix}; B^{u} = \frac{1}{\rho}\tau_{wx}\begin{bmatrix} 0 & 0 & 0 \\ 0 & 1 & 0 \\ 0 & 0 & 0 \end{bmatrix}$$

$$A^{u} = \begin{bmatrix} 0 & 0 & 0 \\ \frac{\beta u_{i-1,j}}{2d\Delta x} + \frac{K_L}{\Delta x^2} & gd\frac{n^2\left(u_{ij}^2 + v_{ij}^2\right)^{1/2}}{d^{1/3}} + \frac{1}{R_u} + \frac{4K_L}{\Delta x^2} & \frac{-\beta u_{i+1,j}}{2d\Delta x} + \frac{-K_L}{\Delta x^2} \\ 0 & 0 & 0 \end{bmatrix}; \quad z^{\mathrm{u}} = 0 \tag{19}$$

- Layer v:

$$A^{vh} = \begin{bmatrix} 0 & \frac{\mathrm{gd}}{2\Delta y} & 0 \\ 0 & 0 & 0 \\ 0 & \frac{-gd}{2\Delta y} & 0 \end{bmatrix}; A^{vu} = \begin{bmatrix} 0 & 0 & 0 \\ \frac{\beta u_{i-1,j}}{2d\Delta x} & 0 & \frac{-\beta u_{i-1,j}}{2d\Delta x} \\ 0 & 0 & 0 \end{bmatrix}; B^{v} = \frac{1}{\rho}\tau_{\mathrm{wy}}\begin{bmatrix} 0 & 0 & 0 \\ 0 & 1 & 0 \\ 0 & 0 & 0 \end{bmatrix}; z^{v} = 0$$

$$A^{v} = \begin{bmatrix} 0 & \frac{\beta v_{i,j+1}}{2d\Delta y} + \frac{K_L}{\Delta y^2} & 0 \\ \frac{K_L}{\Delta y^2} & \mathrm{gd}\frac{n^2\left(u_{i,j}^2 + v_{i,j}^2\right)^2}{d^{1/3}} + \frac{1}{R^v} + \frac{K_L}{\Delta y^2} & \frac{-K_L}{\Delta y^2} \\ 0 & \frac{-\beta v_{i,j+1}}{2d\Delta y} - \frac{K_L}{\Delta y^2} & 0 \end{bmatrix} \tag{20}$$

Step 3: Designing hardware architecture of CNN to solve Navier-Stokes equations

Based on templates found in (18)–(20), we can design an architecture for circuit for CNN chip. It is a three-layered CNN 2D. Then, the arithmetic unit for each layer and links to perform parallel calculation on chip can be made. **Figure 9** shows the architecture of layer h and layer u (the layer v is similar to u).

3.4 Proposed system architecture for MxN CNN

The empirical problems that need a solution is that: firstly, identifying boundary points of whole difference grid (space); secondly, dividing the entire computing space into smaller subspaces. Division and combination of boundary areas need to perform appropriately avoiding incorrect results because of tep time computing time; thirdly, controlling real-time data exchange and combining sequential and parallel computing in a CNN chip. The CNN chip proposed in this chapter has solved similarity in the previous problems [4, 5]. The new issues here are dividing computing space processing dynamic sub-boundary and combining sequential and parallel.

3.4.1 General MxN CNN

Each CNN cell has its own data element and a core that performs the computing function. The CNN has MxN CNN cells in which only (M-2)x(N-2) CNN cells have computing functions, so that the CNN has MxN data elements and (M-2)x(N-2) cores (**Figure 10**).

The Buffer supplies MxN data elements for CNN. Each MxN data element is called as one block of data (**Figure 11**).

The white area is the data element for CNN boundary cells; and the gray part is the data area which requires to be processed by CNN. The CNN arithmetic unit has size of (M-2)x(N-2) cells processing data for the gray area which is inside the input buffer unit.

The Input memory has PxQ blocks of data. It is a true dual port memory.

The Temp memory also has PxQ blocks of data. It is a simple dual port memory. It is used to temporarily store data computed from CNN core and supply data for Boundary updating unit.

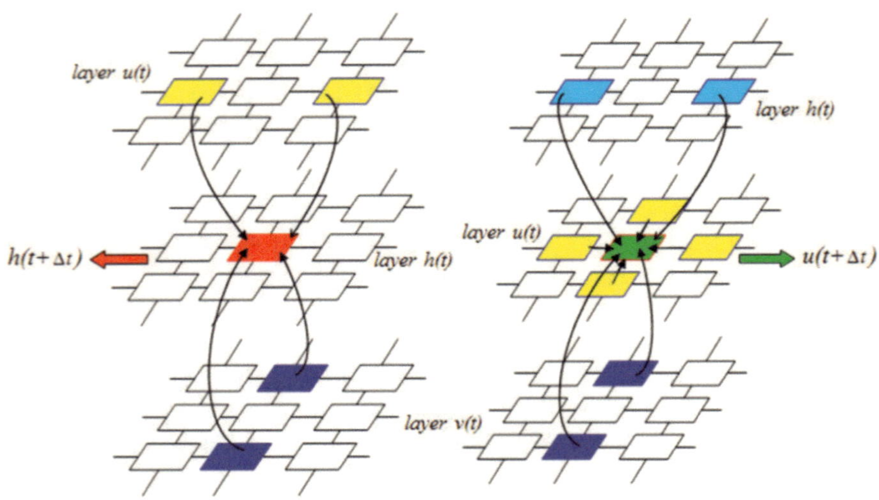

Figure 9.
Logic architecture of cell of h, u.

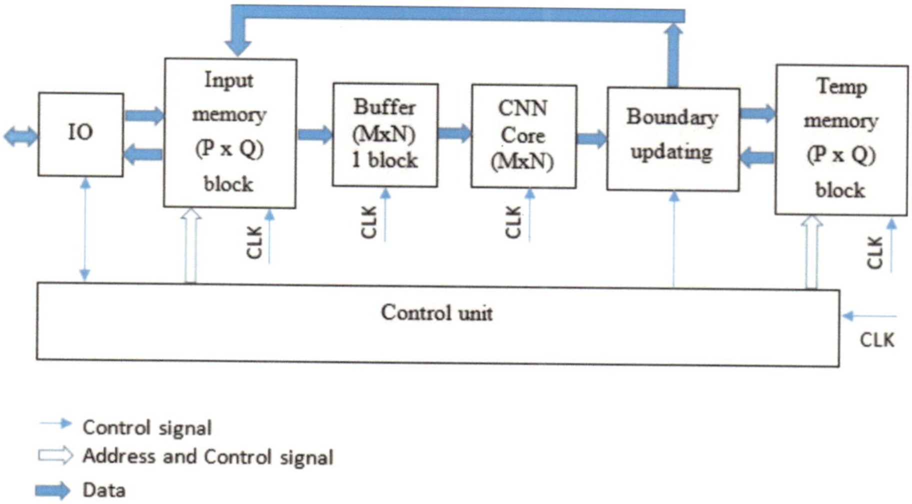

Figure 10.
General architecture of a CNN chip.

0,0	0,1	...	0,N-1
1,0			
...			
M-1,0			

Figure 11.
Buffer (MxN) for CNN core.

Data that need processing sent from PC have the size of mxn (**Figure 12**).

Assume that m = 5, n = 6, M = 3, and N = 4; the white part is boundary and the gray part is the area requiring to be processed. Before the processing data, temporary vertical and horizontal boundaries be need to be added, as in **Figure 13**, column (0,3) and row (3,0).

Temporary vertical and horizontal boundaries are added to the data structure similar to CNN buffer. The data after being added from temporary vertical and horizontal boundaries will be sent to Input memory. The blocks of data in the Input memory unit (in case that mxn = 5x6, MxN = 3x4) are detailed as follows (**Figure 14**).

0,0	0,1	0,2	0,3	...	0,n-1
1,0					
2,0					
...					
m-1,0					

Figure 12.
Computing space with main boundary.

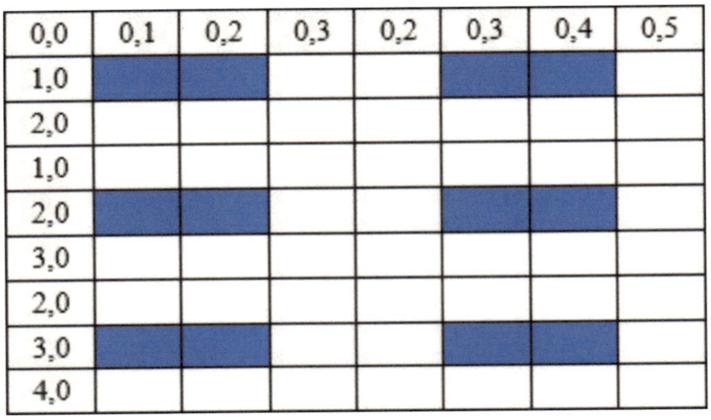

Figure 13.
Divide computing space into subspace with subboundary.

0, 1, 2,.., 6 are the addresses of blocks. In case that mxn = 5x6 and MxN = 3x4, we have P = 3 and Q = 2.

$$PxQ = \frac{m-2}{M-2}x\frac{n-2}{N-2}$$

The Boundary updating unit is in detail structure as follows (in case MxN = 3x4) (**Figure 15**).

The control unit controls the activities of the whole system set by the algorithm which is as follows:

```
At every posedge of clk do
{
        if (has IO event)
                do the IO task;
        else
                buffer = read(Input memory)
                if (finish computing the first block)
                        if (BoundaryUpdating())
                                write(Input memory)
}
```

3.4.2 Proposed CNN architecture when M = 3 (3xN CNN)

The 3xN CNN architecture is similar to the general MxN CNN architecture (M = 3). In order to reduce the memory consumption and simplify the Boundary updating unit, there are some differences (**Figure 16**).

Each block of data in the memory (Input memory or Temp memory) is 1xN data elements. Assume that the data which need processing sent from PC has the size of mxn, m = 5, n = 6, and assume that N = 4. As mention above, the data will be processed after temporary vertical boundaries are added; so that, the Input Memory unit will has 5x2 blocks of data (m = 5, Q = 2) as follow (**Figure 17**).

Each block has size of 1x4 data elements.

The Buffer unit is a Shift up register that has size of 3xN. The input and output have sizes of 1xN and 3xN, respectively. The input is at the bottom.

The Input memory has m rows and Q columns of blocks of data. The control unit reads the blocks in the Input memory by vertical and puts the block of data to the

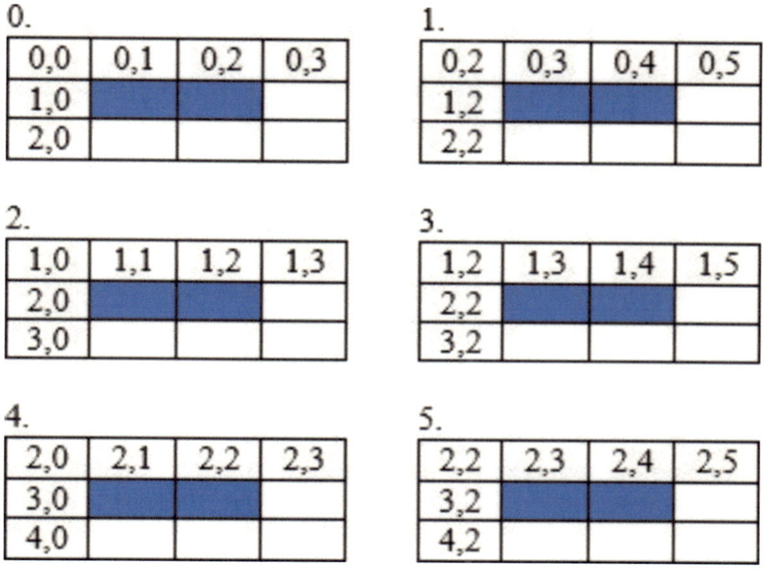

Figure 14.
The blocks of data in the Input memory in case that mxn = 5x6, MxN = 3x4.

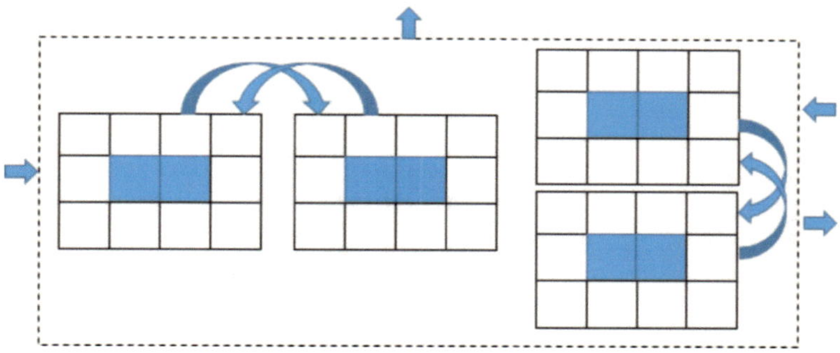

Figure 15.
The Boundary updating structure (MxN = 3x4).

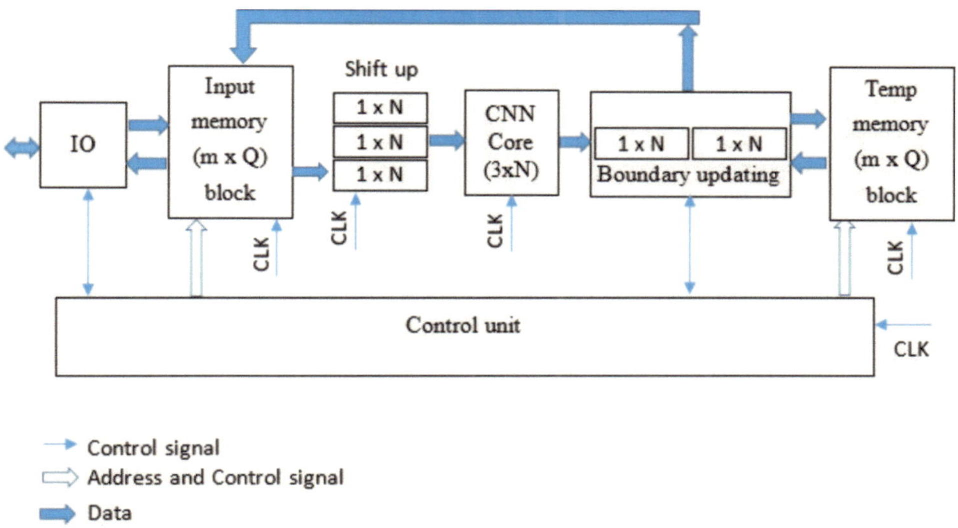

Figure 16.
The architecture of 3xN CNN chip.

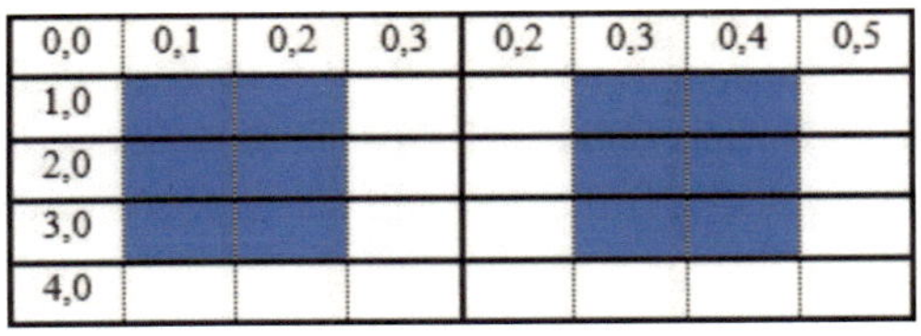

Figure 17.
The memory with 5x2 blocks (m==5, n = 6, N = 4).

input of buffer. The buffer shifts up 1 step. After step 3, the Buffer has 3xN blocks of data to supply to CNN core. After each step, the Buffer has 3xN blocks of data that need to supply to CNN core (**Figure 18**).

The output of CNN core has the size of 1xN.

The Boundary updating unit is shown in **Figure 19**.

The control algorithm for control unit (**Figure 20**).

```
(1) At every posedge of clk do
(2) {
(3)         if (has IO event)
(4)                 do the IO task;
(5)         else
(6)                 buffer = read(Input memory);//read by vertical
(7)                 if (finish computing the first block of column q)
(8)                         if (column_of_current_block==0)
                                    write(Temp memory);
                        else
                                    BoundaryUpdating(CNNoutput,read(Temp
                                    memory));
(9)                                 write(Input memory);
(10) }
```

3.5 Implementation

In this part, we implement the 3xN CNN. Q, m, and N are the parameters that we can configure before compiling and programming to the FPGA chip. For defaulting, we assigned Q = 2, m = 8, and N = 4.

3.5.1 Development environment

For experiencing, the ISE Design Suite software version 14.7 and ML605 evaluation board including chip XCVL240T-1FFG1156 (Virtex 6) are used to implement the schematic of CNN.

First, we use Verilog HDL language to describe the CNN architecture. Then, we use ISim simulator to verify our system. Finally, we program the system to the FPGA chip on ML605 board.

The image of experience system as in **Figure 20** is as follows.

3.5.2 Input data for h, u, v values

The input of CNN to solve the Navier-Stokes Equation has h, u, v values. We use three Input memory units, three Buffer units, and three Temporary memory units to store h, u, v values. The data element is represented in 32-bit floating point

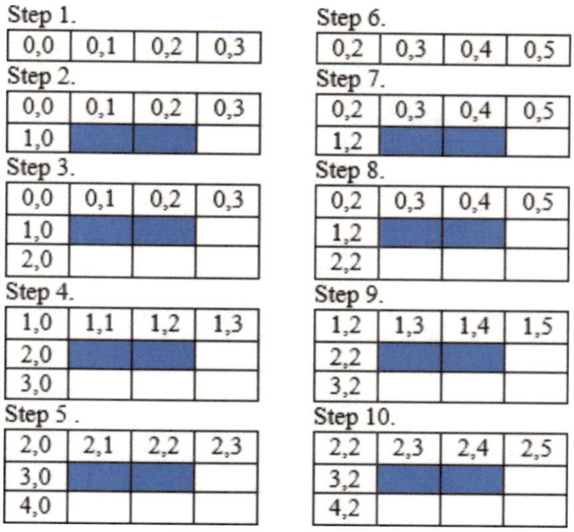

Figure 18.
The Buffer's state after each step (m==5, n = 6, N = 4).

Figure 19.
The output size of CNN core (N = 4).

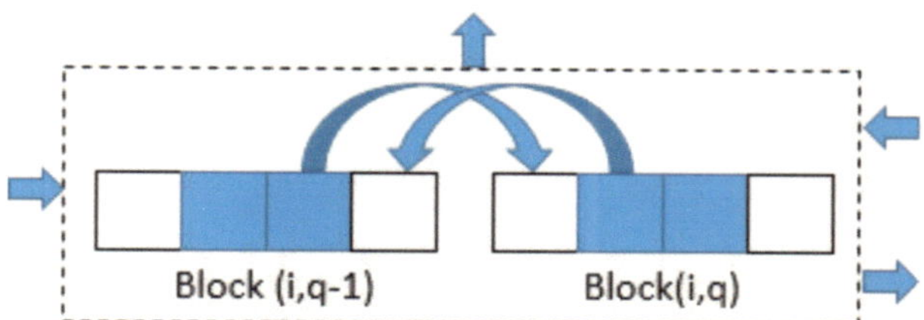

Figure 20.
The Boundary updating structure (N = 4).

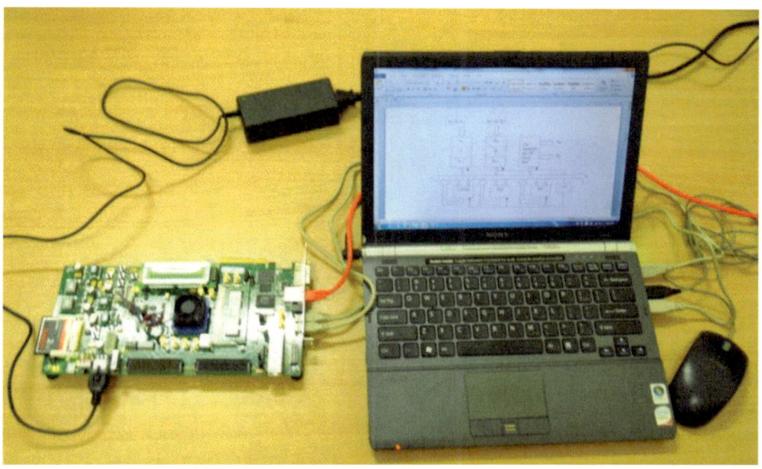

Figure 21.
The chip Virtex 6 (XCVL240T-1FFG1156) connected to PC for configuring to make CNN chip and performing calculation.

real numbers. Data into h, u, v are added with temporary boundaries, detailed as follow (presented in Decimal and Hex of Single-type Floating-point) (**Figure 22**).

The interface of each Input memory, Temporary memory for h, u, v is configurated as same in **Figure 23**. The initial data for the Input memory h, u, v is initialed by COE files. A COE file stores initial values for a memory (**Figure 24**).

0	14	14.5	15	14.5	15	14	0
14	14	14.5	15	14.5	15	14	14
24	24	13.5	13	13.5	13	14	14.5
14	14	13	13	13	13	14	14
20	20	14	13.5	14	13.5	14	14
13	13	13.5	14	13.5	14	13.5	13
14	14	13	13	13	13	14	14
0	13	13.5	14	13.5	14	13.5	0

(a1) Input data values for *h* in decimal

0	41600000	41680000	41700000	41680000	41700000	41600000	0
41600000	41600000	41680000	41700000	41680000	41700000	41600000	41600000
41c00000	41c00000	41580000	41500000	41580000	41500000	41600000	41680000
41600000	41600000	41500000	41500000	41500000	41500000	41600000	41600000
41a00000	41a00000	41600000	41580000	41600000	41580000	41600000	41600000
41500000	41500000	41580000	41600000	41580000	41600000	41580000	41500000
41600000	41600000	41500000	41500000	41500000	41500000	41600000	41600000
0	41500000	41580000	41600000	41580000	41600000	41580000	0

(a2) Input data values for *h* in hex of single-type floating-point

1.3	2	1.8	2.3	1.8	2.3	2.2	2
2	2	1.8	2.3	1.8	2.3	2.2	2
4	4	2.5	3	2.5	3	2.5	2.3
2	2	1.5	1.6	1.5	1.6	2	2.1
2.5	2.5	1.3	1.5	1.3	1.5	2	1.5
1.9	1.9	1.8	1.6	1.8	1.6	3	2
2	2	1.5	1.6	1.5	1.6	2	2.1
1.7	1.9	1.8	1.6	1.8	1.6	3	1.7

(b) Input data values for *u* in decimal

3fa66666	40000000	3fe66666	40133333	3fe66666	40133333	400ccccd	40000000
40000000	40000000	3fe66666	40133333	3fe66666	40133333	400ccccd	40000000
40800000	40800000	40200000	40400000	40200000	40400000	40200000	40133333
40000000	40000000	3fc00000	3fcccccd	3fc00000	3fcccccd	40000000	40066666
40200000	40200000	3fa66666	3fc00000	3fa66666	3fc00000	40000000	3fc00000
3ff33333	3ff33333	3fe66666	3fcccccd	3fe66666	3fcccccd	40400000	40000000
40000000	40000000	3fc00000	3fcccccd	3fc00000	3fcccccd	40000000	40066666
3fd9999a	3ff33333	3fe66666	3fcccccd	3fe66666	3fcccccd	40400000	3fd9999a

(b2) Input data values for *u* in hex of single-type floating-point

2	2.3	2	1	2	1	1.5	1.2
2.3	2.3	2	1	2	1	1.5	2.1
8	8	1.6	2	1.6	2	1.6	2.4
1.6	1.6	2.1	2	2.1	2	2	1.8
10	10	1.5	1.7	1.5	1.7	1.7	1.6
1.6	1.6	2	3	2	3	2	2.5
1.6	1.6	2.1	2	2.1	2	2	1.8
1.8	1.6	2	3	2	3	2	2

(c) Input data values for *v* in decimal

40000000	40133333	40000000	3f800000	40000000	3f800000	3fc00000	3f99999a
40133333	40133333	40000000	3f800000	40000000	3f800000	3fc00000	40066666
41000000	41000000	3fcccccd	40000000	3fcccccd	40000000	3fcccccd	4019999a
3fcccccd	3fcccccd	40066666	40000000	40066666	40000000	40000000	3fe66666
41200000	41200000	3fc00000	3fd9999a	3fc00000	3fd9999a	3fd9999a	3fcccccd
3fcccccd	3fcccccd	40000000	40400000	40000000	40400000	40000000	40200000
3fcccccd	3fcccccd	40066666	40000000	40066666	40000000	40000000	3fe66666
3fe66666	3fcccccd	40000000	40400000	40000000	40400000	40000000	40000000

(c2) Input data values for *v* in hex of single-type floating-point

Figure 22.
Initial data for the Input memory h, u, v.

3.5.3 Shift up register

```
module ShiftUpRegister(clk,din,dout);
parameter M=3;
parameter N=4;
input [32*N-1:0] din;
input clk;
output [32*M*N-1:0] dout;
reg [32*N-1:0] dout1=0;
reg [32*N-1:0] dout2=0;
reg [32*N-1:0] dout3=0;
assign dout={dout3,dout2,dout1};

always @(posedge clk)
begin
     dout3=dout2;
     dout2=dout1;
     dout1=din;
end
endmodule
```

```
module InputBuffer(clk,doutH,doutU,doutV,
          matrixhin,matrixuin,matrixvin);
parameter M=3;
parameter N=4;

input clk;
input[32*N-1:0]doutH,doutU,doutV;
output wire[32*M*N-1:0] matrixhin;
output wire[32*M*N-1:0] matrixuin;
output wire[32*M*N-1:0] matrixvin;

ShiftUpRegister #(M,N) sH(
       .clk(clk),
       .din(doutH),
       .dout(matrixhin));
ShiftUpRegister #(M,N) sU(
       .clk(clk),
       .din(doutU),
       .dout(matrixuin));
ShiftUpRegister #(M,N) sV(
       .clk(clk),
       .din(doutV),
       .dout(matrixvin));
endmodule
```

3.5.4 CNN core

```
module CNNCore(clk,matrixhin,matrixuin,matrixvin,matrixhout,matrixuout,matrixvout);
parameter M=3;
parameter N=4;
parameter PipelineLength=6;

input clk;
input [32*M*N-1:0] matrixhin,matrixuin,matrixvin;
output[32*N-1:0] matrixhout,matrixuout,matrixvout;

ShiftRightN #(PipelineLength) HSR1(clk,matrixhin[32*N+31:32*N],matrixhout[31:0]);
ShiftRightN #(PipelineLength) HSR2(clk,matrixhin[32*(2*N)-1:32*(2*N)-32],matrixhout[32*N-1:32*N-32]);
ShiftRightN #(PipelineLength) HSR3(clk,matrixuin[32*N+31:32*N],matrixuout[31:0]);
ShiftRightN #(PipelineLength) HSR4(clk,matrixuin[32*(2*N)-1:32*(2*N)-32],matrixuout[32*N-1:32*N-32]);
ShiftRightN #(PipelineLength) HSR5(clk,matrixvin[32*N+31:32*N],matrixvout[31:0]);
ShiftRightN #(PipelineLength) HSR6(clk,matrixvin[32*(2*N)-1:32*(2*N)-32],matrixvout[32*N-1:32*N-32]);

generate
   genvar i,j;
   for(i=0;i<M-2;i=i+1)
      for(j=0;j<N-2;j=j+1)
         Cell TB(
            .clk(clk),
            .Hisubljsubl   (matrixhin[32*((i+2)*N+j+2)+31:32*((i+2)*N+j+2)]),
            .Hijsubl       (matrixhin[32*((i+2)*N+j+1)+31:32*((i+2)*N+j+1)]),
            .Hiaddljsubl   (matrixhin[32*((i+2)*N+j+0)+31:32*((i+2)*N+j+0)]),
            .Hisublj       (matrixhin[32*((i+1)*N+j+2)+31:32*((i+1)*N+j+2)]),
            .Hij           (matrixhin[32*((i+1)*N+j+1)+31:32*((i+1)*N+j+1)]),
            .Hiaddlj       (matrixhin[32*((i+1)*N+j+0)+31:32*((i+1)*N+j+0)]),
            .Hisubljaddl   (matrixhin[32*((i+0)*N+j+2)+31:32*((i+0)*N+j+2)]),
            .Hijaddl       (matrixhin[32*((i+0)*N+j+1)+31:32*((i+0)*N+j+1)]),
            .Hiaddljaddl   (matrixhin[32*((i+0)*N+j+0)+31:32*((i+0)*N+j+0)]),
            .Hijnew        (matrixhout[32*(i*N+j+1)+31:32*(i*N+j+1)]),
            .Uisubljsubl   (matrixuin[32*((i+2)*N+j+2)+31:32*((i+2)*N+j+2)]),
            .Uijsubl       (matrixuin[32*((i+2)*N+j+1)+31:32*((i+2)*N+j+1)]),
            .Uiaddljsubl   (matrixuin[32*((i+2)*N+j+0)+31:32*((i+2)*N+j+0)]),
            .Uisublj       (matrixuin[32*((i+1)*N+j+2)+31:32*((i+1)*N+j+2)]),
            .Uij           (matrixuin[32*((i+1)*N+j+1)+31:32*((i+1)*N+j+1)]),
            .Uiaddlj       (matrixuin[32*((i+1)*N+j+0)+31:32*((i+1)*N+j+0)]),
            .Uisubljaddl   (matrixuin[32*((i+0)*N+j+2)+31:32*((i+0)*N+j+2)]),
            .Uijaddl       (matrixuin[32*((i+0)*N+j+1)+31:32*((i+0)*N+j+1)]),
            .Uiaddljaddl   (matrixuin[32*((i+0)*N+j+0)+31:32*((i+0)*N+j+0)]),
            .Uijnew        (matrixuout[32*(i*N+j+1)+31:32*(i*N+j+1)]),
            .Visubljsubl   (matrixvin[32*((i+2)*N+j+2)+31:32*((i+2)*N+j+2)]),
            .Vijsubl       (matrixvin[32*((i+2)*N+j+1)+31:32*((i+2)*N+j+1)]),
            .Viaddljsubl   (matrixvin[32*((i+2)*N+j+0)+31:32*((i+2)*N+j+0)]),
            .Visublj       (matrixvin[32*((i+1)*N+j+2)+31:32*((i+1)*N+j+2)]),
            .Vij           (matrixvin[32*((i+1)*N+j+1)+31:32*((i+1)*N+j+1)]),
            .Viaddlj       (matrixvin[32*((i+1)*N+j+0)+31:32*((i+1)*N+j+0)]),
            .Visubljaddl   (matrixvin[32*((i+0)*N+j+2)+31:32*((i+0)*N+j+2)]),
            .Vijaddl       (matrixvin[32*((i+0)*N+j+1)+31:32*((i+0)*N+j+1)]),
            .Viaddljaddl   (matrixvin[32*((i+0)*N+j+0)+31:32*((i+0)*N+j+0)]),
            .Vijnew        (matrixvout[32*(i*N+j+1)+31:32*(i*N+j+1)])
            );
endgenerate
endmodule
```

3.5.5 Boundary updating

```
module BoundaryUpdating(iEnable,iLeftBoundary,iRightBoundary,oLeftBoundary,oRightBoundary);
parameter N = 4;

    input iEnable;
    input [32*N-1:0] iLeftBoundary;
    input [32*N-1:0] iRightBoundary;
    output reg[32*N-1:0] oLeftBoundary;
    output reg[32*N-1:0] oRightBoundary;
always @(iEnable or iLeftBoundary or iRightBoundary)
   begin
      oLeftBoundary=iLeftBoundary;
      oRightBoundary=iRightBoundary;
      if(iEnable)
         begin
            oRightBoundary[32*N-1:32*N-32]=iLeftBoundary[32*2-1:32*2-32];
            oLeftBoundary[32*1-1:32*1-32]=iRightBoundary[32*(N-1)-1:32*(N-1)-32];
         end
   end
endmodule
```

3.5.6 Control unit

The interface of Control unit is described as follows.

```
module Control(clk,CountCLK,rdaddressHUV,wraddressHUV,wren,
         rdaddressHUVnew,wraddressHUVnew,wrenNew,
         start,EnableBoundaryUpdating,finish,read);
parameter N=4
parameter m_row=8;
parameter Q_Column=2;

input clk;
input start;
input read;

output reg[8:0]rdaddressHUV=0;
output reg[8:0]wraddressHUV=0;
output reg wren=0;
output reg[8:0]rdaddressHUVnew=0;
output reg[8:0]wraddressHUVnew=0;
output reg wrenNew=0;
output reg finish=1;
output reg EnableBoundaryUpdating;
output reg[9:0]CountCLK=0;
```

3.5.7 System scheme

To verify the system, the interface of the top module of the system should include all the signals that we want to verify.

The top module is described as follows.

```
module CNNSystem(clk,start,rdaddr,doutH,doutU,doutV,
              matrixhin,matrixuin,matrixvin,matrixhout,matrixuout,matrixvout,
              wraddrTemp,wrenTemp,HNewTemp,UNewTemp,VNewTemp,
              rdaddrTemp,doutHNewTemp,doutUNewTemp,doutVNewTemp,
              EnableBoundaryUpdating,wraddr,wren,HNew,UNew,VNew,finish,CountCLK);
parameter M=3;
parameter N=4;
input clk;
input start;
//Read Input memory
output wire[5:0]rdaddr;
output wire [32*N-1:0]doutH,doutU,doutV;
//CNN core
output wire[32*M*N-1:0] matrixhin;
output wire[32*M*N-1:0] matrixuin;
output wire[32*M*N-1:0] matrixvin;
output wire[32*N-1:0] matrixhout;
output wire[32*N-1:0] matrixuout;
output wire[32*N-1:0] matrixvout;
//Write Temp memory
output wire[5:0]wraddrTemp;
output wire wrenTemp;
output wire[32*N-1:0]HNewTemp,UNewTemp,VNewTemp;
//Read Temp memory and Boundary updating
output wire[5:0]rdaddrTemp;
output wire [32*N-1:0]doutHNewTemp,doutUNewTemp,doutVNewTemp;
output wire EnableBoundaryUpdating;
//Write results Input memory
output wire[5:0]wraddr;
output wire wren;
output wire[32*N-1:0]HNew,UNew,VNew;
output wire finish;
output wire[9:0]CountCLK;
```

```
Control CU(
        .CountCLK(CountCLK),
        .wraddressHUVTemp(wraddrTemp),
        .rdaddressHUVTemp(rdaddrTemp),
```

```
                    .wrenTemp(wrenTemp),
                    .clk(clk),
                    .wraddressHUV(wraddr),
                    .rdaddressHUV(rdaddr),
                    .wren(wren),
                    .start(start),
                    .EnableBoundaryUpdating(EnableBoundaryUpdating),
                    .finish(finish));
InputMemoryHUV #(N) InputMemory(
                         clk,rdaddr,doutH,doutU,doutV,
                         wraddr,wren,HNew,UNew,VNew);
InputBuffer #(M,N) Buffer(
                         clk,doutH,doutU,doutV,
                         matrixhin,matrixuin,matrixvin);
CNNCore #(M,N) uut(
                         .clk(clk),
                         .matrixhin(matrixhin),
                         .matrixuin(matrixuin),
                         .matrixvin(matrixvin),
                         .matrixhout(matrixhout),
                         .matrixuout(matrixuout),
                         .matrixvout(matrixvout));
BoundaryUpdatingHUV #(N) Boundary(
                         matrixhout,matrixuout,matrixvout,
                         doutHNewTemp,doutUNewTemp,doutVNewTemp,
                         EnableBoundaryUpdating,
                         HNewTemp,UNewTemp,VNewTemp,
                         HNew,UNew,VNew);
TempMemoryHUV #(N) TempMemory(
                    clk,wraddrTemp,wrenTemp,HNewTemp,UNewTemp,
                    VNewTemp,
                    rdaddrTemp,doutHNewTemp,doutUNewTemp,doutVNewTemp);
                    endmodule
```

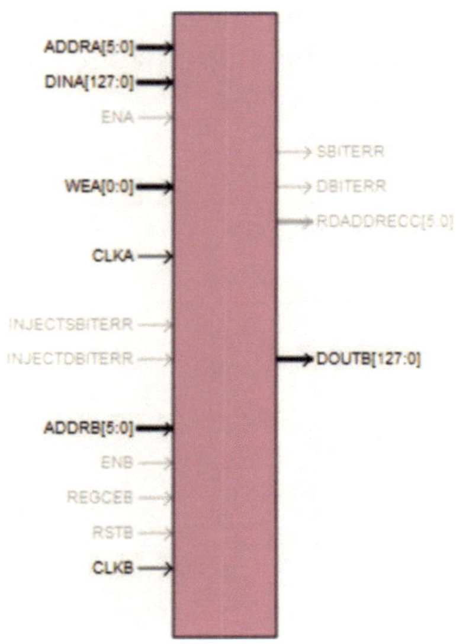

Figure 23.
Interface for Input and Temp memory h, u, v.

```
memory_initialization_radix=16;
memory_initialization_vector=
00000000416000004168000041700000
41680000417000004160000000000000
41600000416000004168000041700000
41680000417000004160000041600000
41c0000041c000004158000041500000
41580000415000004160000041680000
41600000416000004150000041500000
41500000415000004160000041600000
41a0000041a000004160000041580000
41600000415800004160000041600000
41500000415000004158000041600000
41580000416000004158000041500000
41600000416000004150000041500000
41500000415000004160000041600000
00000000415000004158000041600000
41580000416000004158000000000000
```

Figure 24.
An example of h.core file to initial data for the Input memory h.

3.6 Simulation results

The ISE design software shows the device utilization summary as in **Table 1**. **Figures 25–27** show the schematics synthesized by the ISE design software. Comparing the new values of h in **Figure 28i, k** (doutH) with **Figure 29**, we can see that the 3x4 CNN system worked well.

The simulation results show the properness and effectiveness of installation methods. The cost for calculating the first three blocks of 1xN taken from memory units h, u, v is 10 clock pulses, of which 1 clock pulse is for initial reading Input memory, 3 clock pulse is for initial updating buffer to CNN, and 6 clock pulses for initial calculation. Each successive 1xN unit takes only 1 clock pulse to calculate, due to the use of the pipeline mechanism to update buffer to CNN and calculate at CNN arithmetic unit. After finishing reading each column of blocks of data in the Input memory, it needs 2 more clocks for initiating the buffer again. It also takes 1 clk for initial writing Temp memory, 1 clk for initial reading Temp memory, and 1 clk for initial writing result back to Input memory.

Devices used summary (estimated values)			
Logic utilization	**Used**	**Available**	**Utilization**
Number of slice registers	3952	301,440	1%
Number of slice LUTs	16,365	150,720	10%
Number of fully used LUT-FF pairs	1770	18,547	9%
Number of bonded IOBs	3112	600	518%
Number of Block RAM/FIFO	12	416	2%
Number of BUFG/BUFGCTRLs	1	32	3%
Number of DSP48E1s	132	768	17%

Table 1.
Device utilization summary.

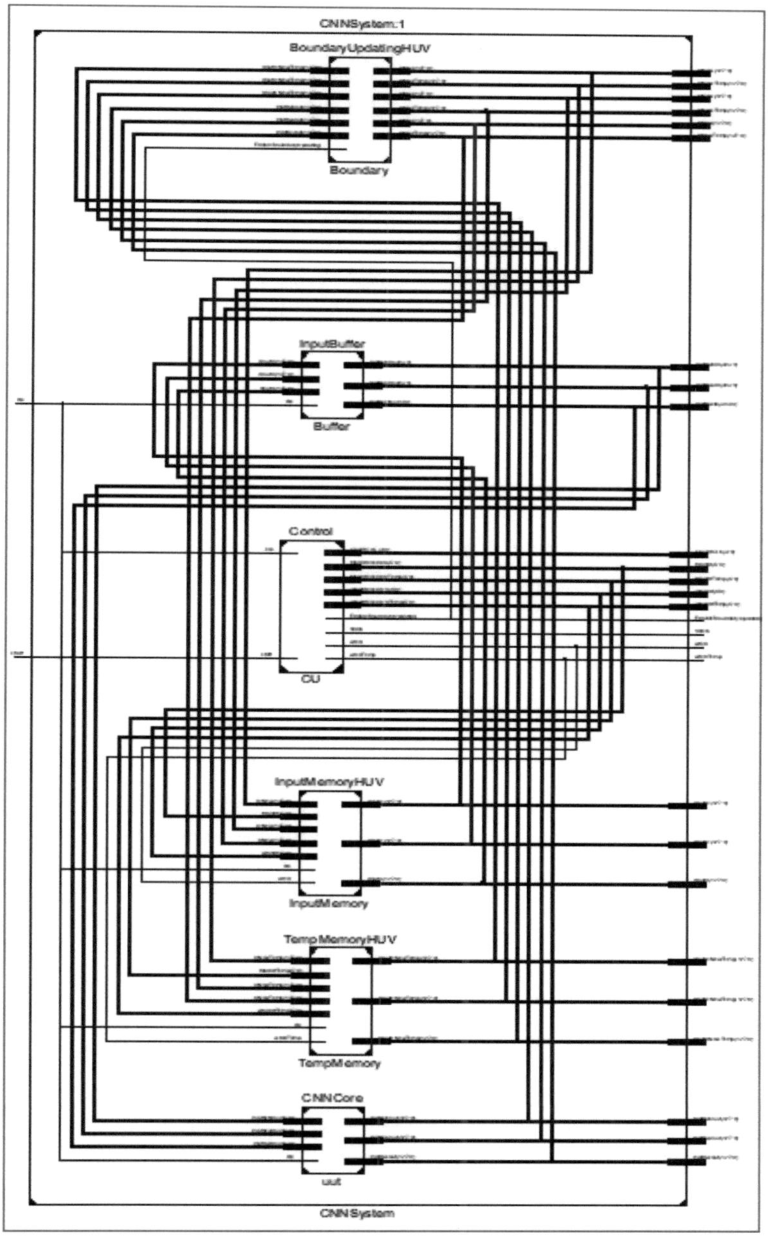

Figure 25.
The architecture of CNN chip.

As a result, the time for one computing cycle is:

$$T = 8 + m(Q + 1)\ (clk)$$

As the above implementation, m = 8, Q = 2, and T = 32 (clk).

4. Conclusion

This chapter gives the solution for configuring CNN chip to solve Navier-Stokes equations, especially concerning to solution in the temporary boundary problem

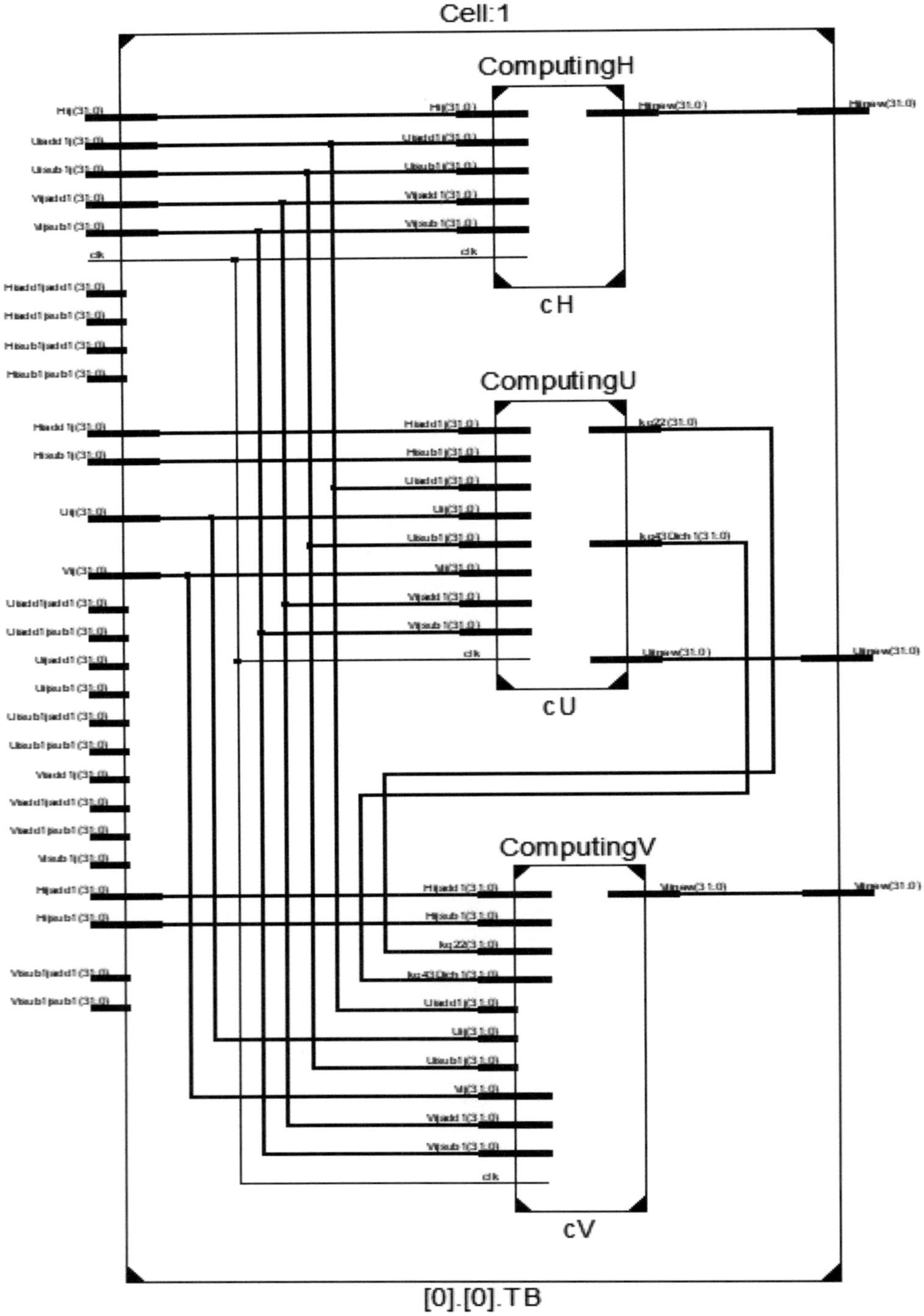

Figure 26.
The architecture of one CNN cell.

when it is required. The purpose is to divide the big data space into many subspaces. The processing of the big data space is based on the calculation of each subdata. With the input data of 32-bit floating point real number and FPGA chip Virtex 6 XCVL240T-1FFG1156, the CNN of 1x12 cells has successfully installed. The installation results show that the effectiveness of this solution mainly lies on the

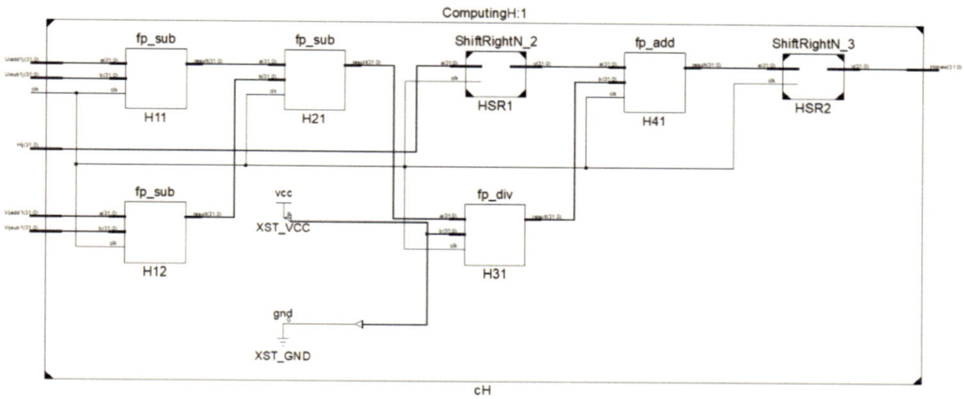

Figure 27.
Inside electronic circuit for h.

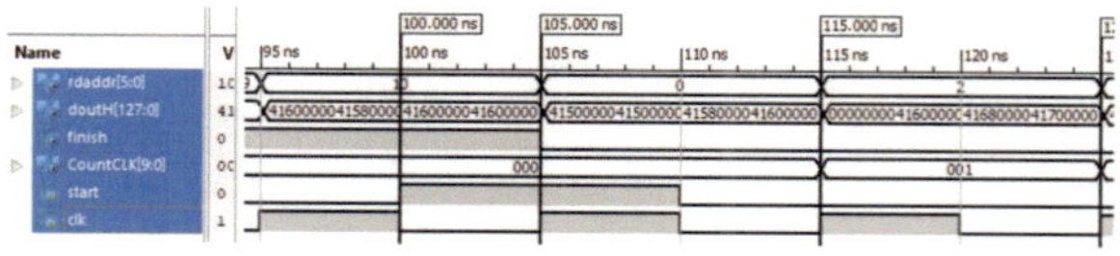

(a)

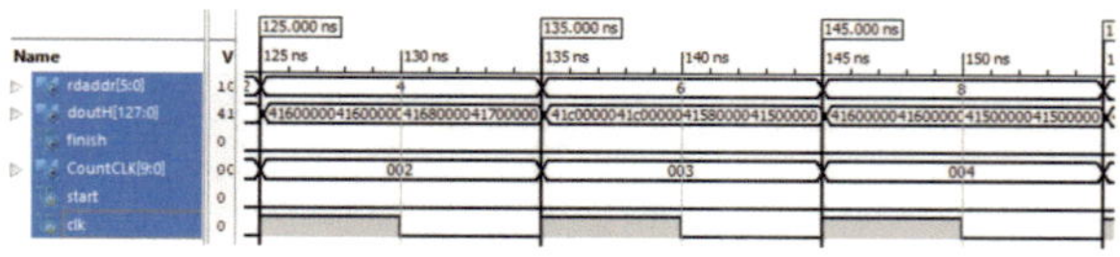

(b)

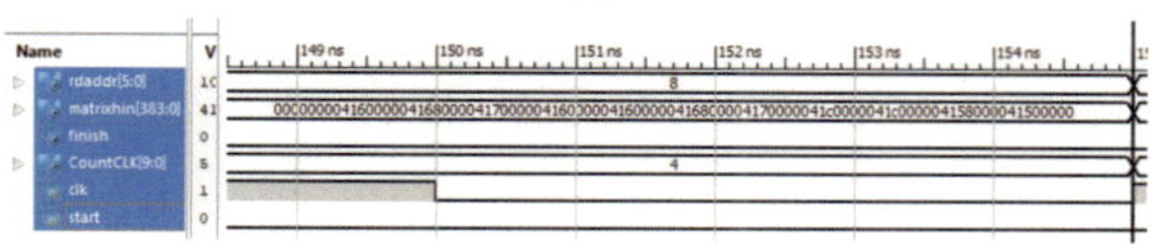

(c)

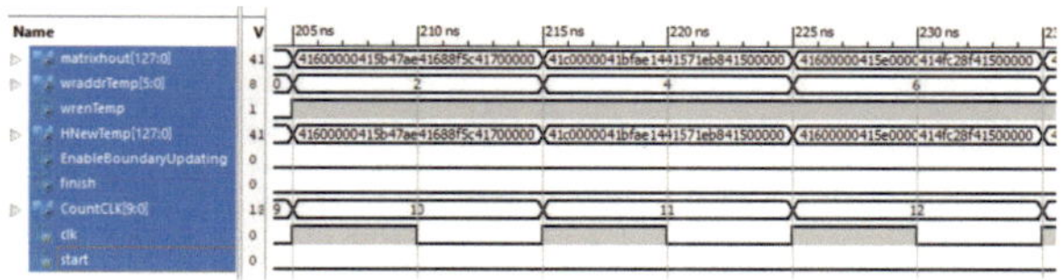

(d)

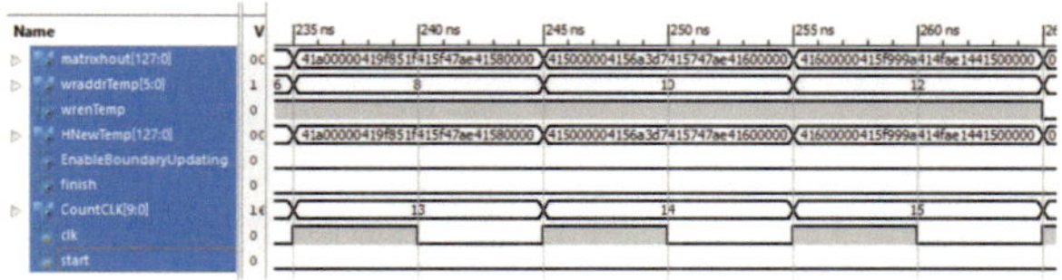

(e)

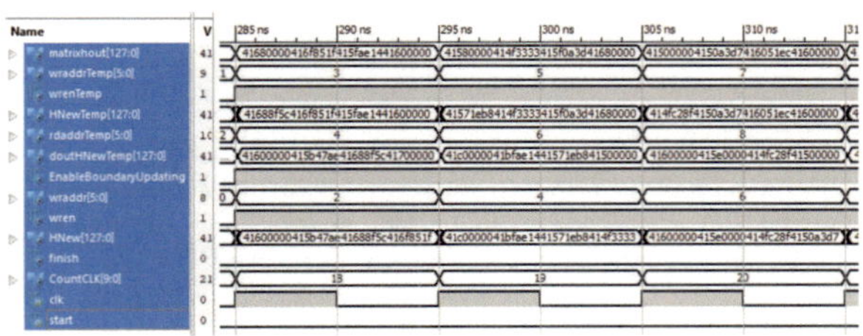

(f)

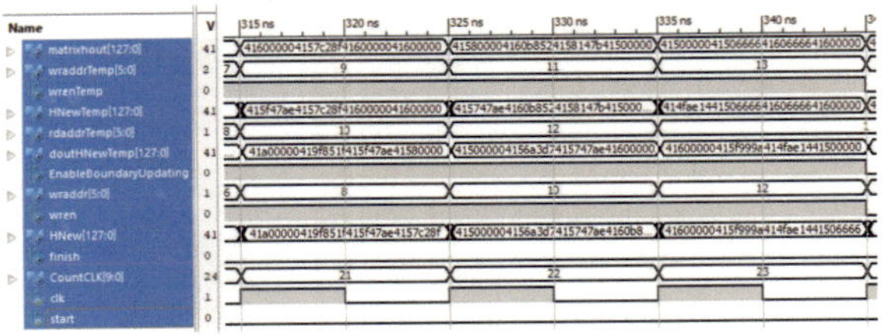

(g)

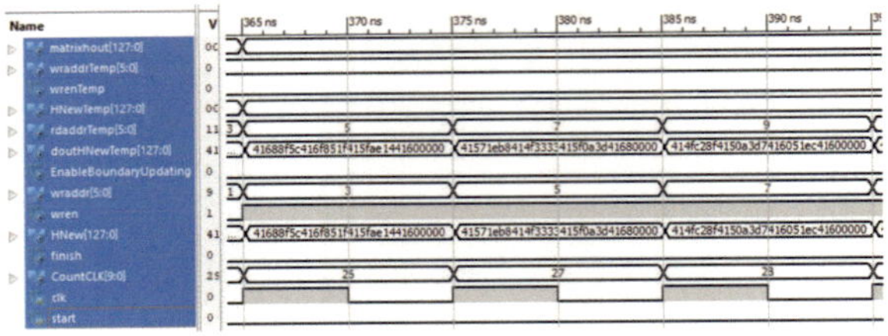

(h)

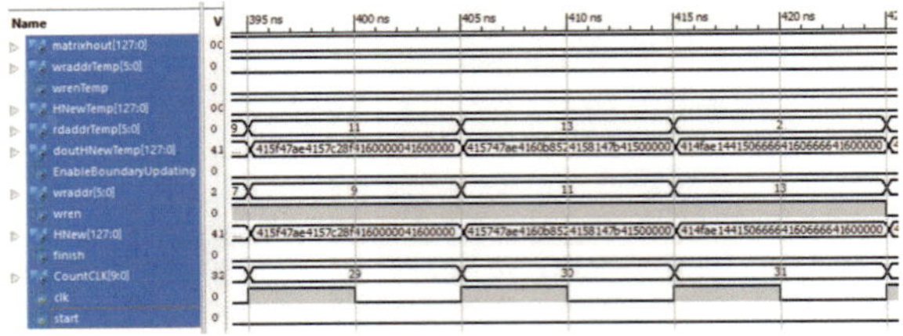

(i)

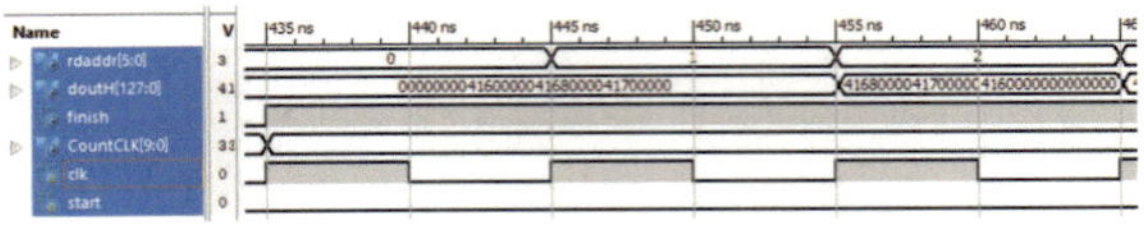

(j)

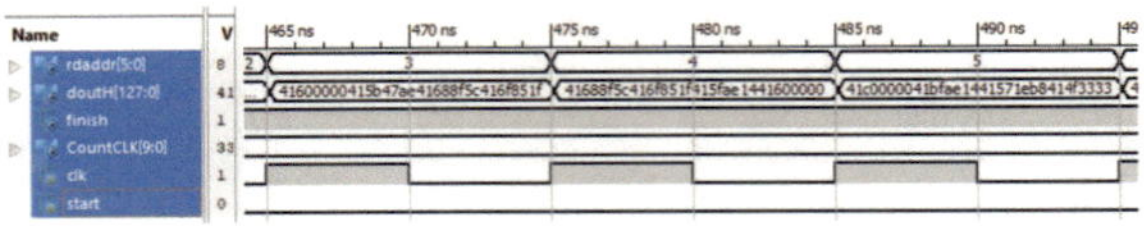

(k)

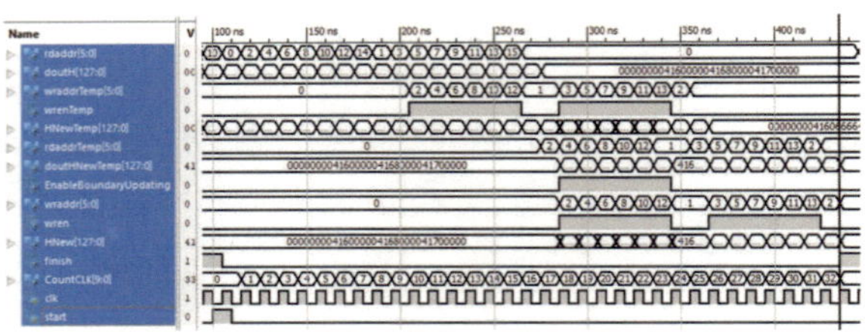

(l)

Figure 28.

Signals operating inside the 3x4 CNN system, m = 8, Q = 2. (a) Starting a computing cycle by setting start = 1. (b) The output of Input memory (doutH). (c) The data outputting from Buffer after 4 clks. (d) The results from CNN core after 10 clks; and start writing the results to Temp memory. (e) The CNN core finish computing the first column of blocks of data at 16 clks; and pause writing the results to Temp memory at 16 clks. (f) The results from CNN core after 18 clks; read Temp memory, start updating boundaries, and write the results to Input memory. (g) Pause updating boundaries from 24 clks. (h) The CNN core finishes computing; read the last column of blocks of data from Temp memory and write to Input memory. (h) Finish writing all results of the first computing cycle to Input memory. (i) The controller sets finish = 1 at 33 clks. (k) The output of Input memory shows the results computed at previous computing cycle. (l) The overview of signals.

0	14.00	14.50	15.00	14.50	15.00	14.00	0
14.00	13.71	14.54	14.97	14.54	14.97	13.98	14.00
24.00	23.96	13.45	12.95	13.45	12.95	13.94	14.50
14.00	13.88	12.99	13.04	12.99	13.04	14.02	14.00
20.00	19.94	13.96	13.49	13.96	13.49	14.00	14.00
13.00	13.42	13.46	14.05	13.46	14.05	13.51	13.00
14.00	13.98	12.98	13.03	12.98	13.03	14.03	14.00
0	13.00	13.50	14.00	13.50	14.00	13.50	0

00000000	41600000	41680000	41700000	41680000	41700000	41600000	00000000
41600000	415b47ae	41688f5c	416f851f	41688f5c	416f851f	415fae14	41600000
41c00000	41bfae14	41571eb8	414f3333	41571eb8	414f3333	415f0a3d	41680000
41600000	415e0000	414fc28f	4150a3d7	414fc28f	4150a3d7	416051ec	41600000
41a00000	419f851f	415f47ae	4157c28f	415f47ae	4157c28f	41600000	41600000
41500000	4156a3d7	415747ae	4160b852	415747ae	4160b852	4158147b	41500000
41600000	415f999a	414fae14	41506666	414fae14	41506666	41606666	41600000
00000000	41500000	41580000	41600000	41580000	41600000	41580000	00000000

Figure 29.
The new values of h computed by excel for the first computing cycle.

expansion of calculation space and resource saving and the accuracy of the calculation acceptable as well. This model can be further developed to feasibly solve similar problems in larger computing space and could be developed for some types of complicated (mixed) boundaries as well.

Acknowledgements

We would like to deeply acknowledge Professor Roska Tamas, the head of the Analogic and Neural Computing Research Laboratory and Chairman of the Scientific Council—Institute of the Hungarian Academy of Sciences; and Associate Professor Pham Thuong Cat, the Head of Automation Laboratory—Institute of Information Technology—Vietnam Academy of Science and Technology, for giving us many important instructions.

Author details

Vu Duc Thai* and Bui Van Tung
Thai Nguyen University, Thai Nguyen, Vietnam

*Address all correspondence to: vdthai@ictu.edu.vn

References

[1] Chua LO, Yang L. Cellular neural networks: Theory. IEEE Transactions on Circuits and Systems. 1988;**35**(10): 1257-1272

[2] Chua LO, Yang L. Cellular neural networks: Application. IEEE Transactions on Circuits and Systems. 1988;**35**:1273-1290

[3] Roska T, Chua LO, Wolf D, Kozek T, Tetzlaff R, Puffer F. Simulating nonlinear waves and partial differential equations via CNN—Part I: Basic techniques. IEEE Transactions on Circuits and Systems. 1995;**42**(10): 807-815

[4] Thai VD, Cat PT. Modeling air pollution problem by cellular neural network. In: Proceeding (ISI) of 10th International Conference on Control, Automation, Robotics and Vision; Hanoi, Vietnam; 2008. pp. 1115-1118

[5] Thai VD, Cat PT. Solving two-dimensional Saint venant equation by using cellular neural network. In: Proceeding of the 7th Asian Control Conference—ASCC2009; Hong Kong; 2009. pp. 1258-1263

[6] Thai VD, Cat PT. Equivalence and stability of two-layer cellular neural network solving Saint venant 1D equation. In: Proceeding (ISI) of 11th International Conference on Control, Automation, Robotics and Vision (ICARCV2010); Singapore; 2010. pp. 704-709

[7] Thai VD, Anh BT, Duong VT. Develop some application of Cyclone—DE2C35 chip. Journal of Science and Technology - Thai Nguyen University. 2015;**10**(140):103-108

[8] Thai VD, Linh LH, Linh NM. Solving Navier-Stokes equation using FPGA cellular neural network chip. In: Proceeding of International Conference on Advances in Information and Communication Technology (ICTA2016). Springer Publishing; 2016. pp. 562-571

[9] Rusin WM. On solution to Navier-Stokes equation in critical spaces [Thesis of Doctor Philosophy]. 2010. Available from: http://conservancy.umn.edu/.../Rusin_umn_0130E_11277.pdf

[10] Hruska J. Intel launches Stratix 10: Altera FPGA combined with ARM CPU, 14nm manufacturing. Extremetech. 2016. Available from: https://www.extremetech.com/computing/237338-intel-launches-stratix-10-altera-fpga-combined-with-arm-cpu-14nm-manufacturing

[11] la Pedus M. Intel-Altera deal to shake up foundry landscape. Chip Design Magazine. 2013. Available from: http://chipdesignmag.com/display.php?articleId=5215

[12] Clive M. The Design Warrior's Guide to FPGAs: Devices, Tools and Flows. Elsevier; 2004. Available from: https://www.eu.elsevierhealth.com/about-us

[13] Clive M. Programmable Logic DesignLine. Xilinx Unveil Revolutionary 65nm FPGA Architecture: The Virtex-5 Family. 2006. Available from: https://www.eetimes.com/document.asp?doc_id=13001899

[14] David WP. Google patent search. Dynamic Data Reprogrammable PLA. 2009

[15] David WP, Peterson LR. Google patent search. Dynamic Data Reprogrammable PLA. 2009

[16] Wisniewski R. Synthesis of Compositional Microprogram Control Units for Programmable Devices.

University of Zielona Góra Press; 2009, (ul. Podgórna 50, 65-246 Zielona Góra. Dane kontaktowe)

[17] Black F, Scholes MS. The pricing of options and corporate liabilities. Journal of Political Economy. 1973;**81**(3):59-637

[18] Chua LO, Roska T. Cellular Neural Networks and Visual Computing. Cambridge University Press; 2000. ISBN: 0-521-65247-2. Available from: https://en.wikipedia.org/wiki/Cambridge_University_Press

[19] FPGA Architecture for the Challenge. Available from: http://www.eecg.toronto.edu/~vaughn/challenge/fpga_arch.html

[20] Intel® FPGAs offer a wide variety of configurable embedded SRAM, high-speed transceivers, high-speed I/Os, logic blocks, and routing. Built-in intellectual property (IP) combined with outstanding software tools lower FPGA development time, power, and cost. Available from: https://www.intel.com/content/www/us/en/products/programmable/fpga.html